THE ANALYSIS OF COMPLEX NONLINEAR MECHANICAL SYSTEMS

A Computer Algebra Assisted Approach

WORLD SCIENTIFIC SERIES ON NONLINEAR SCIENCE - SERIES A

Editor: Leon O. Chua
University of California, Berkeley

Published Titles

Volume 1: From Order to Chaos
L.P. Kadanoff

Volume 6: Stability, Structures and Chaos in Nonlinear Synchronization Networks
V.S. Afraimovich, V.I. Nekorkin, G.V. Osipov, & V.D. Shalfeev
Edited by Academicians *A.V. Gaponov-Grekhov & M.I. Rabinovich*

Volume 7: Smooth Invariant Manifolds and Normal Forms
I.U. Bronstein & A. Ya. Kopanskii

Volume 8: Dynamical Chaos: Models, Experiments, and Applications
V.S. Anishchenko

Volume 12: Attractors of Quasiperiodically Forced Systems
T. Kapitaniak & J. Wojewoda

Volume 14: Impulsive Differential Equations
A.M. Samoilenko & N.A. Perestyuk

Volume 16: Turbulence, Strange Attractors, and Chaos
D. Ruelle

Forthcoming Titles

Volume 9: Frequency Domain Methods for Nonlinear Analysis: Theory and Applications
G.A. Leonov, D.V. Ponomarenko, & V.B. Smirnova

Volume 11: Nonlinear Dynamics of Interacting Populations
A.D. Bazykin

Volume 13: Chaos in Nonlinear Oscillations: Controlling and Synchronization
M. Lakshmanan & K. Murali

Volume 15: One-Dimensional Cellular Automata
B. Voorhees

Volume 18: Wave Propagation in Hydrodynamic Flows
A.L. Fabrikant & Yu. A. Stepanyants

WORLD SCIENTIFIC SERIES ON NONLINEAR SCIENCE Series A Vol. 17
Series Editor: Leon O. Chua

THE ANALYSIS OF COMPLEX NONLINEAR MECHANICAL SYSTEMS

A Computer Algebra Assisted Approach

Martin Lesser
Royal Institute of Technology, Stockholm,
Sweden

World Scientific
Singapore • New Jersey • London • Hong Kong

Published by

World Scientific Publishing Co. Pte. Ltd.

P O Box 128, Farrer Road, Singapore 9128

USA office: Suite 1B, 1060 Main Street, River Edge, NJ 07661

UK office: 57 Shelton Street, Covent Garden, London WC2H 9HE

British Library Cataloguing-in-Publication Data
A catalogue record for this book is available from the British Library.

THE ANALYSIS OF COMPLEX NONLINEAR SYSTEMS: A COMPUTER ALGEBRA ASSISTED APPROACH

ISBN 981-02-2209-2

This book is printed on acid-free paper.

Printed in Singapore by Uto-Print

To Fran and Alice, for being there!

0.1 Preface

During the three centuries since Newton created the subject of mechanics hundreds if not thousands of authors have seen fit to produce texts and treatises. In the light of this perhaps it is not necessary to explain the production of yet another, as it has clearly become a tradition in itself. It is hoped that the present work will be helpful to both students and to those engaged by need or interest in the task of modeling the behavior of mechanical systems. Some novelty is also claimed both in content and manner of presentation. Emphasis is put on a geometrical interpretation of the algorithm developed by Kane for deriving the equations of motion. There is some limited discussion of more traditional methods, such as the Lagrange and Gibbs-Appell equations. Another novel feature of this work is the introduction of a descriptive algorithmical language, 'Sophia' for the formulation and solution of mechanics problems. Sophia is designed to provide a clear and unique encapsulation of the tasks needed to formulate and manipulate theoretical constructs so as to arrive at equations of motion. The test of this clarity of expression is to be seen in the creation of 'interpreters' which translate Sophia statements into equations of motion. Such interpreters have been constructed for two popular computer algebra systems, Maple and Mathematica. The present book uses the Maple version. It is not necessary to have Sophia in its computer embodiment to make use of this text. While it is certainly more interesting to be able to run Sophia as a computer language, the main point is that it provides a detailed prescription for the discussion of particular problems. As different methods are introduced they are 'encapsulated' into algorithms which can be invoked by Sophia statements. This is a currently popular way of looking at problems in the computer science community and it is felt it provides a rich viewpoint for the presentation, understanding of the mechanics of rigid body systems.

The realm treated by this book is sometimes called 'the mechanics of multibody systems'. Most of the literature in this subject concentrates on numerical methods and hence formulations of the subject that will be helpful in writing general purpose computer codes. For truly large systems comprising many bodies this is probably the only reasonable path. In contrast to this the present book is based on the idea that the elusive quality known as 'understanding' is best achieved by only modeling the most essential features of a problem. In fact the very problem of determining just what the essential features are is at the root of achieving understanding. While the techniques treated here may also be useful for large scale numerical modeling the main thrust is towards dealing with moderately complex systems. Here it is possible, using computer algebra, to obtain constraint free representations of the equations of motion in forms that can be readily integrated. It is known that such forms frequently lead to the most efficient numerical integration procedures, but it is also hoped that they can help the investigator achieve analytical understanding. Some discussion of this is included in the last chapter of this book, which covers methods of approximation.

Notes, on which this book is based, have been used in two advanced mechanics courses given at the Royal Institute. For the most part students have come from the engineering physics, vehicle technology and mechanical engineering departments. The course consisted of some 60 contact hours, including a number of active computer demonstrations. The students have had good access to the Maple system and have been able to make considerable use of the Sophia programs.

Finally a word about the Sophia programs. Appendix A describes how to use the attached disks. They are also available by FTP over the internet at ftp.mech.kth.se in the directory *sophia*. Included are some tutorial files as well as a somewhat more primitive version of Sophia in Mathematica. The present main form of the programs are intended to run with MapleV release 3. There are also interface programs for placing equations derived with Sophia into Matlab format.

(Matlab, Maple and Mathematica are trade marks of their respective developers.)

I have had the great benefit of advice and encouragement from a number of my colleagues, both at the Royal Institute and other institutions. In particular I would like to thank Dr. Hanno Essén for his careful reading of the text, numerous discussions in which he set me 'on the right path' and toleration if not acceptance of some notational heresy. Dr. Arne Nordmark, a former student who in many ways has become my teacher, showed me the need to respect the difference between tangent and cotangent spaces, though the conversion was not complete. Professors Sören Andersson and Lennart Karlsson and their students helped put Sophia and the geometrical interpretation of Kane's equations into practice on 'real' problems. Professor Håkan Gustavsson, though busy with his own research programs found time to support my work through his activites with the Swedish Technical Research Council (TFR). Support has also come from the Swedish Industrial Research Council (STU) and the Volvo Research Foundation. I also am indebted to the Royal Institute of Technology, and the Mechanics department for both a wonderful research atmosphere, some of the best students in the world and a sabbatical leave which allowed me to finish the task of preparing this book. Especial thanks go to Claes Tissell and Annika Stensson for their work on applications. My own graduate student, Anders Lennartsson, has made major contributions to the utility of the Sophia programs as well as developing interface tools for moving Sophia output to Matlab and C. None of this work would have been carried out if I had not come across the works of Thomas Kane. The reading of his works has given me both new insight and interest in this classical subject and I thank him for this. Professor Leon Chua, the editor of this series, encouraged me to produce this version of the work. As editor of the Journal of Bifurcation and Chaos his belief that the community of dynamical systems researchers will obtain benefit from the ability to easily deal with complex mechanical systems has been a great encouragement to me.

The preparation of this work by one person was made possible by the modern computer tools we now are taking for granted. In particular I thank the authors of TeX and LaTeX, as well as Adobe Dimension and Adobe Illustrator. The latter made it possible for a draftsman of modest talent, myself, to produce the illustrations of the present text.

Martin Lesser
Stockholm
November, 1994.

It's so simple, So very simple, That only a child can do it!
New Math, Tom Lehrer

Contents

Chapter 1

The Problems of Mechanics

Classical Mechanics provides the basic grammar of physical science. It is a traditional practice for scientists and engineers to begin their educational training with courses in the fundamentals of mechanics. It is assumed that the root vocabulary of the subject, together with at least a passing familiarity with the associated concepts, is known to the reader. The goal is to provide an understanding of powerful techniques that will help formulate mechanical models of pertinent aspects of the physical world. A prime contribution of the twentieth century's work in this classical subject has been in the development of tools so that the concepts of mechanics can be applied to the design and understanding of complex systems. Standard introductions and even advanced works do not provide these tools. The formulation of equations of motion for what appear to be simple mechanical systems composed of only a few interconnected rigid bodies require considerable algebraic manipulation. Therefore the modern tool of computer algebra must be included in any serious modeling of mechanical systems.

The study of grammar and spelling will not make one a great author, but they do provide the creative writer with the technical tools of his trade. The material in this book is designed to provide the technical tools that will assist creativity and physical intuition in formulating appropriate mechanical models of the physical world. This introductory chapter supplies some motivation and a brief survey of the territory to be covered. At first we concentrate on the task of describing position and motion, the subject known as kinematics. Then we focus on the task of formulating equations that govern the behavior of complex systems of particles and rigid bodies according to the laws of Newtonian Mechanics. Finally we examine the structure of mechanics and seek a qualitative understanding of how mechanical systems must behave.

1.1 Newtonian Principles

In the context of how natural science was practiced in the seventeenth century, the contribution of Isaac Newton's work is without peer. After three hundred years of development and refinement, many of Newton's conceptual advances have become so much part of our civilization's world view that it is easy to see them as obvious and even trivial or worse not to see them at all. The study of mechanics begins with some deceptively simple looking statements, the understanding of which depends on the reader already being a party to the cultural and intellectual heritage created by Newton and his successors.

The ideas needed for the development of mechanics may be summarized in the form of a few *working principles.* These are not meant to be axiomatic or mathematically rigorous, but are to be taken as a way of expressing some of the basic notions of our subject.

The first principle provides information as to the form of the equations of motion. The statement assumes that one 'understands' such terms as coordinate and geometric configuration. Considerable effort will be spent in an examination of these deceptively simple concepts. The first of these working principles is:

Principle 1 *The motion of a mechanical system is governed by second order differential equations which describe the values of the independent coordinates needed to determine the geometric configuration of the system as a function of time.*

The rates of change of the above noted coordinates are related to the *velocities* needed to describe the state of the system's motion. The rates of change of these velocities define the *accelerations.* The first principle allows the conclusion that the accelerations, which are related to second time derivatives of the coordinates, are equal to functions of the coordinates and velocities. The second working principle is:

Principle 2 *The accelerations are proportional to functions of the current coordinates and their rates of change. These functions can be understood as describing* forces *applied to various parts of the mechanical system.*

Finally the Newtonian model assumes that these functions, representing different forces follow a principle of superposition:

Principle 3 *The forces acting on bodies are linearly additive, i.e. they can be superposed.*

The discovery and invention of reasonable force laws, that is suitable functions of position and acceleration, give substance to mechanics. The principle of superposition of forces *does not* imply that the equations of motion are linear, only that the forces from different causes are additive. Thus the gravitational force proposed by Newton is inversely proportional to the inverse square of the distance between the mass centers of bodies, which is certainly a nonlinear function of the distance. On the other hand the principle implies that the gravitational forces acting on a body from several other bodies is the sum of forces caused by each body.

Acceptance of these working principles implies acceptance of the idea of *Newtonian Determinism*, i.e. knowledge of the coordinates and their rates of change at any instant determines *all* past and future states of the mechanical system. The world view implicit in this idea still dominates our culture and patterns of thought. Many volumes have been written on the meaning of Newton's laws and their philosophical implications. In this work a practical working view of these matters is assumed, the main interest being on how mechanics may be applied to the problems of science and technology.

Instead of making further generalizations at this introductory stage, it is helpful to examine some concrete examples. It is assumed that the reader has some familiarity with mechanics, as expressed in the formulation of free body diagrams, the solution of elementary statics problems, and the study of rectilinear motion in a uniform gravitational field.

• Illustration

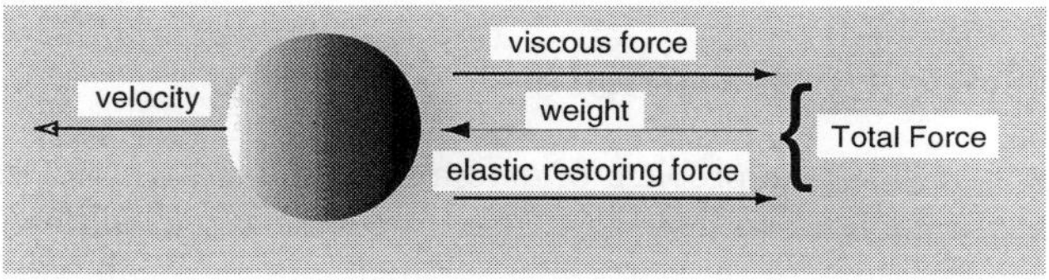

Figure 1.1: Superposition of Point Particle Forces

About the simplest mechanical system one can imagine is a single point particle confined to motion along a straight line. Label the particle position by x. The principle of Newtonian Determinism says that in general the motion will be governed by a second order differential equation, with time as the independent variable. That is an equation of the form:

$$\frac{d^2x}{dt^2} = G(x, \frac{dx}{dt}, t). \tag{1.1}$$

Assumptions about the homogeneity and isotropic structure of space and about the invariance of the form of the equations of motion under uniform velocity transformations, set restrictions on the form of the function G. A careful mathematically oriented discussion of this is given in the text by Arnold. For example the explicit dependence of G on time t is only justifiable if we consider the system at hand to be part of a larger closed system, the other part of the system being only 'weakly' influenced by the part under study. The superposition principle requires that force laws are additive, i.e. G can be composed of a sum of terms each of which represent some other aspect of the larger part of the system's influence. Examples are weight, viscous friction and elastic response. This suggests the existence of a constant parameter, which characterizes the particle, such that when divided into each of the additive terms will provide a quantity with the appropriate dimensions of length over the square of time, i.e. acceleration. Thus if $R(x, \frac{dx}{dt}, t)$, represents the sum of forces acting on the particle, define $1/m$, as the constant parameter, where m is called the mass, and the equation takes the form:

$$\frac{d^2x}{dt^2} = \frac{1}{m} R(x, \frac{dx}{dt}, t). \tag{1.2}$$

Examples of useful force expressions that are commonly used in engineering systems are the linear spring, $-kx$, the viscous damper, $-r\frac{dx}{dt}$, and weight, mg. All these are linear, leading to relatively simple explicit solutions of the equation of motion. The linear spring might also be considered as a first approximation to a more complex elastic force law such as, $f(x) = kx + \gamma x^3$. Time dependent forcing might be used to approximate the connection of the simple particle system to a massive particle moving under the influence of another linear spring and giving rise to a force of the form: $A\cos(\omega t)$. Readers of this text have most likely spent a large part of their educational experience dealing with these various possibilities.

1.2 A Specific Example

As an example, consider the problem of determining the motion of a small particle of matter which is confined to move in a given two dimensional surface as shown in Figure 1.1. One can think of the particle as approximating a small ball bearing, and the surface as being formed from two thin glass forms which fit around the ball. Also assume that the ball is subject to forces which are more or less controllable. For example electromagnetic forces induced by the investigator, and gravitational forces due to the experiment being at the Earth's surface. Let $\mathbf{a}$ represent the vectorial acceleration of the particle and let $\mathbf{R}$ denote the vectorial *force* that drives the motion. The working principles above imply that this mechanical system may be described by a mathematical expression of the form:

$$\mathbf{R} - m\mathbf{a} = 0, \tag{1.3}$$

which is simply the common modern form of the Newtonian Law of motion for a particle. From the viewpoint of technique, the question of notation must be taken with some care. The way we represent things has a profound influence on how we think about them. For the moment the aim is to set forth the basic problems faced in formulating equations of motion, hence somewhat casual use is made of notational structures that are assumed familiar to the reader. Thus the bold face quantities represent vectors, which are in some sense independent of coordinate systems. Likewise $\boldsymbol{r}(t)$ is the vectorial position of the particle, with respect to some given reference point in space. The reference point must be considered fixed in a so called *inertial reference frame*, which is a reference frame in which a particle's motion is described by the Newtonian Law. Later more precise consideration must be applied to such

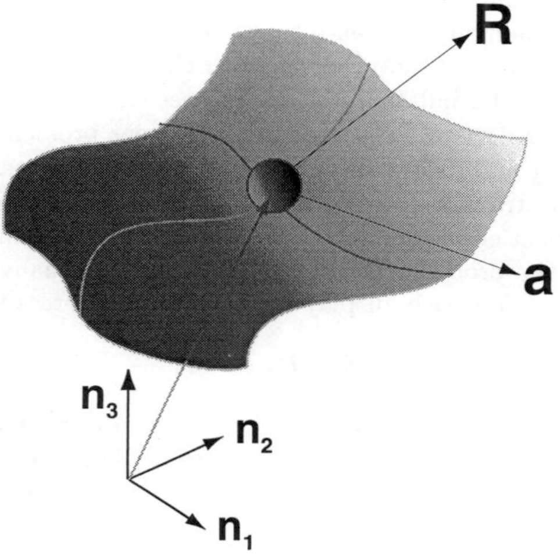

Figure 1.2: Particle moving in a confining surface

questions. In terms of the position, $\boldsymbol{r}(t)$, the acceleration is given by:

$$\boldsymbol{r} = \frac{d^2\boldsymbol{r}}{dt^2}. \tag{1.4}$$

It may appear that three second order differential equations are required to represent the vectorial form. The existence of the constraint reduces the dimensionality of the problem, so that in scalar form the motion is described by two second order differential equations, and thus, a fully posed problem requires knowledge of four initial conditions. These can be provided by the position and velocity of the particle at some given time. Since the particle is confined to move in the given two dimensional surface, four quantities must be known to determine the motion for all time.

Also, it might appear that equation 1.3 together with knowledge of position and velocity at any given time and an expression for the applied force provide all the information needed for a fully determined mathematical description of the problem. To see that this is not so, look more carefully at the way force is applied to the particle. Thus the force, $\boldsymbol{R}$, is composed of several parts. Gravitational force might take the form $\boldsymbol{R}_a = m\boldsymbol{g}$, where $\boldsymbol{g}$ is a constant vector pointing in the direction of the gravitational attraction and m is the mass of the particle. The subscript on $\boldsymbol{R}_a$ is intended to indicate that this is an *applied* force. If this were the only force on the particle, it would not stay confined in the given surface. For the particle's motion to be so confined, the surface must exert forces on it. These forces depend on the motion itself. Thus the problem as stated specifies a detail of the allowed motion based on the assumption that the material composing the surface will in fact constrain the motion as indicated. It is a common occurrence that information is available in regard to the possible motions of a mechanism. Dealing with the *unknown* forces of constraint is a vital part of problem formulation. In many cases the determination of constraint forces are the prime objective for study. In the present problem it might be known that the constraining glass surfaces would fracture under certain loading, hence a knowledge of the constraining forces would be an imperative. In any event they are unknown and one must either develop equations for their determination or somehow remove them from the problem. The applied forces are usually known in terms of position and velocity. To clearly display the constraining forces write:

$$\boldsymbol{R} = \boldsymbol{R}_a + \boldsymbol{R}_c, \tag{1.5}$$

where $\boldsymbol{R}_c$ now represents the constraint forces. To proceed further assumptions are needed about the behavior of the constraint forces. These assumptions are *in addition* to the basic principles of mechanics, and must ultimately be judged on their usefulness in the explanation of mechanical phenomena. There can certainly exist circumstances in which they are not appropriate. It is also assumed that the surface is very smooth so that frictional forces opposing the motion of the particle inside the glass sheets may be neglected. The surface may be moving as a rigid body or even deforming. The physical mechanism for the development of constraint forces will arise from electromagnetic interactions connected with the atomic structure of the moving mass and the constraining surface. With friction ignored, this will take the form of a force at right angles to the surface. If at a given time the particle is located at some specific point, this implies that the constraining force will be in the same direction as the vector normal to the constraining surface at that point. Thus the geometry of the constraining surface provides information about the direction of the unknown constraining force. Therefore suitable mathematical apparatus is required for the description of the constraint surface's geometrical features. The relation:

$$\phi(\boldsymbol{r}, t) = 0, \tag{1.6}$$

provides an implicit statement that defines a two dimensional changing surface in three dimensional space. The position vector

$$\mathbf{r} = r_1\mathbf{n}_1 + r_2\mathbf{n}_2 + r_3\mathbf{n}_3 \tag{1.7}$$

represents a location in space in terms of a triad of orthogonal unit vectors $\boldsymbol{n}_j$, where the index j runs from 1 to 3. Any $\boldsymbol{r}$ which satisfies 1.6 at a time t, represents a point on the geometric surface at that time.

Another convenient means of representing the surface makes use of parametric equations for the surface coordinates. Thus at any fixed time, points on the two dimensional surface embedded in three dimensional space, can be represented by three equations of the form:

$$r_j = r_j(q_1, q_2, t), \tag{1.8}$$

where again j takes on the values 1,2 or 3. If we fix time and one of the surface parameters, q_j, the equation 1.8 represents a curve in three dimensional space. The derivatives of the three functions with respect to q_j then represent the components of a vector which is tangent to this curve, and hence also tangent to the surface. Therefore the vectors:

$$\boldsymbol{\tau}_j = \frac{\partial}{\partial q_j}\boldsymbol{r}, \tag{1.9}$$

are tangent to the surface. It can be shown that if the surface is properly described by 1.8, the vectors $\boldsymbol{\tau}_j$ must be independent, i.e. not collinear. Therefore their vector cross product is non-zero and perpendicular to the surface. The vector with components:

$$\frac{\partial}{\partial r_j}\phi \tag{1.10}$$

is also perpendicular to the surface. To show this insert the relations 1.8 into the implicit description of the surface given by 1.6 and take the derivative of the result with respect to q_k. Application of the chain rule in evaluating the partial derivatives then gives the result:

$$\sum_{i=1}^{3} \frac{\partial \phi}{\partial r_i}\frac{\partial r_i}{\partial q_k} = 0 \tag{1.11}$$

with $k = 1, 2$. The vector gradient of ϕ is:

$$\nabla\phi = \sum_{i=1}^{3} \frac{\partial \phi}{\partial r_i}\boldsymbol{n}_i \tag{1.12}$$

which shows that

$$\nabla\phi \cdot \boldsymbol{\tau}_k = 0. \tag{1.13}$$

This result demonstrates that the gradient vector, $\boldsymbol{N} = \nabla\phi$ is perpendicular to any vector tangent to the surface given by $\phi = 0$.

At any point on the surface an *arbitrary* vector which is tangent to the surface can be represented using the *coordinate* tangent vectors, $\boldsymbol{\tau}_k$ with an expression of the form:

$$\boldsymbol{\tau} = \beta_1\boldsymbol{\tau}_1 + \beta_2\boldsymbol{\tau}_2 \tag{1.14}$$

where the β_k are *expansion coefficients*, and $\boldsymbol{N} \cdot \boldsymbol{\tau} = 0$. In general $\boldsymbol{\tau}$ or $\boldsymbol{N}$ will *not* be unit vectors, and $\boldsymbol{\tau}_1 \cdot \boldsymbol{\tau}_2$ will be non-zero. Sometimes the q_k are called surface coordinates, and the vectors $\boldsymbol{\tau}_k$,*coordinate tangent vectors.*

The physical assumption is that for a *smooth* surface at a fixed time, the constraint force that the surface exerts on a particle, which is assumed to stay on the surface, is normal to the surface. In terms of the mathematical representations of surfaces discussed above, this implies that:

$$\boldsymbol{R}_c \cdot \boldsymbol{\tau} = 0 \tag{1.15}$$

and that for some suitable λ:

$$\boldsymbol{R}_c = \lambda \boldsymbol{N} \tag{1.16}$$

The Newtonian law of motion for the particle is:

$$\boldsymbol{R}_a + \boldsymbol{R}_c - m\boldsymbol{a} = 0. \tag{1.17}$$

To eliminate the constraint force, $\boldsymbol{R}_c$, from 1.17, dot an arbitrary vector tangent to the surface into the equation. Thus if $\boldsymbol{\tau}$ is such a vector we see that

$$(\boldsymbol{R}_a + \boldsymbol{R}_c - m\boldsymbol{a}) \cdot \boldsymbol{\tau} = 0 \tag{1.18}$$

so that with 1.15

$$(\boldsymbol{R}_a - m\boldsymbol{a}) \cdot (\beta_1 \boldsymbol{\tau}_1 + \beta_2 \boldsymbol{\tau}_2) = 0. \tag{1.19}$$

As β_k are arbitrary, this equation implies the two *independent* scalar equations:

$$(\boldsymbol{R}_a - m\boldsymbol{a}) \cdot \boldsymbol{\tau}_k = 0 \tag{1.20}$$

The equations 1.20 are entirely expressed in terms of the applied forces, which have a known dependence on position and possibly velocity. The constraint force is found by returning to 1.17, replacing $\boldsymbol{R}_c$ by $\lambda\boldsymbol{N}$ and taking the dot product of the result with $\boldsymbol{N}$. This provides a scalar equation for λ. The technical solution of the problem also involves the ability to express the acceleration in terms of the surface coordinates, q_k, however for the moment the main interest is in the overall approach to elimination of constraint forces from the equations of motion.

The above shows that the crucial step in performing this task is to obtain vectors that are *tangent* to the directions in which motion is permitted by the instantaneous constraints. For the simple case above, this involved the use of vectors tangent to the physical surface in question. The treatment of more complex systems, involving many particles and rigid bodies with interconnections requires some more technical tools, but the basic idea of some kind of tangency still provides the crucial step in the elimination procedure.

•Illustration

A particle of mass 'm' is allowed to move in the surface of a paraboloid of revolution. The paraboloid opens up in the direction of positive x_3, and the direction of the uniform gravitational force is towards the base, as shown in Figure 1.3. The problem is to write equations of motion which are independent of the constraint forces that keep the particle confined to the surface. We will assume that there is no friction, i.e. that the particle slides smoothly within the surface. Following the previous discussion of tangent vectors, find suitable vectors which are tangent to the surface. To accomplish this task, first construct a parametric representation of the surface. Thus if the equation for the surface is

$$x_3 = k(x_1^2 + x_2^2), \tag{1.21}$$

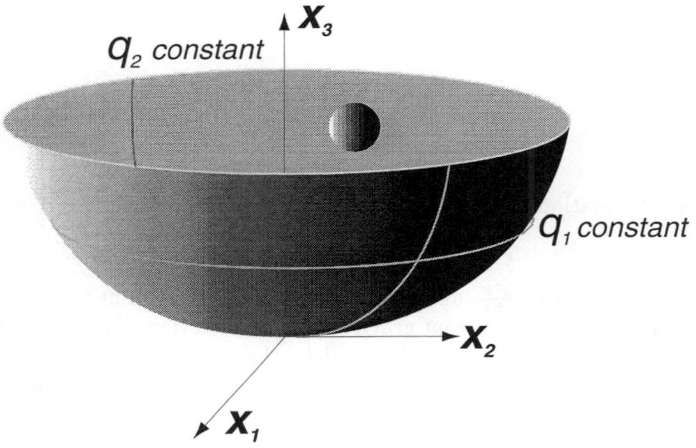

Figure 1.3: Particle moving in a paraboloid of revolution

where k is a constant, obtain a parametric representation in terms of the parameters q_1 and q_2. A convenient one is given by:

$$x_1 = q_1 \cos q_2, \tag{1.22}$$

$$x_2 = q_1 \sin q_2 \tag{1.23}$$

and

$$x_3 = k{q_1}^2. \tag{1.24}$$

If q_1 is held constant, these equations describe circles which are the intersection of the parabolic surface with planes orthogonal to the axis of the paraboloid. The curves for q_2 held constant are simply parabolas formed by the intersection of a plane passing through the axis with the surface. Thus in this simple case, geometric insight shows that the tangent vectors to these curves will be tangents to the circles and parabolic curves and that at each point of the surface, these tangent vectors will be orthogonal. Also in specifying the initial position of the particle, it is only needed to specify the values of q_j that fix the x_α coordinates. For the present discussion the convention is adopted that Latin subscripts vary between 1 and 2, while Greek subscripts go from 1 to 3. To formally obtain expressions for the tangent vectors, let $\boldsymbol{e}_\alpha$ denote three orthogonal unit vectors, fixed in space and aligned with the axis of x_α. The position of an arbitrary point on the surface is then given by the expression:

$$\boldsymbol{r} = \sum_\alpha x_\alpha(q_1, q_2)\boldsymbol{e}_\alpha. \tag{1.25}$$

The tangent vectors to the coordinate lines in the surface are:

$$\boldsymbol{\tau}_j = \frac{d\boldsymbol{r}}{dq_j}, \tag{1.26}$$

hence

$$\boldsymbol{\tau}_1 = \cos q_2 \boldsymbol{e}_1 + \sin q_2 \boldsymbol{e}_2 + 2kq_1 \boldsymbol{e}_3, \tag{1.27}$$

and

$$\boldsymbol{\tau}_2 = -q_1 \sin q_2 \boldsymbol{e}_1 + q_1 \cos q_2 \boldsymbol{e}_2. \tag{1.28}$$

The Newtonian equation of motion for the particle is

$$\boldsymbol{R} - m\boldsymbol{a} = 0, \tag{1.29}$$

where $\boldsymbol{a}$ is the acceleration $\frac{d^2\boldsymbol{r}}{dt^2}$, and $\boldsymbol{R}$ is the sum of the constraint forces $\boldsymbol{R}_c$, and the weight $-mg\boldsymbol{e}_3$, g being the constant gravitational acceleration. The basic assumption is that the constraint forces are orthogonal to the tangent vectors $\boldsymbol{\tau}_j$, hence that $\boldsymbol{\tau}_j \cdot \boldsymbol{R}_c = 0$. Therefore obtain two second order differential equations for the coordinates q_j, in the form:

$$(g\boldsymbol{e}_3 + \boldsymbol{a}) \cdot \boldsymbol{\tau}_j = 0. \tag{1.30}$$

The acceleration is obtained by differentiation of the expression for $\boldsymbol{r}$, which is carried out in two steps, first obtaining the particle velocity, $\boldsymbol{v} = \frac{d\boldsymbol{r}}{dt} = \sum_\alpha v_\alpha \boldsymbol{e}_\alpha$ as

$$v_1 = \frac{dx_1}{dt} = \dot{q}_1 \cos q_2 - q_1 \dot{q}_2 \sin q_2, \tag{1.31}$$

$$v_2 = \frac{dx_2}{dt} = \dot{q}_1 \sin q_2 + q_1 \dot{q}_2 \cos q_2 \tag{1.32}$$

and

$$v_3 = \frac{dx_3}{dt} = 2kq_1\dot{q}_1. \tag{1.33}$$

Further differentiation provides the acceleration components, thus:

$$a_1 = \ddot{q}_1 \cos q_2 - 2\dot{q}_1\dot{q}_2 \sin q_2 - q_1\ddot{q}_2 \sin q_2 - q_1{\dot{q}_2}^2 \cos q_2, \tag{1.34}$$

$$a_2 = \ddot{q}_1 \sin q_2 + 2\dot{q}_1\dot{q}_2 \cos q_2 + q_1\ddot{q}_2 \cos q_2 - q_1{\dot{q}_2}^2 \sin q_2, \tag{1.35}$$

and

$$a_3 = 2k{\dot{q}_1}^2 + 2kq_1\ddot{q}_1. \tag{1.36}$$

Taking the appropriate dot products with the tangent vectors leads to the two equations:

$$2gkq_1 + a_1 \cos q_2 + a_2 \sin q_2 + 2ka_3 q_1 = 0, \tag{1.37}$$

and

$$-a_1 q_1 \sin q_2 + a_2 q_1 \cos q_2 = 0. \tag{1.38}$$

The constraint force can also be found by finding the component of the equation of motion in the direction of the normal to the surface. Thus if $\boldsymbol{n} = \boldsymbol{\tau}_1 \times \boldsymbol{\tau}_2$ the constraint force can be computed from the relation:

$$\boldsymbol{n} \cdot (\boldsymbol{R}_c - mg\boldsymbol{e}_3 - m\boldsymbol{a}) = 0. \tag{1.39}$$

This example is simpler than the general case treated in the text as the position of the surface is fixed. A somewhat more interesting case in which the surface oscillates in the e_3 direction can be used to illustrate the point that, in elimination of the constraint force, it is the instantaneous direction of the tangent vectors that are normal to the constraint at any given instant. •

The main theme of this text is that much of mechanics can be understood in terms of a suitable generalization of this situation. This requires techniques for the precise description of the configuration of complex systems of interconnected bodies, the extension of the notion of tangency to higher dimensional spaces and the connection of these ideas with the classical notions of mechanical science.

1.3 Mechanics and Geometry

The theme of this book is to closely coordinate mechanics and geometry in both an abstract and a concrete sense. The concrete aspect involves the description of the possible motions that a constrained mechanical system may undergo. The abstract part involves a geometrical interpretation of the structure of mechanical theory. The following overall plan of this work is intended to give the reader some idea of what follows. Many of the specific terms used in the plan will most likely be unknown to most readers, but the 'flavor' of the work may be previewed by reading it.

Chapter two develops notation for dealing with the description of rotation and for comparing rates of change of vectors as measured by observers stationed on bodies which are in relative motion. The mathematical apparatus is the theory of orthogonal transformations. A notation is developed which should help to avoid problems in keeping track of the transformations.. This is done by embedding the properties of the transformations into logical rules for constructing the symbols representing them. The concept of angular velocity is introduced as an operator for relating rates of change of vectors as seen by different observers. Dyadic forms and projection operators, which play an important role in the rest of the text, are discussed.

The next chapter examines the kinematics of motion in detail. A geometrical object, called the Kvector, is employed to describe the configuration of constrained multibody mechanisms. Using the Kvector, one obtains an explicit representation of the many dimensional instantaneous configuration hypersurface of a mechanism. Tangent Kvectors, which can be used to eliminate constraint forces from the equations of motion, are arrived at by considering so called test motions. These latter are finite extensions of what are commonly called virtual displacements in the traditional literature of mechanics. Arbitrary linear combinations of the tangent Kvectors lead to the idea of *non coordinate* tangent basis vectors for the configuration hypersurface and generalized speed parameters. These quantities are used in the sequel to simplify the derivation of motion equations for particular systems.

The many point particle mechanism is discussed in chapter 4. The word mechanism is used to indicate a definite scheme for representing both the configuration and motion of a system composed of point mass particles, which may or may not be subject to constraints. Tangent Kvectors are used to eliminate constraints and to derive the so called Kane's Equations for the mechanism. In addition two forms of Lagrange's equations are derived from the Kvector representations. The last form uses the concept of potential energy, which is also discussed at some length for the general point mechanism.

Chapter 5 introduces the mechanics of rigid bodies from the point of view of a mechanism with a continuous mass distribution function. The applied forces are also taken in the form of a continuous distribution over the body. Based on Euler's extension of the Newtonian equations of motion, these distributions are used to develop equations for the center of mass and attitude motion of a single rigid body. A natural result of the rigid body constraint is that applied force distributions are equivalent to each other if they only differ by what is called a null force system. This is used to develop the theory of equipollent force systems in a well motivated manner. The moment of inertia dyad is seen to be a natural means for describing the way a mass distribution effects the attitude motion of the rigid body. The theory of the single constrained rigid body is developed using suitable tangent vectors in the space of Kvectors composed of the linear and angular velocity. Finally this is extended to systems composed of multiple rigid bodies.

1.4 Mechanics and Computer Algebra

Until the middle of the twentieth century tedious hand calculations aided by slide rules and mechanical adding machines were the rule for carrying out actual calculations in mechanics. Today it is taken for granted that even extremely complex numerical algorithms can be programmed and run on readily available personal computers. Even so it is still common to exert considerable effort in carrying out algebraic tasks associated with the development and analysis of suitable equations of motion. One method of avoiding such work is to use numerical methods to directly eliminate constraint forces from the equations of motion For really complex problems this will most likely continue to be the case for the immediate future. Another option is to use software dedicated to the symbolic manipulation of mathematical constructs. Software of this type has been around since the early days of computing but it is only in the last decade that it has become easily usable on hardware accessible to students as well as practicing engineers and scientists. In this text the student will be introduced to this subject and to its applications in classical mechanics. As noted in the introduction it is not necessary to have a computer algebra system available to benefit from this material, though of course it is certainly desirable. Sections of each chapter will be devoted to a tutorial in computer algebra and the use of a set of special programmes for the solution of mechanics problems.

Mechanics is enduring! The last generation's mechanics text may appear slightly odd in terms of its examples and the emphasis of its presentation, but its basic content is as meaningful today as it was at the time of writing. In contrast to this we are living in a time of rapid change and development when it comes to the way we approach computing. Today's graphical interfaces and software tools would have the air of magic to the intended audience of that old mechanics book. Therefore as soon as we bring computer algebra into our presentation of mechanics it becomes a certainty that the material will be rapidly outdated. One possible way around this problem would be to treat system in some ideal fashion, i.e. to construct a kind of pseudo-language or code which the reader could translate into the computer algebra system of choice. The choice made here is that it is better to live with the problem of rapid outdating and to use a particular computer algebra system which is now easily available and is able to deal with the typical mechanics problem on modest hardware. To this end we will use the language Maple and we will explain enough of Maple so that it can also be thought of as a means for the general description of the needed algorithms.

One way of using a computer algebra system is as a kind of symbolic calculator. Thus commands and expressions are presented to the system which then carries out appropriate transformations and simplifications. Another approach is to use the languages programming features to write code for the solution of specific problems. A modification of this approach, which fits very nicely into the philosophy of symbolic manipulation systems, is to write programs which can be used as new commands in the language. In effect we extend the language, one might even say we construct a new language designed for the some specific problem domain. This is the viewpoint we shall take, treating Maple as a kind of general language from which we construct a specific language that is suitable for treating problems in mechanics.

A convenient approach is to first see how we go about solving a problem in 'calculator mode', and to then use this experience to construct a suitable set of new commands that make solving the problem a simpler task. In this first chapter we will not write new commands but simply use basic Maple in 'symbolic calculator' mode. If you are fortunate enough to have a running Maple or other computer algebra system you should follow along with the examples modifying them so as to use your particular system. If not you can still follow the logic and general style. In order to help distinguish what is entered by a user, input will be shown in typewriter style, e.g. `this is input` Note that the last sentence does not end with a period. This odd grammatical practice is to make sure that the reader does not mix punctuation with computer input!

1.4.1 The Paraboloid Revisited

Let us now return to the problem of the ball moving on the surface of a paraboloid of revolution. If you have Maple running on your computer you may type in an expression such as:

```
> x1 := q1*cos(q2);
```

$$q1 \cos(q2)$$

The above is the typical result of interaction with the Maple system. The > symbol is the Maple prompt, i.e. the signal that Maple is waiting for input. The input expression follows the standard form used in languages such as Pascal. The symbol `x1` will be replaced by the symbols `q1*cos(q2)`. There is one big difference from what occurs in a programming language oriented toward numerical computation. To appreciate this note that the above is not a code fragment but is something which is actually input into the system and evaluated. In a traditional language we would certainly get some sort of error message informing us that `q1` and `q2` were not assigned, i.e. they have no value. In Maple as in most languages designed for symbolic computation a symbol can evaluate to itself. Indeed this is the essence of symbolic computation. The assignment symbol, := should not be confused with equality =. Thus while we are using the above statement to represent one of the parametric equations for the paraboloid it is not itself an equation but simply means that whenever we use the left side in an expression it will evaluate to the right side. The line below the input line is the actual output that Maple sends to the screen, i.e. the value of the symbol `x1`. Note that the semicolon does not appear in the output line. The reason is that the semicolon is a reserved symbol which tells Maple to evaluate the input and show the result after the user presses the enter key.

The remaining part of the parametric definition is typed in as follows:

```
> x2 := q1*sin(q2):
```

and

```
> x3 := k*q1^2:
```

The alert reader may ask why no output line is shown for these terms. The answer to this is to note that a colon rather than a semicolon was typed in at the end of the input line. The rule is that when an input line is terminated by a colon no output is explicitly shown. On the other hand if we type:

```
> x2;
```

$$q1\ \sin(q2)$$

as expected when a line ends with a semicolon!

To obtain the tangent vectors we use Maples facility to carry out symbolic differentiation. The symbols t11, t12, t13 will be used to store the components of the tangent vector along the direction of the curves obtained from keeping q2 constant. The q1 constant tangent vector's components will be designated as t21, t22 and t23. The differentiation operator takes two arguments, the expression to be differentiated and the symbol of the differentiation variable. This proceeds as follows:

```
>t11 := diff(x1,q1);
```

$$\cos(q2)$$

```
>t12 := diff(x2,q1);
```

$$\sin(q2)$$

```
>t13 := diff(x3,q1);
```

$$2\,k\,q1$$

The components of the other tangent vector are obtained by differentiating with respect to q2. To construct the equations of motion we need the velocity and acceleration components in terms of the time derivatives of the generalized coordinates. Here we run into a slight problem as we have not indicated that q1 and q2 depend on anything. Designate time by the symbol 't' and attempt to apply the differentiation operator:

```
>diff(q1,t);
```

$$0$$

Therefore some way of indicating that q1 depends on the variable t is needed. The standard way of indicating this dependence is to write q1(t) rather than simply q1. We could go back to the beginning of our calculation and do this. Instead we introduce another Maple function, subs, which is used for substitution of a symbol or expression for a specific symbol in an expression. First let us see how this works:

```
>subs(q1=q1(t),x1);
```

$$q1(t)\cos(q2)$$

It is important to note that in the substitution operation we use the equality sign '=' not the assignment operator ':='. To get what we want we must also substitute for the symbol q2. Maple recognizes the double quote symbol as evaluating to the last output expression in a work session. Therefore we can write:

```
>subs(q2=q2(t),");
```

$$q1(t)\cos(q2(t))$$

giving the desired form. This can now be differentiated to give the velocity and acceleration components, e.g.

```
>v1 := diff(",t);
```

$$\left(\frac{d}{dt}q1(t)\right)\cos(q2(t)) - q1(t)\sin(q2(t))\frac{d}{dt}q2(t)$$

Instead of carrying out two separate applications of the substitution operation we can use another feature of Maple, the set. Maple has three constructs that we will make constant use of, the sequence, the set and the list. A sequence is a group of elements separated by commas, thus:

```
> q1, q2, w=r
```

is a sequence with three elements, q1, q2 and the equality w=r. A set in Maple is a sequence enclosed with curly brackets, e.g.

```
>{q1,q2,w=r}
```

The important thing to know about Maple sets is that neither order or duplication is maintained. To see the latter type in a case with duplicate elements:

```
>{x,x,x};
```

$$\{x\}$$

If order and or duplication is important Maple provides the list structure, which is a sequence surrounded by square brackets:

```
>[x,x,x];
```

$$[x, x, x]$$

The important point for now is that the subs operation accepts a sequence, list or set of substitutions instead of a single substitution. When given a sequence of substitutions it proceeds to carry them out from left to right. This means that if a symbol is substituted in one substitution it can be effected by a following substitution in the sequence. When a list or set is used all substitutions are made together, at least logically. To keep track of what we are doing it is reasonable to define a set of substitutions which converts q1 and q2 to q1(t) and q2(t):

```
toTimeFunction := {q1=q1(t),q2=q2(t)};
```

$$\{q1 = q1(t), q2 = q2(t)\}$$

Now it is an easy matter to carry out the differentiation using the command

```
diff( subs(toTimeFunction,x1),t);
```

$$\left(\frac{d}{dt}q1(t)\right)\cos(q2(t)) - q1(t)\sin(q2(t))\frac{d}{dt}q2(t)$$

It can be difficult to read expressions containing explicit functional dependences. Therefore it is useful to provide a means to convert back to the notation which does not show the explicit time dependence. To do this define another substitution set. Because mechanics involves velocities and accelerations it is necessary to include expressions for first and second time derivatives. Indicate a first derivative by appending a t to the q and a second derivative by appending tt. This leads to the substitution set:

```
toTimeExpression :=
{q1(t)=q1,diff(q1(t),t)=q1t,diff(q1(t),t,t)=q1tt,
q2(t)=q2,diff(q2(t),t)=q2t,diff(q2(t),t,t)=q2tt}:
```

Note the use of the diff operator to provide the symbol for the differentiated variable and the use of diff(q1(t),t,t) for the second derivative with respect to t. To complete the arrangements for substitution we now modify the definition of the toTimeFunction expression, thus:

```
>toTimeFunction := {q1=q1(t),q1t=diff(q1(t),t),
q1tt=diff(q1(t),t,t),q2=q2(t),q2t=diff(q2(t),t),
 q2tt=diff(q2(t),t,t)};
```

With this arrangement it is simply a matter of properly arranging the arguments to obtain the desired result, thus for the velocity components:

```
>v1 := subs(toTimeExpression,diff(subs(toTimeFunction,x1),t));
```

$$q1t\,\cos(q2) - q1\,\sin(q2)\,q2t$$

```
>v2 := subs(toTimeExpression,diff(subs(toTimeFunction,x2),t));
```

$$q1t\,\sin(q2) + q1\,\cos(q2)\,q2t$$

```
>v3 := subs(toTimeExpression,diff(subs(toTimeFunction,x3),t));
```

$$2\,kq1\;q1t$$

The same pattern can be followed to obtain the acceleration components, however the reader might like to have some means of avoiding all the repetition involved in the previous calculations. Maple does offer a number of techniques for this and we will explore one of them now. To do this we need two new Maple constructs, the dot operator and the 'for' statement. If we write a symbol 's' followed by a dot followed by another symbol the second symbol is evaluated and the result appended to the first to form a new symbol.

```
> h := 3:
> s := w:
> s.h;
```

$$s3$$

Note that s was not evaluated. If it was evaluated it would have been replaced by w. The symbol h however was evaluated to its value 3. Now that we have the dot operator we can use it in an iterative calculation:

```
>for j from 1 to 3 do a.j:=
subs(toTimeExpression,diff(subs(toTimeFunction,v.j),t)) od:
```

The key words in the construct are for, from, to, do and od. The do-od pair is used to set off the series of statements that are iterated. The iterated index, indicated by j in this case takes on a range starting with 1 and ending with 3. If the step is different from 1 the key word 'by' is used, e.g. for j from 1 by 2 to 10. Output has been suppressed by use of the colon terminator. We can also use 'for' to see the results:

```
>for j from 1 to 3 do a.j od;
```

$$q1tt\;\cos(q2) - 2\,q1t\;\sin(q2)q2t - q1\;\cos(q2)q2t^2 - q1\;\sin(q2)q2tt$$

$$q1tt\;\sin(q2) + 2\,q1t\;\cos(q2)q2t - q1\;\sin(q2)q2t^2 + q1\;\cos(q2)q2tt$$

$$2\,kq1t^2 + 2\,kq1\;q1tt$$

It is left as an exercise to complete the derivation of the equations of motion using the tangent vectors and the applied gravitational force.

If you are using Maple in your work you should now explore the system using the online tutorial and the help facility. If you want help on any particular topic you simply type ?topic, e.g. ?diff or ?for. If possible you should also read the introduction and first chapter in the book 'First Leaves: A Tutorial Introduction To Maple V' by Char and Geddes et. al. The second group of problems below provide some practice in using the techniques that have been discussed above.

1.4.2 Using Lists

The previous example treated each vectorial component as a separate object. It is also possible to use the Maple list structure as a representation for vectors. In fact Maple has another data structure, the array, which is more suitable for this purpose. Lists are very important in symbolic processing and it is worthwhile to examine an application that uses them. Consider the situation shown in figure 1.4, where a mass is suspended at the end of a light rod of length l attached to the edge of a disk of radius s. The rod is pinned so that it rotates in a plane perpendicular to and passing through the center of the disk. The two parametric coordinates that describe the

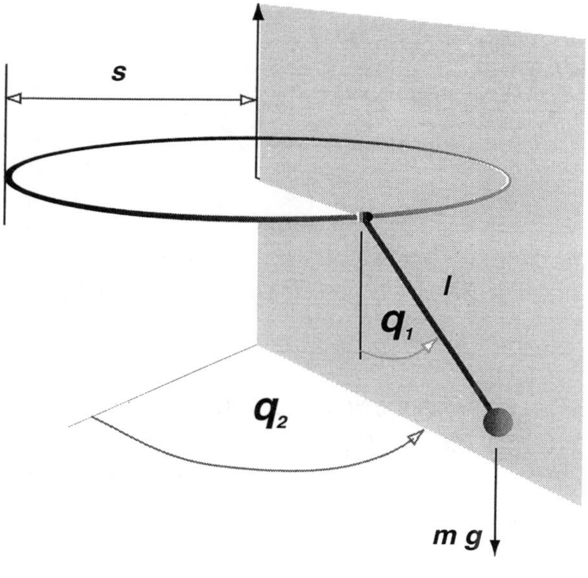

Figure 1.4: Pendulum on Circular Support

possible configurations are taken as the rotation angle of the disk q_1 and the angle of the rod with the vertical q_2, as shown in the figure. Take the origin of a coordinate system fixed in space as the center of the disk, so the coordinates of the mass point are input as:

```
>x := (s+l*sin(q2))*cos(q1):
>y := (s+l*sin(q2))*sin(q1):
>z := -l*cos(q2):
```

Again note that no output is shown since the statements terminate with a colon. A Maple list is simply a sequence surrounded by square brackets, hence the vector from the origin to the mass point can be represented as:

```
>r := [x,y,z]:
```

Following the ideas of the previous section we again set up two substitution sets for the purpose of differentiation and simplified representation of results. In fact the same sets as used above for a set of two parameters can be used for this purpose. The differentiation operator is designed to operate on a single scalar entity. Because operations on structures made up of component parts occur often in Maple a special function has been built into the system for this purpose. It is called *map*. Here we see it in action using the differentiation function to obtain the velocity and acceleration:

```
>rs := subs(toTimeFunction,r):
>vs := map(diff,rs,t):
>v  := subs(toTimeExpression,vs):
>as := map(diff,vs,t):
>a  := subs(toTimeExpression,as):
```

To see the last result explicitly:

```
>a[1];
```

$$l\cos(q2)\,q2tt\,\cos(q1) - l\sin(q2)\,q2t^2\cos(q1) - 2\,l\cos(q2)\,q2t\,\sin(q1)\,q1t \\ -(s + l\sin(q2))\cos(q1)\,q1t^2 - (s + l\sin(q2))\sin(q1)\,q1tt$$

```
>a[2];
```

$$l\cos(q2)\,q2tt\,\sin(q1) - l\sin(q2)\,q2t^2\sin(q1) \\ +2\,l\cos(q2)\,q2t\,\cos(q1)\,q1t - (s + l\sin(q2))\sin(q1)\,q1t^2 \\ +(s + l\sin(q2))\cos(q1)\,q1tt$$

```
>a[3]
```

$$l\cos(q2)\,q2t^2 + l\sin(q2)\,q2tt$$

The selection operation of attaching a number in square brackets to the symbol representing the list to extract an ordered component has been used in the above. Another useful Maple function is the sequence operator, seq. For example the sequence consisting of the first ten odd integers is obtained as follows:

```
>seq(2*'j'+1, 'j'=0..9);
```

$$1, 3, 5, 7, 9, 11, 13, 15, 17, 19$$

To form the equations of motion we need a list of the acceleration components multiplied by the mass. The sequence function can be used to do this as follows:

```
>pt := [ seq(m*a[j],j=1..3)]:
>pt[3]
```

$$m\left(l\cos(q2)\,q2t^2 + l\sin(q2)\,q2tt\right)$$

A somewhat more elegant approach can be carried out using the map function. For this task we need to use one of Maple's techniques to define new functions. Later we will see how to write programs in Maple but for the moment we only use the 'arrow' method of function definition. For example if we want to define a function called square, simply assign the the mapping $x \rightarrow x^2$ to the symbol square, thus:

```
>square := x -> x^2:
>square(b);
```

$$b^2$$

In some cases it is not necessary or desirable to assign the function to a symbol, but simply to use it. An example of this is with map. Consider the following:

```
map(x-> sin(x),[f,g,h]);
```

$$[\sin(f), \sin(g), \sin(h)]$$

This idea shows how we can deal with the acceleration list as an entity. The operation to obtain pt is now:

```
>pt := map(x->m*x,a):
```

It is also convenient to use map to obtain the tangent vectors:

```
>tau1 := map(diff,r,q1);
```

$$[-(s + l\sin(q2))\sin(q1), (s + l\sin(q2))\cos(q1), 0]$$

```
>tau2 := map(diff,r,q2);
```

$$[l\cos(q2)\cos(q1), l\cos(q2)\sin(q1), l\sin(q2)]$$

In order to form the equations of motion we must take the scalar product of the tangent vectors with pt, the rate of change of the momentum vector. We will also call the later the inertial force vector. This can be done in a number of ways, but we will use a method that introduces you to some more features of Maple. The quote symbol, ', is very important in that it prevents evaluation. It provides a way of removing an assignment. For example in the above application of 'for' and 'seq' the index variable j may be left with an assigned value. If used again as an index this will result in an error as the function expects a symbol and if j was assigned a number it will get a number. To avoid this problem one 'unassigns' the symbol by just giving it itself as a value. Thus the statement j:='j' means that j will only evaluate to itself. In the following we will use a combination of the 'sum' function and the 'zip' function. The later is a generalization of the map function. It takes three arguments. The first is a function and the other two are structures such as two equal length lists. The function takes two arguments, one from each list and produces a third list of the results. See if you can follow the code for carrying out the inner product:

```
>j:='j': Fs1:=sum( zip((x,y)->x*y,pt,tau1)[j],j=1..3):
>j:='j': Fs2:=sum( zip((x,y)->x*y,pt,tau2)[j],j=1..3):
```

Now consider another more direct approach to obtain the inner product of the tangent vectors with the applied gravitational force:

```
>j:='j': F1 := sum(Rg[j]*tau1[j],j=1..3):
>j:='j': F2 := sum(Rg[j]*tau2[j],j=1..3):
```

The equations of motion are now given by:

```
>Eq1 := Fs1 = F1:
>Eq2 := Fs2 = F2:
```

Before leaving this problem it is useful to note another Maple function, used for solving equations. The above equations contain acceleration terms involving both coordinate parameters. To use standard Runge-Kutta integration schemes it is helpful to put the equations in the form of each acceleration parameter equaling an expression in the velocity and coordinate parameters. This is accomplished by using the solve function. The arguments are a set of the equations and a set of the parameters which are to be solved for. In the present case this is:

```
>Eqs := solve({Eq1,Eq2},{q1tt,q2tt}):
```

Typical packages for solving differential equations expect the equation or equations to be expressed as a set of first order equations in the form:

$$\frac{dy_j}{dt} = f_j(y_j, t).$$

In agreement with the notation that will be used in the remainder of this text we introduce the *generalized speed* parameters u_j which will be linear combinations of derivatives of the coordinate parameters. In the present case let:

$$\frac{dq_j}{dt} = u_j.$$

Equations that relate the generalized speeds to coordinate derivatives will be call *kinematic differential equations*. The generalized speeds can now be substituted into the two force balance equations so that we arrive at a system of four first order differential equations in a standard form:

```
>Eqs := subs({q1tt=u1t,q2tt=u2t,q1t=u1,q2t=u2},Eqs);
```

$$\begin{aligned} u1t &= -\frac{2\,l\cos(q2)u2\;u1}{s+l\sin(q2)} \\ u2t &= n/d \end{aligned}$$

where

$$\begin{aligned} n = {} & l\cos(q2)\cos(q1)^2\sin(q2)u2^2 + \cos(q2)\cos(q1)^2 u1^2 s \\ & +l\cos(q2)\cos(q1)^2 u1^2\sin(q2) - l\cos(q2)u2^2\sin(q2) \\ & -g\sin(q2) + l\cos(q2)\sin(q1)^2\sin(q2)u2^2 \\ & +\cos(q2)\sin(q1)^2 u1^2 s + l\cos(q2)\sin(q1)^2 u1^2\sin(q2) \end{aligned}$$

$$d = l\left(\cos(q2)^2\sin(q1)^2 + \cos(q2)^2\cos(q1)^2 + \sin(q2)^2\right)$$

This last equation is not in the form that Maple outputs, but has been adjusted for easier reading by defining the numerator and denominator of the first equation as n and d.

If you have access to Maple or an equivalent system you should now consult your manuals and try to repeat the above steps on your own. If you are using another system than Maple it would be an excellent exercise to attempt the above calculation with your particular system. Even if you do not have a computer algebra system it is still worthwhile to examine the above for its algorithmic character. Aside from the practical value of actually using a computer algebra system the expression of a problem's solution in computer algebra terms provides a clear description of just what steps are needed. The solution expressed in this manner must be free of ambiguity as it has to be interpreted by the computer algebra system. In the end the computer algebra system is just a means of implementing these steps and of packaging well know or often repeated components of a calculation. While it may be tedious it is always possible in principle for a trained individual to carry out these steps. On the other hand even the simple calculations shown above indicate that it is quite hard to avoid the introduction of errors in a hand calculation. It is a fact of life that even simple mechanical systems can lead to quite lengthy expressions. It is also true that one should not be too optimistic about the capabilities of computer algebra systems. They are quite complex in themselves and one should always check results against common sense and the examination of simple limiting cases. This problem of verification is of course common to hand calculation and numerical oriented programming as well.

1.5 Problems

•Problem 1.1

A rigid wire is placed in the x, y-plane at an angle θ to the x axis. A bead of mass m is attached to the wire and free to move along it without frictional resistance. There is a uniform gravitational acceleration in the direction of the negative y axis. The intercept point of the wire with the y axis oscillates, so that with y_i denoting the distance of the intercept point from the origin:

$$y_i = y_0 \sin \gamma t.$$

Using the idea of tangency for the elimination of constraint forces, find the equation of motion for the bead. Find the constraint force required to maintain the bead on the wire.

•Problem 1.2

A ball of mass m and radius l is dropped from a large height in a uniform gravitational field of acceleration g. Fluid mechanical theory and experiment indicate that if the ball is moving so that the Reynolds number (a parameter which characterizes the fluid flow) is above 100, the resistance force generated by surrounding fluid will be proportional to $\rho v^2 A$, where ρ is the fluid density, v the relative velocity of the ball to the fluid and A the frontal cross sectional area of the ball. In most cases the proportionality constant is close to one. If the Reynolds number is one or less, the resistance force is given by $6\pi\mu l v$, where μ is the viscosity of the fluid. The Reynolds number is defined as the non-dimensional quantity $\rho v l/\mu$. Discuss the problem of what happens to the ball given the above information. Support the discussion by formulating, and as far as possible solving, suitable equations of motion.

•Problem 1.3

A parabolic shaped rigid wire is forced to oscillate in the $x - y$ plane. At $t = 0$ the equation for the shape of the wire is given by $y = 4x^2$. The y coordinate of the lowest point on the wire is given by $l \sin 5t$. A bead of mass m is allowed to slide freely on the wire, which is in a gravitational field of strength g. Formulate an equation of motion for the bead and discuss the situation in regard to the constraint forces and suitable tangent and normal vectors.

1.5.1 Computer Algebra Problems

•Problem 1.4

Complete the derivation of the equations of motion for the particle in a parabolic bowl using computer algebra techniques. Note that you have to specify another tangent vector and that you must include the gravitational force on the particle.

•Problem 1.5 Carry out the calculations for problems 1.1-1.3 using computer algebra.

•Problem 1.6 Use computer algebra to find an expression for the kinetic and potential energy of the particle in a parabolic bowl using the generalized coordinates introduced in the text. If you know about Lagrange's equations derive the equations of motion from the Lagrangian.

Chapter 2

The Description of Motion

To survive in the 'real world' we must carry a 'model' of it in our minds. Our everyday language reflects this model, which we acquired both as a gift of our evolutionary heritage and from the experience of growing up in our culture. The initial study of mechanics makes some use of this intuitive model's concepts, but at the same time attempts to set up a more restrictive set of meanings for everyday words that we use in describing geometry and motion. The real excitement of the subject is its predictive power in describing the action of forces on the behavior of bodies. For this reason in a first introduction to mechanics one is led quickly past the precise development of concepts such as *space, body, reference frame* and *observer.*

The ability to deal with complex mechanical systems depends on having a very clear idea as to how to describe the configuration of such a system, which in turn requires some attention to the basic concepts used to structure our thoughts about such matters. Clear concepts are not sufficient. We must also be able to express facts about the configuration and motion of mechanical systems in mathematical terms which we can manipulate in convenient ways. The role of proper notation for such tasks can not be overestimated. Therefore a special effort is made to provide the student with a notation for the description of motion that is suggestive and which provides a maximum amount of help in formulating relations between different systems of reference. Patience is needed, as the rewards of seeing how these concepts relate to the computation of motions due to the action of forces on bodies will at times seem distant.

2.1 Space Organization

The mechanics of a *single body* placed in an infinite and empty space might have some interest for a theologian, but very little for a mechanical analyst as no basis would exist to describe the change of the mechanical state of such a *pure* system. Two bodies placed in space leads to a more interesting situation for which it is useful to define some precise terminology. At the start a somewhat formal manner of expressing basic notions is adopted, however this will frequently be relaxed in favor of a more informal mode of description.

Space is a set of points which can be enumerated by use of n-tuples of real numbers. Thus three dimensional space consists of *points* which correspond to triples of the form $(\mathcal{P}_1, \mathcal{P}_2, \mathcal{P}_3)$. This gives space the property of *continuity*. The notion of distance is imparted by adding the structure of 'Euclidian Geometry.' This is done by providing a distance function or metric. If $\mathcal{P}$ denotes a point, the distance between two points will be given by $\| \mathcal{P} - \mathcal{Q} \|$, where for an n-tuple:

$$\| \mathcal{P} - \mathcal{Q} \| = \sqrt{\sum_{k=1}^{n} (\mathcal{P}_k - \mathcal{Q}_k)^2} \tag{2.1}$$

A *body* is a connected set of points in three dimensional space, that is any point in the body can be reached from any other point in the body by traversing a continuous path made up only of body points. At this stage nothing of a physical or mechanical nature is assumed about a body. The question is simply which set of points is to be identified as a body. Time and the related idea of change does not arise at this level of the theories structure.

Given any body, construct an *extended body* consisting of the entire space of points, and call such an extended body an *observer*. Now suppose there are two bodies in space, which are labeled as body A and B. Extend A into an observer. It is now possible to define the motion of B relative to the observer A. To do this introduce a scalar parameter, which in some cases may be the time, denoted by t. Other parameters such as angles may also be used. To emphasize this 'neutrality of function the symbol q is used for this purpose. At some fixed q, it is possible to describe the configuration of B to the observer A as a connected subset of the extended points of A. A *motion* then consists of a continuous mapping, as a function of q, of this subset of the observer A into another subset of the observer A. Thus the observer A characterizes the motion of B by the changing the set of points traversed in the extended body.

For the most part classical mechanics is concerned with a special class of bodies which are either single points or so called *rigid* bodies. The latter are bodies for which the distance between *any two points* in the body is invariant, i.e. preserved, during a motion. Observers are now assumed to be based on rigid bodies.

The idea here is that an observer considers itself to be a fixed entity watching the activity of the bodies that move about it. To quantify this activity the observer must employ measurement tools. The abstract model for the organization of the world requires mathematical constructs to represent these tools. To this end use is made of vectors. It is assumed that the reader is familiar with vectors, and operations such as dot and cross products, as working tools from basic mechanics courses.

A *reference triad* consists of three non co-planer vectors and a rigid body. The directions, but not the point of origin of these vectors are assumed fixed with respect to the rigid body. Thus the vectors comprising the triad are *free*. If a particular one of the vectors is represented by a line from a point $\mathcal{P}$ to a point $\mathcal{Q}$ for an observer A, any vector with the same distance $\|\mathcal{P}-\mathcal{Q}\|$ and parallel to that line will be considered equivalent. The *combination of a fixed point in A and a triad is called a reference frame.* For the treatment of the motions of point and rigid body masses it is frequently convenient to use orthonormal vectors as a reference triad. Such a triad is a *standard* reference triad, and a reference frame using a standard triad is a *standard* reference frame.

A special class of standard reference frames, inertial frames, are needed for the equations of Newtonian motion. In discussing the basic organization of space it is desirable to retain total equivalence among standard reference frames. None is better than another! A *standard observer* is taken as an observer based on any rigid body and equipped with a standard reference frame. A mathematical framework is needed for communicating information between observers based on different frames.

The basic problem is how to relate the observations between standard observers. These observations will involve the motions of bodies as noted by each standard observer. For example suppose standard observer A notes a *fixed point* $\mathcal{Q}$ in his standard reference frame. In terms of an orthonormal set of vectors, $\boldsymbol{a}_k$ and a fixed point $\mathcal{O}$ which comprise the standard reference frame of A, such a point would be given by a vector:

$$\boldsymbol{r}^{\mathcal{OQ}} = r_1\mathbf{a}_1 + r_2\mathbf{a}_2 + r_3\mathbf{a}_3. \tag{2.2}$$

Since it is taken that the point $\mathcal{Q}$ is fixed in standard reference frame A, the quantities r_k will be fixed numbers, hence their derivatives with respect to time or some other parameter will vanish. An observer based on a body B will have a standard reference frame based on a triad $\boldsymbol{b}_k$. For this observer:

$$\boldsymbol{r}^{\mathcal{OQ}} = s_1\mathbf{b}_1 + s_2\mathbf{b}_2 + s_3\mathbf{b}_3. \tag{2.3}$$

In general observer B will consider observer A to be in motion, hence the quantities s_k will not be constants. Therefore the derivative of the vector $\boldsymbol{r}^{\mathcal{OQ}}$ will depend on which observer is calculating it! A special point of this has been made by Kane, who suggests using a notation that indicates that the derivative of a vector depends on the observer. Thus the following definition is used for the derivative of a vector quantity $\boldsymbol{w}$:

$$\frac{{}^A d\boldsymbol{w}}{dq} = \frac{dw_1}{dq}\mathbf{a}_1 + \frac{dw_2}{dq}\mathbf{a}_2 + \frac{dw_3}{dq}\mathbf{a}_3. \tag{2.4}$$

Notice that taking a derivative of a vector requires expressing that vector in a standard reference frame. This is indicated by the left superscript on the derivative symbol. If the quantities involved are considered to be functions of more than one parameter, partial differential symbols should be used as in:

$$\frac{{}^A \partial\boldsymbol{w}}{\partial q} = \frac{\partial w_1}{\partial q}\mathbf{a}_1 + \frac{\partial w_2}{\partial q}\mathbf{a}_2 + \frac{\partial w_3}{\partial q}\mathbf{a}_3. \tag{2.5}$$

It is left as a problem for the reader to show that if $\boldsymbol{a}_k$ and $\hat{\boldsymbol{a}}_k$ are reference triads attached to the *same body* then:

$$\frac{dw_1}{dq}\mathbf{a}_1 + \frac{dw_2}{dq}\mathbf{a}_2 + \frac{dw_3}{dq}\mathbf{a}_3 = \frac{dw_1}{dq}\hat{\mathbf{a}}_1 + \frac{dw_2}{dq}\hat{\mathbf{a}}_2 + \frac{dw_3}{dq}\hat{\mathbf{a}}_3, \tag{2.6}$$

which permits using the same symbol as a left superscript for all reference triads attached to A. The computation of the rate of change of the distance vector between two points fixed in one body as viewed by an observer attached to another body shows that in general:

$$\frac{{}^{A}d\boldsymbol{w}}{dq} \neq \frac{{}^{B}d\boldsymbol{w}}{dq}. \tag{2.7}$$

The interesting question is what is the correct relation between derivatives of a vector taken with respect to different standard reference frames. To answer this additional mathematical tools are needed.

•Illustration

Before continuing with the development of a formal apparatus for the taking of derivatives of vectors in different reference frames, it is worthwhile to examine a simple situation. Our illustration shows that in the Euclidian space of classical mechanics the need for this apparatus is connected with the concept of 'rotation'. For simplicity we look at a two dimensional situation. An observer A has a reference set of orthonormal vectors $\boldsymbol{a}_1$ and $\boldsymbol{a}_2$. Sketches containing these vectors must display

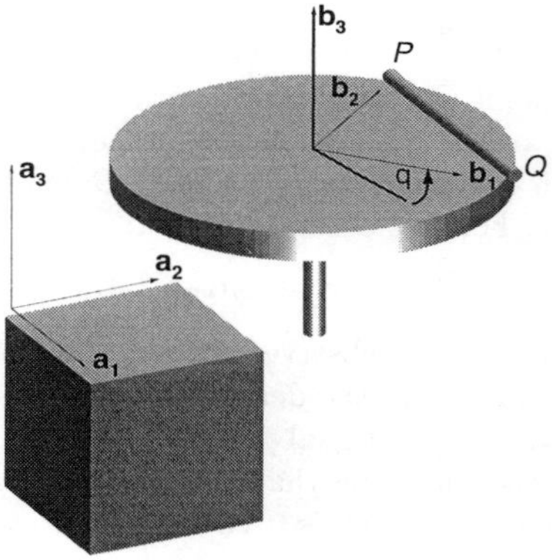

Figure 2.1: Derivatives of vectors in different frames.

them at some particular place. It is conventional to choose a point which is associated as the origin of some standard reference frame. In the theory these vectors are *free*, i.e. they can be anywhere in space, and need not be together. The idea of a distinguished point or origin is distinct from the concept of these triad vectors! These are shown together at the origin of observer A. Another observer, B, is attached to a circular table which is free to rotate with respect to observer A. The angle of rotation q is measured as the angle going from the vector $\boldsymbol{a}_1$ to the vector $\boldsymbol{b}_1$. The vectors $\boldsymbol{b}_1$ and $\boldsymbol{b}_2$ are orthonormal, and possess fixed orientations in the frame of observer B. The points $\mathcal{P}$ and $\mathcal{Q}$ are taken as fixed in the table and located at unit distance from its center and in the directions of $\boldsymbol{b}_2$ and $\boldsymbol{b}_1$ respectively. Therefore the vector representing the displacement between these points can be expressed as:

$$\boldsymbol{r}^{\mathcal{PQ}} = \boldsymbol{b}_1 - \boldsymbol{b}_2. \tag{2.8}$$

It is clear that this vector does not depend on the angle q, and hence its derivative with respect to q will certainly vanish. This is expressed in more formal notation by the statement:

$$\frac{{}^{B}d\boldsymbol{r}^{\mathcal{PQ}}}{dq} = 0. \tag{2.9}$$

The rule for computing the rate of change of $\boldsymbol{r}^{\mathcal{PQ}}$ with respect to q for observer A requires expressing this vector in terms of the frame vectors $\boldsymbol{a}_j$. This is accomplished by finding expressions for both the vectors $\boldsymbol{b}_j$ in terms of $\boldsymbol{a}_j$ and substituting into the expression for $\boldsymbol{r}^{\mathcal{PQ}}$ given in terms of $\boldsymbol{b}_j$. To visualize this computation it is useful to draw all the frame vectors as emerging from a common point so as to easily see the angular relationships. From such a diagram:

$$\boldsymbol{b}_1 = \cos q \boldsymbol{a}_1 + \sin q \boldsymbol{a}_2, \tag{2.10}$$

and

$$\boldsymbol{b}_2 = -\sin q \boldsymbol{a}_1 + \cos q \boldsymbol{a}_2. \tag{2.11}$$

It is common practice in mechanics to abbreviate terms such as $\sin q$ and $\cos q$ by s and c. When several variables are involved, such as q_1 and q_2 one uses abbreviations such as s_1 and s_2. Substituting these expressions for $\boldsymbol{b}_j$ in terms of $\boldsymbol{a}_j$ shows that:

$$\boldsymbol{r}^{\mathcal{PQ}} = \boldsymbol{b}_1 - \boldsymbol{b}_2 = (c\boldsymbol{a}_1 + s\boldsymbol{a}_2) - (-s\boldsymbol{a}_1 + c\boldsymbol{a}_2) = (s+c)\boldsymbol{a}_1 + (s-c)\boldsymbol{a}_2. \tag{2.12}$$

It is now possible to compute the derivative of this vector with respect to q referred to frame A. Thus:

$$\frac{{}^{A}d\boldsymbol{r}^{\mathcal{PQ}}}{dq} = (c-s)\boldsymbol{a}_1 + (c+s)\boldsymbol{a}_2. \tag{2.13}$$

It is often necessary to express the rate of change of a vector referred to a frame such as A in terms of the frame vectors of B. This is easily done if one realizes that all that is needed is to now express the frame vectors $\boldsymbol{a}_j$ in terms of the frame vectors $\boldsymbol{b}_j$. Thus :

$$\boldsymbol{a}_1 = c\boldsymbol{b}_1 - s\boldsymbol{b}_2 \tag{2.14}$$

and

$$\boldsymbol{a}_2 = s\boldsymbol{b}_1 + c\boldsymbol{b}_2. \tag{2.15}$$

Using these relations we see that:

$$\frac{{}^{A}d\boldsymbol{r}^{\mathcal{PQ}}}{dq} = (c-s)(c\boldsymbol{b}_1 - s\boldsymbol{b}_2) + (c+s)(s\boldsymbol{b}_1 + c\boldsymbol{b}_2) = \boldsymbol{b}_1 + \boldsymbol{b}_2, \tag{2.16}$$

where the dependence on q is implicit. One might hope for computational techniques that would eliminate many of the above steps. In fact there are such, which are developed in the remainder of this chapter. •

2.2 Standard Triads

The crucial step in solving this problem is to realize that any given reference triad can be expressed in terms of any other reference triad. As a reference triad consists of *free vectors* this relation will only involve the angles between vectors. Thus consider two reference triads $\boldsymbol{a}_k$ and $\boldsymbol{b}_k$, and recall that each one is orthonormal, providing the relation:

$$\boldsymbol{a}_i \cdot \boldsymbol{a}_j = \delta_{ij} \tag{2.17}$$

with $\delta_{ij} = 0$ or 1 as $i \neq j$ or $i = j$. A similar relation also holds for $\boldsymbol{b}_k$. Any given vector $\boldsymbol{w}$ can be expressed as a linear combination of the unit vectors, that is as:

$$\mathbf{w} = w_1\mathbf{a}_1 + w_2\mathbf{a}_2 + w_3\mathbf{a}_3. \tag{2.18}$$

The orthonormality of the standard triad makes it easy to find the coefficients in this expansion. Thus dot any one of the triad vectors into both sides of the above equation and use orthonormality to show that

$$w_j = \boldsymbol{a}_j \cdot \boldsymbol{w}. \tag{2.19}$$

Therefore in terms of a standard triad we have the important relation:

$$\boldsymbol{w} = (\boldsymbol{a}_1\boldsymbol{a}_1 + \boldsymbol{a}_2\boldsymbol{a}_2 + \boldsymbol{a}_3\boldsymbol{a}_3) \cdot \boldsymbol{w} \tag{2.20}$$

We can use the terms in parentheses on the right side of equation 2.20 to suggest a useful definition of the product of two vectors. Thus the *dyadic* or *outer* product of two vectors will simply be indicated by placing the two vectors side by side with the understanding that taking a dot product with another vector from the left or right will produce a vector. For example:

$$\boldsymbol{u} \cdot (\boldsymbol{v}\boldsymbol{w}) = (\boldsymbol{u} \cdot \boldsymbol{v})\boldsymbol{w} \tag{2.21}$$

and

$$(\boldsymbol{v}\boldsymbol{w}) \cdot \boldsymbol{u} = (\boldsymbol{u} \cdot \boldsymbol{w})\boldsymbol{v}. \tag{2.22}$$

With this understanding we see that if $\boldsymbol{a}_k$ are a set of orthonormal vectors then

$$\boldsymbol{a}_1\boldsymbol{a}_1 + \boldsymbol{a}_2\boldsymbol{a}_2 + \boldsymbol{a}_3\boldsymbol{a}_3 \tag{2.23}$$

when dotted into any vector produces an expanded form of the same vector, i.e. this *unit dyad* acts as an identity. Dyads act on vectors by taking the dot product with the vector. This operation can take place with the vector on the left or right side of the dyad, and in general the results will differ, i.e. if $\boldsymbol{D}$ is a dyad then

$$\boldsymbol{u} \cdot \boldsymbol{D} \neq \boldsymbol{D} \cdot \boldsymbol{u} \tag{2.24}$$

The general dyadic in three dimensional Euclidian space can be expanded using a standard triad in the form:

$$\boldsymbol{D} = \sum_{i,j=1}^{i,j=3} D_{ij} \boldsymbol{a}_i \boldsymbol{a}_j. \tag{2.25}$$

Note our convention that vectors are represented as bold lower case letters, dyads as bold capital letters.

The above shows that once a standard triad is selected, one may obtain representations of vectors and dyads in terms of expansion coefficients. Often it is convenient to present these representations in the form of matrices, e.g.

$$\boldsymbol{D} \doteq \begin{bmatrix} D_{11} & D_{12} & D_{13} \\ D_{21} & D_{22} & D_{23} \\ D_{31} & D_{32} & D_{33} \end{bmatrix}. \tag{2.26}$$

The symbol $\doteq$ is used to indicate that the matrix on the right is a *representation* of the dyad and that the right side is not literally equal to the dyad. Many mechanics texts deal with this situation by defining a dyad as the class of all possible arrays that represent it under orthogonal transformations. Taking the dot product from the left and right with the standard reference frame vectors $\boldsymbol{a}_k$ and $\boldsymbol{a}_m$ shows that

$$D_{km} = \boldsymbol{a}_k \cdot \boldsymbol{D} \cdot \boldsymbol{a}_m. \tag{2.27}$$

The components of vectors with respect to a standard triad can also be represented in terms of matrix notation. The convention is adopted of using a column matrix in the form:

$$\boldsymbol{w} \doteq \begin{bmatrix} w_1 \\ w_2 \\ w_3 \end{bmatrix} \tag{2.28}$$

to represent the components of a vector with respect to a reference triad.

As there can be many representations for the same vector or dyad, they must be distinguished. If it is necessary to specify the standard triad for a representation it is done by using an upper left sided super script. Thus

$$\boldsymbol{w} \doteq \begin{bmatrix} {}^a w_1 \\ {}^a w_2 \\ {}^a w_3 \end{bmatrix} \tag{2.29}$$

stands for the column matrix representation of the vector $\boldsymbol{w}$ in the standard triad $\boldsymbol{a}_k$. In a similar manner the dyad, $\boldsymbol{D}$ would have the representation, ${}^a D_{ij}$. As far as possible this convention is respected. If a capital Latin letter, such as A, represents a body, the same lower case letter represents a standard triad. Different triads fixed in the same body may be indicated by primes, i.e. a, a', a'' represent triads fixed in the body A.

A very useful notation can be developed by the use of matrices with vector entries. To keep the notation from becoming too complicated indicate such a matrix representing a frame triad by Latin lower case letters. Thus the standard triad a indicates the *column* matrix

$$a = \begin{bmatrix} \boldsymbol{a}_1 \\ \boldsymbol{a}_2 \\ \boldsymbol{a}_3 \end{bmatrix}. \tag{2.30}$$

A right superscript T will indicate the *transpose*, so that the row form is given by

$$a^T = \begin{bmatrix} \boldsymbol{a}_1 & \boldsymbol{a}_2 & \boldsymbol{a}_3 \end{bmatrix}. \tag{2.31}$$

This notation, together with the normal rules of matrix multiplication, gives us a convenient way of writing vector equations. A vector $\boldsymbol{w}$ now can be given in terms of a standard triad in the two forms:

$$\boldsymbol{w} = {}^a w^T a = a^T\, {}^a w, \tag{2.32}$$

both of which give

$$\boldsymbol{w} = {}^a w_1 \mathbf{a}_1 + {}^a w_2 \mathbf{a}_2 + {}^a w_3 \mathbf{a}_3. \tag{2.33}$$

The unit dyad can be expressed as

$$\boldsymbol{U} = a^T a \tag{2.34}$$

and a general dyad has the form

$$\boldsymbol{D} = a^T\, {}^a D a, \tag{2.35}$$

which the reader should verify.

•Illustration

A situation that frequently arises in mechanics is the need to project a vector into parts parallel and perpendicular to a given plane. A simple example of this is if we are given a vector $\boldsymbol{w}$ in terms of a standard triad, a. Then the vector $w_1\boldsymbol{a}_1 + w_2\boldsymbol{a}_2$, which is the projection of the vector onto the plane parallel to the vectors $\boldsymbol{a}_1$ and $\boldsymbol{a}_2$, provides one part of the decomposition. The other part is of course the vector $w_3\boldsymbol{a}_3$. We can think of the dyad:

$$\Pi_{\boldsymbol{a}_1\boldsymbol{a}_2} = \boldsymbol{a}_1\boldsymbol{a}_1 + \boldsymbol{a}_2\boldsymbol{a}_2 \tag{2.36}$$

as a *projection* operator onto the plane of $\boldsymbol{a}_1$ and $\boldsymbol{a}_2$. Now suppose we have two arbitrary but non-coplanar vectors, $\boldsymbol{p}_1$ and $\boldsymbol{p}_2$, and we wish to find the projection of a third arbitrary vector, $\boldsymbol{w}$ onto the plane of these two vectors. Again we can form a projection operator as a dyad. Here is one way to proceed. Take one of the vectors, $\boldsymbol{p}_1$, and form a unit vector:

$$\boldsymbol{k}_1 = \boldsymbol{p}_1/|\boldsymbol{p}_1|, \tag{2.37}$$

where $|\boldsymbol{p}_1| = \sqrt{\boldsymbol{p}_1 \cdot \boldsymbol{p}_1}$. We can obtain a unit vector that is orthogonal to $\boldsymbol{k}_1$ and in the plane of $\boldsymbol{p}_1$ and $\boldsymbol{p}_2$, by subtracting the component of $\boldsymbol{p}_2$ in the direction of $\boldsymbol{k}_1$ and dividing the result by its magnitude. Thus let:

$$\boldsymbol{k}_2 = (\boldsymbol{p}_2 - (\boldsymbol{p}_2 \cdot \boldsymbol{k}_1)\boldsymbol{k}_1)/|(\boldsymbol{p}_2 - (\boldsymbol{p}_2 \cdot \boldsymbol{k}_1)\boldsymbol{k}_1)|. \tag{2.38}$$

The dyad

$$\Pi_{\boldsymbol{k}_1\boldsymbol{k}_2} = \boldsymbol{k}_1\boldsymbol{k}_1 + \boldsymbol{k}_2\boldsymbol{k}_2 \tag{2.39}$$

provides the desired projection operator. Note the easily verified property of projection operators that if:

$$\Pi \cdot \boldsymbol{w} = \boldsymbol{w}_{\|}, \tag{2.40}$$

then

$$\Pi \cdot \boldsymbol{w}_{\|} = \boldsymbol{w}_{\|}. \tag{2.41}$$

In three dimensions the vector cross product operation may be used to find projections onto a plane perpendicular to a given vector. The above approach has the advantage of generalization to spaces of higher dimensions. Projection operators are used in the discussion of equivalent force systems, moments of inertia of rigid bodies and methods for explicit determination of constraint forces. •

2.2.1 Rotation Dyads

A linear transformation of a vector which preserves the vector's magnitude is called a *rotation*. Dyads provide a convenient means of characterizing such transformations. A conventional approach to this is to develop the consequences of magnitude invariance on a general linear transformation. Instead of this we will develop the subject from our intuitive grasp of rotation.

The simplest situation of any interest is the rotation of a vector constrained to a plane. Assume we have a vector $\boldsymbol{w}$ which is parallel to the plane of the triad vectors $\boldsymbol{a}_1$ and $\boldsymbol{a}_2$. Now obtain a new vector by rotating the direction of this vector about the axis determined by $\boldsymbol{a}_3$. Let the angle of rotation be θ and the direction be positive, i.e. in the sense of the right hands fingers when wrapped about the axis with the thumb pointing in the positive direction. Our task is to find a dyad which carries out this transformation. To set the stage for what is to follow we define the rotation by introducing a new triad $b^T = [\boldsymbol{b}_1, \boldsymbol{b}_2, \boldsymbol{b}_3]$, where $\boldsymbol{a}_3 = \boldsymbol{b}_3$. The vector $\boldsymbol{b}_1$ is taken to be at a positive angle θ to $\boldsymbol{a}_1$. Therefore the orthonormal unit vectors $\boldsymbol{b}_1$ and $\boldsymbol{b}_2$ are given in terms of the unrotated frame vectors $\boldsymbol{a}_j$ by the relations

$$\boldsymbol{b}_1 = \boldsymbol{a}_1 \cos\theta + \boldsymbol{a}_2 \sin\theta \tag{2.42}$$

$$\boldsymbol{b}_2 = -\boldsymbol{a}_1 \sin\theta + \boldsymbol{a}_2 \cos\theta. \tag{2.43}$$

With a little thought you should see that the rotation operation is the same as assigning the vectors components with respect to the a triad to the b triad. Therefore

$$\boldsymbol{R}_\theta \boldsymbol{w} = \boldsymbol{b}_1 {}^a w_1 + \boldsymbol{b}_2 {}^a w_2, \tag{2.44}$$

with

$$^{a}w_j = \boldsymbol{a}_j \cdot \boldsymbol{w}. \tag{2.45}$$

This allows us to express the rotation as the dyad

$$\boldsymbol{R}_\theta = \boldsymbol{b}_1\boldsymbol{a}_1 + \boldsymbol{b}_2\boldsymbol{a}_2. \tag{2.46}$$

As the components are the same as in the a triad, it should be evident that the magnitude of the vector obtained by this transformation is unchanged, fitting the original requirement for a rotation! Substitution of the above expressions for $\boldsymbol{b}_j$ in terms of $\boldsymbol{a}_j$ provides an expression for the dyad in the a representation, thus

$$\boldsymbol{R}_\theta = (\boldsymbol{a}_1\boldsymbol{a}_1 + \boldsymbol{a}_2\boldsymbol{a}_2)\cos\theta + (\boldsymbol{a}_2\boldsymbol{a}_1 - \boldsymbol{a}_1\boldsymbol{a}_2)\sin\theta \tag{2.47}$$

This result is considerably more general then it appears here. The general rotation operator can be expressed in the form:

$$\boldsymbol{R}_{a\to b} = \sum_{j=1}^{3} \boldsymbol{b}_j\boldsymbol{a}_j, \tag{2.48}$$

which represents a rotation into the frame of triad b, i.e. the components of the new vector in frame b are the same as the components of the original vector in frame a. It should be obvious that the magnitude is preserved, since the components are the same in the new orthogonal system! The above case appears special in that the rotation was about the $\boldsymbol{a}_3$ axis. Intuitively one might expect that a general rotation could be represented by a rotation of some angle about some axis. Now let this axis be identified with the vector $\boldsymbol{a}_3$. To emphasize that we are dealing with a more general situation it is convenient to also identify this unit vector with a vector $\boldsymbol{\lambda}$, called the rotation axis vector. If the rotation is about this axis with an angle θ we can use the above formula and the fact that $\boldsymbol{a}_3 = \boldsymbol{b}_3 = \boldsymbol{\lambda}$. This gives the rotation dyad as

$$\boldsymbol{R}_{a\to b} = \boldsymbol{b}_1\boldsymbol{a}_1 + \boldsymbol{b}_2\boldsymbol{a}_2 + \boldsymbol{b}_3\boldsymbol{a}_3, \tag{2.49}$$

which can now be written as

$$\boldsymbol{R}_{a\to b} = (\boldsymbol{a}_1\boldsymbol{a}_1 + \boldsymbol{a}_2\boldsymbol{a}_2)\cos\theta + (\boldsymbol{a}_2\boldsymbol{a}_1 - \boldsymbol{a}_1\boldsymbol{a}_2)\sin\theta + \boldsymbol{a}_3\boldsymbol{a}_3. \tag{2.50}$$

The coefficient of the $\cos\theta$ term is almost in the form of a unit dyad. Therefore it is natural to add and subtract the term $\boldsymbol{a}_3\boldsymbol{a}_3\cos\theta$ from the equation. This allows us to write

$$\boldsymbol{R}_{a\to b} = \boldsymbol{U}\cos\theta + (1-\cos\theta)\boldsymbol{\lambda\lambda} + (\boldsymbol{a}_2\boldsymbol{a}_1 - \boldsymbol{a}_1\boldsymbol{a}_2)\sin\theta. \tag{2.51}$$

This latter expression is almost in a form which is in terms of an arbitrary unit rotation direction vector $\boldsymbol{\lambda}$, the rotation angle and the unit dyad. To obtain a fully frame invariant form for the rotation dyad it is necessary to exam the effect of the last term on an arbitrary vector $\boldsymbol{v}$. Thus note that

$$(\boldsymbol{a}_2\boldsymbol{a}_1 - \boldsymbol{a}_1\boldsymbol{a}_2)\cdot\boldsymbol{v} = \boldsymbol{a}_2\,{}^{a}v_1 - \boldsymbol{a}_1\,{}^{a}v_2. \tag{2.52}$$

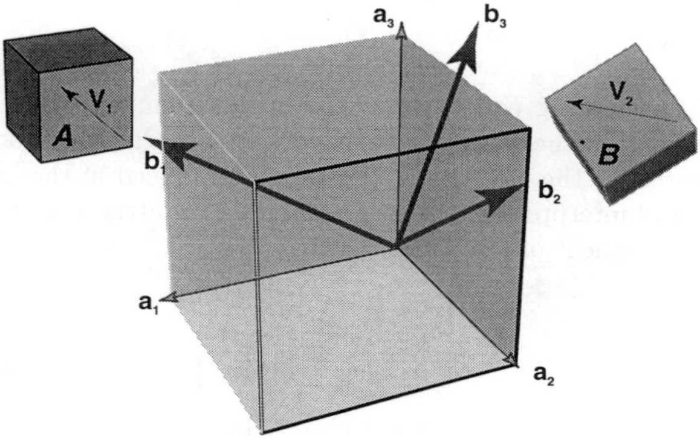

(b1 a1 + b2 a2 + b3 a3) • V1 ⟶ V2

Figure 2.2: Rotation Operator

It is easy to verify that this is the same as the vector cross product $\boldsymbol{\lambda} \times \boldsymbol{v}$. We can use the unit dyad to put this into an operator form, thus

$$\boldsymbol{\lambda} \times \boldsymbol{v} = \boldsymbol{\lambda} \times (\boldsymbol{U} \cdot \boldsymbol{v}) = (\boldsymbol{\lambda} \times \boldsymbol{U}) \cdot \boldsymbol{v}. \tag{2.53}$$

The cross product acts on the dyad by taking the cross product with each left hand vector in the dyadic representation, e.g.

$$\boldsymbol{\lambda} \times (\boldsymbol{u}\boldsymbol{v}) = (\boldsymbol{\lambda} \times \boldsymbol{u})\boldsymbol{v}. \tag{2.54}$$

Therefore we can now write the rotation dyad as

$$\boldsymbol{R}_{a \to b} = \boldsymbol{U} \cos\theta + (1 - \cos\theta)\boldsymbol{\lambda}\boldsymbol{\lambda} + \boldsymbol{\lambda} \times \boldsymbol{U} \sin\theta. \tag{2.55}$$

This is a form that depends only on the vector λ and the rotation angle θ. A representation with respect to any triad can easily be obtained by expressing $\boldsymbol{\lambda}$ in terms of that triad. Therefore it is reasonable to also use the notation $\boldsymbol{R}_{(\lambda,\theta)}$ to represent the rotation dyad.

Finally the reader should verify that $\boldsymbol{R}_{(\lambda,\theta)} \cdot \boldsymbol{\lambda} = \boldsymbol{\lambda}$. This demonstrates that the rotation operation leaves any vector parallel to the rotation axis in its original orientation. In other words $\boldsymbol{\lambda}$ is an *eigenvector* of the rotation operator with eigenvalue 1. Examination of the matrix representation of the rotation operator in a frame with $\boldsymbol{\lambda}$ aligned in the direction of one of the triad vectors shows that this is the only real eigenvalue.

The rotation dyad provides all of the important properties of a rotation in space in a relatively transparent form. It is directly related to another, somewhat more common, representation of rotations introduced in the following section.

2.3 Observer Relations and Direction Cosine Matrices

The process of formulating equations of motion for complex systems requires the ability to relate motions between observers attached to different bodies. This task is greatly facilitated by the notational structures introduced in the last section. For what follows we will interpret the effect of a Dyad on a matrix of vectors as the Dyad operating on each element of the matrix. Thus

$$\boldsymbol{D} \cdot \begin{bmatrix} \boldsymbol{a}_1 \\ \boldsymbol{a}_2 \\ \boldsymbol{a}_3 \end{bmatrix} = \begin{bmatrix} \boldsymbol{D} \cdot \boldsymbol{a}_1 \\ \boldsymbol{D} \cdot \boldsymbol{a}_2 \\ \boldsymbol{D} \cdot \boldsymbol{a}_3 \end{bmatrix} \tag{2.56}$$

Consider two standard triads, a and b, attached to bodies A and B respectively. Using the properties of the unit dyad we obtain the relations:

$$b = b \cdot \boldsymbol{U} = b \cdot (a^T a) = (b \cdot a^T) a \tag{2.57}$$

where the last grouping of parentheses indicates an *outer* matrix product, i.e. the product of column and row results in a 3 by 3 matrix. The elements, as indicated, consist of vector dot products, hence are scalar quantities, thus:

$$b \cdot a^T = \begin{bmatrix} \boldsymbol{b}_1 \\ \boldsymbol{b}_2 \\ \boldsymbol{b}_3 \end{bmatrix} \cdot \begin{bmatrix} \boldsymbol{a}_1 & \boldsymbol{a}_2 & \boldsymbol{a}_3 \end{bmatrix} = \begin{bmatrix} \boldsymbol{b}_1 \cdot \boldsymbol{a}_1 & \boldsymbol{b}_1 \cdot \boldsymbol{a}_2 & \boldsymbol{b}_1 \cdot \boldsymbol{a}_3 \\ \boldsymbol{b}_2 \cdot \boldsymbol{a}_1 & \boldsymbol{b}_2 \cdot \boldsymbol{a}_2 & \boldsymbol{b}_2 \cdot \boldsymbol{a}_3 \\ \boldsymbol{b}_3 \cdot \boldsymbol{a}_1 & \boldsymbol{b}_3 \cdot \boldsymbol{a}_2 & \boldsymbol{b}_3 \cdot \boldsymbol{a}_3 \end{bmatrix} \tag{2.58}$$

This formal calculation has a very simple geometric interpretation in that each row of equation 2.58 represents the components of one of the b triad vectors in terms of the a triad. This *direction cosine matrix* will be indicated by the notation:

$$R_{ba} = b \cdot a^T, \tag{2.59}$$

where the subscripts match up with the way the transformation runs, thus R_{ba} represents the b triad in terms of the a triad. Clearly we also have the relation

$$R_{ab} = a \cdot b^T \tag{2.60}$$

representing the a triad in terms of the b triad. These relations are easy to recall, simply note the placement of the subscripts on R and observe that, to obtain a matrix result from the product, it is necessary to have the form 'column' · 'row'.

In what follows extensive use is made of the property of matrix products that

$$(AB)^T = B^T A^T. \tag{2.61}$$

An important property of R_{ab} is seen by taking its transpose:

$$R_{ab}^T = (a \cdot b^T)^T = b \cdot a^T = R_{ba}. \tag{2.62}$$

Thus the transpose of R simply represents the inverse transformation between representations! We also have the relation:

$$R_{ab}R_{ba} = a \cdot b^T b \cdot a^T = a \cdot (b^T b) \cdot a^T. \tag{2.63}$$

The central grouping in the above represents the unit dyad, hence

$$R_{ab}R_{ba} = a \cdot \boldsymbol{U} \cdot a^T = a \cdot a^T = \begin{bmatrix} 1 & 0 & 0 \\ 0 & 1 & 0 \\ 0 & 0 & 1 \end{bmatrix} \tag{2.64}$$

Representing the unit matrix as $\mathbf{1}$ we have the result that

$$RR^T = \mathbf{1} \tag{2.65}$$

for *any* direction cosine matrix between standard triads. This shows that

$$R^{-1} = R^T, \tag{2.66}$$

with R^{-1} indicating the inverse matrix. Matrices with the property $RR^T = \mathbf{1}$ are called orthogonal. The above shows that direction cosine matrices are orthogonal, so that the inverse equals the transpose, or in other words the product of such a matrix with its transpose yields the unit matrix.

We are now in a position to easily relate the representations of vectors and dyads relative to different standard reference triads. For any vector:

$$\boldsymbol{w} = {}^a w^T a = {}^b w^T b. \tag{2.67}$$

Now use the relation that $b = R_{ba}a$ to show that

$$\boldsymbol{w} = {}^b w^T R_{ba} a \tag{2.68}$$

from which we can see by inspection that

$${}^a w^T = {}^b w^T R_{ba} \tag{2.69}$$

or from the transpose relation

$${}^a w = R_{ab} {}^b w. \tag{2.70}$$

It is left as an exercise to show that for dyads one has the relation

$${}^a D = R_{ab} {}^b D R_{ba}. \tag{2.71}$$

All of these relations are quite easy to recall if one notes the logical ordering of subscripts.

The direction cosine matrix is simply related to the rotation operator introduced in the last section. This is shown by an easy calculation using the notation introduced above. First express the rotation dyad in the a representation, thus:

$$\boldsymbol{R}_{a\to b} = a^T ({}^aR_{a\to b})a. \tag{2.72}$$

If we dot both sides of this equation from the left by a and from the right by a^T we obtain an expression for the matrix representing the rotation dyad in the a representation:

$$a \cdot \boldsymbol{R}_{a\to b} \cdot a^T = a \cdot a^T ({}^aR_{a\to b})a \cdot a^T. \tag{2.73}$$

Using the property that $a \cdot a^T$ is a unit matrix it is clear that

$${}^aR_{a\to b} = a \cdot \boldsymbol{R}_{a\to b} \cdot a^T. \tag{2.74}$$

Now simply substitute the definition of the rotation dyad from the last section, i.e. that it is given by the expression $b^T a$:

$${}^aR_{a\to b} = a \cdot b^T a \cdot a^T = a \cdot b^T. \tag{2.75}$$

Therefore we have the important result that

$${}^aR_{a\to b} = R_{ab}. \tag{2.76}$$

Proceeding in the same manner it is possible to obtain the representation of the rotation dyad in the b frame. The result may be somewhat surprising at first sight, thus

$${}^bR_{a\to b} = b \cdot b^T a \cdot b^T = a \cdot b^T. \tag{2.77}$$

Which implies that

$${}^aR_{a\to b} = {}^bR_{a\to b} = R_{ab}, \tag{2.78}$$

i.e. the rotation dyad that rotates vectors from the a to the b frame has the same representation in both frames. This is not so astonishing if one recalls the representation of the dyad in terms of the rotation direction vector λ and the rotation angle θ. Clearly the rotation vector is the same for both frames, i.e. it is an eigenvector of the transformation.

To complete the discussion, derive the rotation dyad's explicit matrix representation in terms of the rotation vector and angle. In particular compare it with the direction cosine matrix and use the well known fact that the trace of a matrix is invariant under orthogonal transformations. This exercise provides explicit relations between the components of the rotation vector, the rotation angle and the direction cosine components.

The components of the rotation matrix are given by the relation:

$${}^aR_{a\to b} = a \cdot (\cos\theta \boldsymbol{U} + (1 - \cos\theta)\boldsymbol{\lambda\lambda} + \sin\theta \boldsymbol{\lambda} \times \boldsymbol{U}) \cdot a^T. \tag{2.79}$$

The calculation naturally divides into three parts, one for each term in the rotation dyad. The result will of course be the same for frame a or b hence the frame name is omitted from λ and R components. The result is that

$$R_{ij} = \delta_{ij}\cos\theta + (1-\cos\theta)\lambda_i\lambda_j - \sin\theta\epsilon_{ijk}\lambda_k. \tag{2.80}$$

As the components are the same in both frames the trace can be evaluated as convenient. Its invariance provides the equation

$$\cos\theta = \frac{1}{2}(Trace(R_{ab}) - 1), \tag{2.81}$$

providing a formula for the rotation angle. It is also a routine matter to evaluate the direction cosine components of the rotation vector from these formula or from the property that it is an eigenvector of R_{ab}. The rotation dyad has one real and two conjugate complex eigenvalues. The meaning of the complex eigenvalues can be seen from the following calculation.

Take $\boldsymbol{\kappa}_1$ and $\boldsymbol{\kappa}_2$ as a set of two orthonormal vectors which are also both normal to the rotation direction vector. Assume they have been chosen so that $\boldsymbol{\lambda} \times \boldsymbol{\kappa}_2 = \boldsymbol{\kappa}_1$. This implies that $\boldsymbol{\kappa}_1, \boldsymbol{\lambda}, \boldsymbol{\kappa}_2$ forms a right handed orthonormal triad. Define a *complex* vector:

$$\boldsymbol{\kappa} = \boldsymbol{\kappa}_1 + i\boldsymbol{\kappa}_2 \tag{2.82}$$

Using the orthonormality properties operate on this expression with the rotation dyad. It is thus seen that

$$\boldsymbol{R}_{a\to b} \cdot \boldsymbol{\kappa} = e^{i\theta}\boldsymbol{\kappa}. \tag{2.83}$$

The complex conjugate of this equation is also satisfied, hence $e^{\pm i\theta}$ are complex conjugate eigenvalues of the rotation operator associated with the indicated complex conjugate eigenvectors. Thus the rotation operators complex conjugate eigenvalues determine the rotation angle about the axis given by the eigenvector associated with the real eigenvalue.

•Illustration

In this illustration we examine how the above notation can be used to describe geometrical relationships for several connected rigid bodies. Three square plates of side l are connected as indicated in the figure. A standard triad is fixed in plate A, with the frame vector $\boldsymbol{a}_1$ aligned in the direction $\mathcal{P}_1 - \mathcal{P}_0$, and $\boldsymbol{a}_2$ along the direction $\mathcal{P}_1 - \mathcal{Q}_1$. The vector $\boldsymbol{a}_3 = \boldsymbol{a}_1 \times \boldsymbol{a}_2$. A second plate, B, is connected to A by a hinge along the line $\mathcal{P}_1 - \mathcal{Q}_1$. The standard triad attached to B is such that $\boldsymbol{b}_1$ is in the direction $\mathcal{P}_2 - \mathcal{P}_1$, and $\boldsymbol{b}_2$ is in the direction $\mathcal{P}_1 - \mathcal{Q}_1$, with $\boldsymbol{b}_3 = \boldsymbol{b}_1 \times \boldsymbol{b}_2$. A third plate, C, is connected along the line $\mathcal{P}_1 - \mathcal{P}_2$. Frame vectors $\boldsymbol{c}_1, \boldsymbol{c}_2$ and $\boldsymbol{c}_3$ aligned in the directions $\mathcal{P}_2 - \mathcal{P}_1$, $\mathcal{P}_2 - \mathcal{P}_3$ and the cross product. The lines $\mathcal{P}_0 - \mathcal{P}_1$ and $\mathcal{P}_1 - \mathcal{P}_2$ are at an angle q_1. The angle $\mathcal{Q}_2\mathcal{P}_2\mathcal{P}_3$ is denoted by q_2. The view of this system shown in the figure is intended to help find the relationships among the various frame vectors. The convention is adopted that all angles are positive in the displayed configuration.

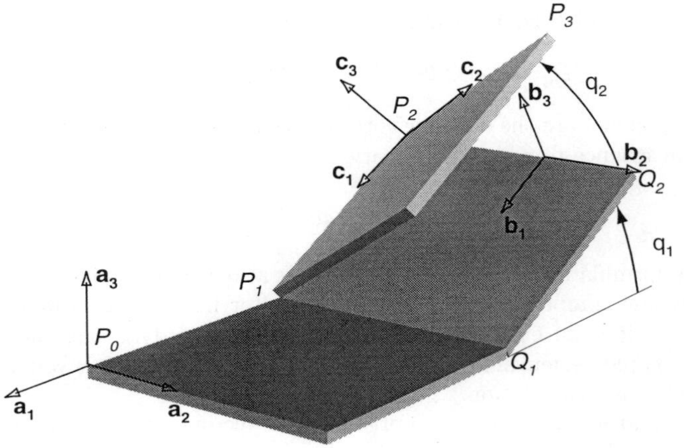

Figure 2.3: Geometric relationships among hinged plates

In the notation developed above the direction cosine matrix relating b to a is given by $R_{ba} = b \cdot a^T$. From the figure it should be clear that:

$$\begin{aligned} \boldsymbol{b}_1 \cdot \boldsymbol{a}_1 = c_1, \quad \boldsymbol{b}_1 \cdot \boldsymbol{a}_2 = 0, \quad \boldsymbol{b}_1 \cdot \boldsymbol{a}_3 = -s_1, \\ \boldsymbol{b}_2 \cdot \boldsymbol{a}_1 = 0, \quad \boldsymbol{b}_2 \cdot \boldsymbol{a}_2 = 1, \quad \boldsymbol{b}_2 \cdot \boldsymbol{a}_3 = 0, \\ \boldsymbol{b}_3 \cdot \boldsymbol{a}_1 = s_1, \quad \boldsymbol{b}_3 \cdot \boldsymbol{a}_2 = 0, \quad \boldsymbol{b}_3 \cdot \boldsymbol{a}_3 = c_1, \end{aligned}$$

hence the direction cosine matrix is:

$$R_{ba} = \begin{bmatrix} c_1 & 0 & -s_1 \\ 0 & 1 & 0 \\ s_1 & 0 & c_1 \end{bmatrix}. \tag{2.84}$$

In a similar manner we find the matrix:

$$R_{cb} = \begin{bmatrix} 1 & 0 & 0 \\ 0 & c_2 & s_2 \\ 0 & -s_2 & c_2 \end{bmatrix}. \tag{2.85}$$

The direction cosine matrix relating frame a to frame c is obtained by matrix multiplication, thus:

$$R_{ca} = R_{cb} R_{ba} = \begin{bmatrix} c_1 & 0 & -s_1 \\ s_1 s_2 & c_2 & c_1 s_2 \\ s_1 c_2 & -s_2 & c_1 c_2 \end{bmatrix}. \tag{2.86}$$

To see how these can be applied to the derivation of geometric relations consider the problem of determining an analytic relation for the displacement between points $\mathcal{P}_0$ and $\mathcal{P}_3$. We have the relationship:

$$\boldsymbol{r}^{\mathcal{P}_0\mathcal{P}_3} = \boldsymbol{r}^{\mathcal{P}_0\mathcal{P}_1} + \boldsymbol{r}^{\mathcal{P}_1\mathcal{P}_2} + \boldsymbol{r}^{\mathcal{P}_2\mathcal{P}_3} = l(-\boldsymbol{a}_1 - \boldsymbol{b}_1 + \boldsymbol{c}_2), \tag{2.87}$$

which in terms of our matrix containing vectors notation can be expressed as:

$$\frac{1}{l}\boldsymbol{r}^{\mathcal{P}_0\mathcal{P}_3} = (-1,0,0)a + (-1,0,0)b + (0,1,0)c. \tag{2.88}$$

Using the above direction cosine matrices we can put this in terms of any of the reference frames, so choosing the frame a, and carrying out the indicated matrix products we have:

$$\begin{aligned} \frac{1}{l}\boldsymbol{r}^{\mathcal{P}_0\mathcal{P}_3} &= [(-1,0,0) + (-1,0,0)R_{ba} + (0,1,0)R_{ca}]a \\ &= (-1 - c_1 + s_1 s_2, c_2, s_1 + c_1 s_2)a. \end{aligned} \tag{2.89}$$

In the notation introduced in the above discussion this last result would be denoted by $({}^a r^{\mathcal{P}_0\mathcal{P}_3})^T$, indicating the row matrix of components of the displacement vector expressed in the a system of frame vectors. •

2.4 More About Dyads

The main advantage of dyads is their invariant character. Thus a relationship derived in terms of dyads and vectors can be expressed in terms of any desired standard triad. A number of relationships involving dyads are useful in the development of rigid body mechanics.

A *symmetric dyad* is defined in terms of operations on vectors by the relationship

$$\boldsymbol{D} \cdot \boldsymbol{w} = \boldsymbol{w} \cdot \boldsymbol{D}, \tag{2.90}$$

and an *antisymmetric dyad* by

$$\boldsymbol{D} \cdot \boldsymbol{w} = -\boldsymbol{w} \cdot \boldsymbol{D}. \tag{2.91}$$

To see the consequences of these definitions in a particular reference triad observe that:

$$\boldsymbol{D} = a^T D a, \tag{2.92}$$

$$\boldsymbol{w} = a^T w, \tag{2.93}$$

where, as only the triad a is involved, a left superscript is not used on the matrix D or the column w that represent the dyad and vector in the a representation. Recalling the fact that $a \cdot a^T = \mathbf{1}$ compute:

$$\boldsymbol{D} \cdot \boldsymbol{w} = a^T D a \cdot a^T w = a^T D w \tag{2.94}$$

and

$$\boldsymbol{w} \cdot \boldsymbol{D} = a^T w \cdot a^T Da = w^T a \cdot a^T Da = w^T Da. \tag{2.95}$$

The relation, $a^T w = w^T a$, was used in the above. In the case where D is symmetric both the right hand sides are equal, hence:

$$a^T Dw = w^T Da \tag{2.96}$$

but $a^T Dw = (Dw)^T a = w^T D^T a$, therefore if $\boldsymbol{D}$ is symmetric its matrix representation satisfies the relation:

$$D^T = D. \tag{2.97}$$

Clearly by similar reasoning the result for the antisymmetric case is that:

$$D^T = -D. \tag{2.98}$$

The reader should be able to show both that any dyad can be decomposed into the sum of a symmetric and antisymmetric part and what this implies for the matrix representation of the dyad.

If $\boldsymbol{D}$ is antisymmetric its matrix representation in any standard triad must have the form:

$$\boldsymbol{D} \doteq \begin{bmatrix} 0 & z_1 & z_2 \\ -z_1 & 0 & z_3 \\ -z_2 & -z_3 & 0 \end{bmatrix} \tag{2.99}$$

so that the matrix representation of $\boldsymbol{D} \cdot \boldsymbol{w}$ will be:

$$\begin{bmatrix} z_1 w_2 + z_2 w_3 \\ -z_1 w_1 + z_3 w_3 \\ -z_2 w_1 - z_3 w_2 \end{bmatrix}. \tag{2.100}$$

Now consider the vector cross product of some vector $\boldsymbol{s}$ with $\boldsymbol{w}$:

$$\boldsymbol{s} \times \boldsymbol{w} \doteq \begin{bmatrix} -s_3 w_2 + s_2 w_3 \\ s_3 w_1 - s_1 w_3 \\ -s_2 w_1 + s_1 w_2 \end{bmatrix}. \tag{2.101}$$

Comparing these results we see that if we form a vector from the three independent components of $\boldsymbol{D}$ with the matrix column representation

$$\boldsymbol{vect}[\boldsymbol{D}] \doteq \begin{bmatrix} D_{32} \\ D_{13} \\ D_{21} \end{bmatrix}, \tag{2.102}$$

then the operation of an antisymmetric dyad on a vector can be replaced by the cross product operation, i.e.

$$\boldsymbol{D} \cdot \boldsymbol{w} = \boldsymbol{vect}[\boldsymbol{D}] \times \boldsymbol{w}. \tag{2.103}$$

The formation of $\boldsymbol{vect}[\boldsymbol{D}]$ in any particular representation is easy to recall as the subscript grouping in the column representation can be obtained by taking 321321 to 32 13 21.

Equipped with these results about standard triads and dyad relations we can return to the problem of comparing rates of change as calculated by different observers.

•Illustration

The triple cross product $\boldsymbol{n} \times (\boldsymbol{w} \times \boldsymbol{n})$, where $\boldsymbol{n}$ is a unit vector, is clearly a vector perpendicular to the plane determined by the vectors $\boldsymbol{n}$ and $\boldsymbol{w} \times \boldsymbol{n}$. Consider this product in a representation a, so that $\boldsymbol{a}_i \cdot \boldsymbol{n} = n_i$ and $\boldsymbol{a}_i \cdot \boldsymbol{w} = w_i$. The equivalence between the dot product of an antisymmetric dyad and a vector and the vector cross product allows us to write the triple product as:

$$\boldsymbol{n} \times (\boldsymbol{w} \times \boldsymbol{n}) \doteq \begin{bmatrix} 0 & -n_3 & n_2 \\ n_3 & 0 & -n_1 \\ -n_2 & n_1 & 0 \end{bmatrix} \begin{bmatrix} 0 & -w_3 & w_2 \\ w_3 & 0 & -w_1 \\ -w_2 & w_1 & 0 \end{bmatrix} \begin{bmatrix} n_1 \\ n_2 \\ n_3 \end{bmatrix}. \tag{2.104}$$

Multiplying the indicated matrices gives the result:

$$\boldsymbol{n} \times (\boldsymbol{w} \times \boldsymbol{n}) \doteq \begin{bmatrix} w_1 - n_1(n_1w_1 + n_2w_2 + n_3w_3) \\ w_2 - n_2(n_1w_1 + n_2w_2 + n_3w_3) \\ w_3 - n_3(n_1w_1 + n_2w_2 + n_3w_3) \end{bmatrix}, \tag{2.105}$$

which, translating back to vector notation, can be written as

$$\boldsymbol{n} \times (\boldsymbol{w} \times \boldsymbol{n}) = \boldsymbol{w} - \boldsymbol{n}(\boldsymbol{n} \cdot \boldsymbol{w}). \tag{2.106}$$

The interpretation of this last result is that the component of the vector $\boldsymbol{w}$ in the direction of $\boldsymbol{n}$ is subtracted from $\boldsymbol{w}$, hence the resultant vector can be considered as the projection of $\boldsymbol{w}$ onto a unit vector which is perpendicular to the plane determined by $\boldsymbol{n}$ and $\boldsymbol{w}$. This type of orthogonal projection will be needed in several areas of our development of mechanics. The calculation with antisymmetric matrices is more laborious than a purely vectorial approach, however the role played by the antisymmetric matrices can be generalized to higher dimensional situations. The correspondence between cross product and vector in three dimensions must be considered a happy coincidence, as only in three dimensions are the number of independent components of an antisymmetric dyad equal to the dimension of the space! •

2.5 Rates and Observers

The rate of change of a vector depends on the observer. Thus an observer sees a fixed vector in its standard reference frame as unchanging. Other observers that are in motion with respect to it will not agree with this assessment. Given the tools developed above it is quite easy to calculate the relationship between derivatives of vectors as obtained by different observers. Thus if t, representing time, is the differentiation parameter:

$$\frac{{}^{A}d\boldsymbol{w}}{dt} = \frac{{}^{A}d}{dt}(b^T\,{}^{b}\!w). \tag{2.107}$$

The key point in the above is that the vector $\boldsymbol{w}$ is resolved into components with respect to a standard triad, b, that is fixed in the body B, while the derivative with respect to t is taken from the viewpoint of an observer attached to body A. Expanding the right hand side and rearrangement shows that

$$\frac{{}^{A}d}{dt}\boldsymbol{w} - \frac{{}^{B}d}{dt}\boldsymbol{w} = (\frac{{}^{A}d}{dt}b^T)\,{}^{b}\!w. \tag{2.108}$$

To express this result in an invariant form, i.e. eliminate the explicit reference to the component array ${}^{b}\!w$. A simple and useful trick for doing this is to note that $b \cdot b^T$ produces the unit matrix, hence that $b \cdot b^T = \mathbf{1}$. Therefore the above equation can written as:

$$\frac{{}^{A}d}{dt}\boldsymbol{w} - \frac{{}^{B}d}{dt}\boldsymbol{w} = ((\frac{{}^{A}d}{dt}b^T)b) \cdot (b^T\,{}^{b}\!w) = ((\frac{{}^{A}d}{dt}b^T)b) \cdot \boldsymbol{w}. \tag{2.109}$$

This expresses the difference in the derivatives of $\boldsymbol{w}$ as calculated by the two observers as a dyad dotted into the vector. This dyad plays an extremely important role in mechanics as it provides the tool for calculating velocities of objects as seen by different observers. This important quantity deserves a name and symbol of its own. The symbol should reflect the fact that two observers are involved. As the dyad is an invariant object, any standard triad fixed in B can be used in the calculation. It will soon be evident that the dyad in question is related to the concept of angular velocity as understood in introductory mechanics courses. Therefore it is reasonable to define the *angular velocity dyad* as:

$${}^{A}\boldsymbol{\Omega}^{B} = (\frac{{}^{A}d}{dt}b^T)b. \tag{2.110}$$

Because of its importance it is worth examining 2.110 in expanded form, thus:

$${}^{A}\boldsymbol{\Omega}^{B} = \begin{bmatrix} \frac{{}^{A}d\boldsymbol{b}_1}{dt} & \frac{{}^{A}d\boldsymbol{b}_2}{dt} & \frac{{}^{A}d\boldsymbol{b}_3}{dt} \end{bmatrix} \begin{bmatrix} \boldsymbol{b}_1 \\ \boldsymbol{b}_2 \\ \boldsymbol{b}_3 \end{bmatrix} \tag{2.111}$$

which gives

$${}^{A}\boldsymbol{\Omega}^{B} = \frac{{}^{A}d\boldsymbol{b}_1}{dt}\boldsymbol{b}_1 + \frac{{}^{A}d\boldsymbol{b}_2}{dt}\boldsymbol{b}_2 + \frac{{}^{A}d\boldsymbol{b}_3}{dt}\boldsymbol{b}_3. \tag{2.112}$$

This last form is almost obvious if one notes that operating with it on a vector takes the components of the vector in the b representation and multiplies them by the *rates of change* of the b vectors with respect to the a representation.

•Illustration

The direct calculation of the angular velocity dyad, using the results of this section, can be extremely tedious, even in very simple cases. Once one carries out a few such calculations it becomes obvious from the relative simplicity of the results that there

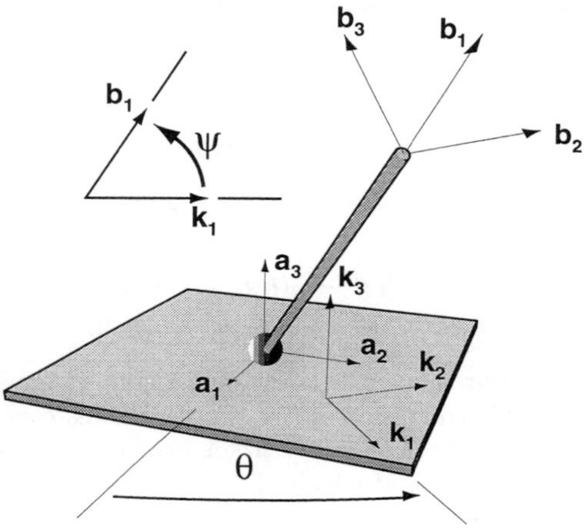

Figure 2.4: Angular position of pivoted rod.

might be a simpler way. To see this, consider the case shown in Figure 2.4, which shows a rod which can pivot about the point $\mathcal{O}$. The configuration is defined by two angles, designated as θ and ψ, which are themselves functions of the time, t. A standard observer, with triad a, is used as reference. An arbitrary configuration can be specified in terms of two auxiliary triads, k and b. The vectors $\boldsymbol{a}_3$ and $\boldsymbol{k}_3$ coincide and $\boldsymbol{k}_1 \cdot \boldsymbol{a}_1 = \cos\theta$, defining the rotation of the rod about the $\boldsymbol{a}_3$ axis. The system b is defined so as to maintain equality between $\boldsymbol{b}_2$ and $\boldsymbol{k}_2$, with $\boldsymbol{b}_1 \cdot \boldsymbol{k}_1 = \cos\psi$. Thus ψ describes the angle made by the rod with the plane defined by $\boldsymbol{a}_1$ and $\boldsymbol{a}_2$. The notation that, for example $c_2 = \cos\psi$ and$s_1 = \sin\theta$, will be used in what follows. With this information it is possible to construct the direction cosine matrices needed for the problem, thus

$$R_{ka} = \begin{bmatrix} c_1 & s_1 & 0 \\ -s_1 & c_1 & 0 \\ 0 & 0 & 1 \end{bmatrix}, \tag{2.113}$$

$$R_{bk} = \begin{bmatrix} c_2 & 0 & s_2 \\ 0 & 1 & 0 \\ -s_2 & 0 & c_2 \end{bmatrix}, \tag{2.114}$$

$$R_{ba} = R_{bk}R_{ka} = \begin{bmatrix} c_1c_2 & s_1c_2 & s_2 \\ -s_1 & c_1 & 0 \\ -c_1s_2 & -s_1s_2 & c_2 \end{bmatrix}. \tag{2.115}$$

The equation for ${}^{A}\boldsymbol{\Omega}^{B}$ requires the calculation of the total time derivative of each of the vectors $\boldsymbol{b}_i$ with respect to the triad a, hence going back to the definition of the derivative each $\boldsymbol{b}_i$ must be represented in terms of its expansion in the triad a, before taking the derivatives. The result, after some calculation, is

$$\frac{^A d}{dt}\boldsymbol{b}_1 = c_2\dot{\theta}\boldsymbol{b}_2 + \dot{\psi}\boldsymbol{b}_3, \tag{2.116}$$

$$\frac{^A d}{dt}\boldsymbol{b}_2 = -\dot{\theta}c_2\boldsymbol{b}_1 + \dot{\theta}s_2\boldsymbol{b}_3, \tag{2.117}$$

$$\frac{^A d}{dt}\boldsymbol{b}_3 = -\dot{\psi}\boldsymbol{b}_1 - \dot{\theta}s_2\boldsymbol{b}_2. \tag{2.118}$$

Inserting these results into the formula for the angular velocity dyad and carrying out some simplifications gives the result:

$$\begin{aligned} {}^A\boldsymbol{\Omega}^B = {} & \dot{\psi}(\boldsymbol{b}_3\boldsymbol{b}_1 - \boldsymbol{b}_1\boldsymbol{b}_3) + \\ & \dot{\theta}[s_2(\boldsymbol{b}_3\boldsymbol{b}_2 - \boldsymbol{b}_2\boldsymbol{b}_3) + c_2(\boldsymbol{b}_2\boldsymbol{b}_1 - \boldsymbol{b}_1\boldsymbol{b}_2)]. \end{aligned} \tag{2.119}$$

The final result lends encouragement to a search for simpler methods. It shows that the angular velocity dyad is composed of two simpler dyads, each expressed in terms of one of the angles and one of the standard triads used in the problem. The next two sections go on to develop a method which allows an even simpler form of the result to be found by inspection! •

2.6 Angular Velocity As Dyad, Matrix or Vector

Depending on the problem at hand various representations and properties of the angular velocity dyad can prove useful. This section is devoted to deriving some of them and to justifying the name angular velocity.

For concrete calculation we need the representation of the angular velocity dyad in terms of a standard triad. Choose the triad b for this task to obtain the *matrix representation of the dyad for the standard triad.* Following the previous work with dyads this is seen to be:

$$^b\Omega_{AB} = b \cdot {}^A\boldsymbol{\Omega}^B \cdot b^T = \begin{bmatrix} \boldsymbol{b}_1 \cdot {}^A\boldsymbol{\Omega}^B \cdot \boldsymbol{b}_1 & \boldsymbol{b}_1 \cdot {}^A\boldsymbol{\Omega}^B \cdot \boldsymbol{b}_2 & \boldsymbol{b}_1 \cdot {}^A\boldsymbol{\Omega}^B \cdot \boldsymbol{b}_3 \\ \boldsymbol{b}_2 \cdot {}^A\boldsymbol{\Omega}^B \cdot \boldsymbol{b}_1 & \boldsymbol{b}_2 \cdot {}^A\boldsymbol{\Omega}^B \cdot \boldsymbol{b}_2 & \boldsymbol{b}_2 \cdot {}^A\boldsymbol{\Omega}^B \cdot \boldsymbol{b}_3 \\ \boldsymbol{b}_3 \cdot {}^A\boldsymbol{\Omega}^B \cdot \boldsymbol{b}_1 & \boldsymbol{b}_3 \cdot {}^A\boldsymbol{\Omega}^B \cdot \boldsymbol{b}_2 & \boldsymbol{b}_3 \cdot {}^A\boldsymbol{\Omega}^B \cdot \boldsymbol{b}_3 \end{bmatrix} \tag{2.120}$$

Note that three objects come into the computation of the angular velocity matrix, the two bodies A and B and the standard triad b chosen for the representation.

Closer examination of the components of the angular velocity matrix reveal some interesting facts. For example

$$\boldsymbol{b}_1 \cdot {}^A\boldsymbol{\Omega}^B \cdot \boldsymbol{b}_1 = \boldsymbol{b}_1 \cdot \frac{^A d\boldsymbol{b}_1}{dt} = \frac{1}{2}\frac{^A d\boldsymbol{b}_1 \cdot \boldsymbol{b}_1}{dt} = \frac{1}{2}\frac{^A d}{dt}(1) = 0, \tag{2.121}$$

with a similar result for the other diagonal components, that is all the diagonal components vanish, at least in the b representation. Now consider a typical off diagonal component $^b\Omega_{AB_{12}}$, thus:

$$\boldsymbol{b}_1 \cdot {}^A\boldsymbol{\Omega}^B \cdot \boldsymbol{b}_2 = \boldsymbol{b}_1 \cdot \frac{^A d\boldsymbol{b}_2}{dt} \tag{2.122}$$

and note that

$$\boldsymbol{b}_1 \cdot \boldsymbol{b}_2 = 0 \tag{2.123}$$

which implies that

$$\boldsymbol{b}_1 \cdot \frac{{}^A d\boldsymbol{b}_2}{dt} = -\boldsymbol{b}_2 \cdot \frac{{}^A d\boldsymbol{b}_1}{dt}. \tag{2.124}$$

Similar results obtain for the other off diagonal components, hence in the b representation ${}^b\Omega_{AB}$ is an antisymmetric matrix. It can be shown that antisymmetry is preserved under transformation between standard triads, hence the angular velocity dyad itself must be antisymmetric! This is seen directly in terms of the dyadic definition of antisymmetry. The implication of the antisymmetry is that the dyadic operation is replaceable with a vector cross product operation by using the results of equation 2.103, so that.

$${}^A\boldsymbol{\Omega}^B \cdot \boldsymbol{w} = \boldsymbol{vect}[{}^A\boldsymbol{\Omega}^B] \times \boldsymbol{w}. \tag{2.125}$$

The angular velocity vector is thus defined as:

$${}^A\boldsymbol{\omega}^B = \boldsymbol{vect}[{}^A\boldsymbol{\Omega}^B], \tag{2.126}$$

which using the definition of $\boldsymbol{vect}$ is seen to be equivalent to:

$${}^A\boldsymbol{\omega}^B = \boldsymbol{b}_1(\frac{{}^A d\boldsymbol{b}_2}{dt} \cdot \boldsymbol{b}_3) + \boldsymbol{b}_2(\frac{{}^A d\boldsymbol{b}_3}{dt} \cdot \boldsymbol{b}_1) + \boldsymbol{b}_3(\frac{{}^A d\boldsymbol{b}_1}{dt} \cdot \boldsymbol{b}_2). \tag{2.127}$$

The angular velocity vector now provides a vectorial way to express the difference in rates of change as calculated by observers attached to different bodies, i.e.

$$\frac{{}^A d\boldsymbol{w}}{dt} = \frac{{}^B d\boldsymbol{w}}{dt} + {}^A\boldsymbol{\omega}^B \times \boldsymbol{w}. \tag{2.128}$$

The pair of equations 2.127 and 2.128 provide a convenient bases for the velocity and acceleration analysis of mechanical systems. Some authors, such as Kane and Levinson, take equation 2.127 as the definition of angular velocity, deriving 2.128 as a consequence. The vectorial, matrix and dyadic forms of angular velocity prove useful both in theory and explicit calculation. As the alert reader may have noted, there is a relationship between the angular velocity matrix and the direction cosine matrix. Returning to the expression:

$$\frac{{}^A d\boldsymbol{w}}{dt} - \frac{{}^B d\boldsymbol{w}}{dt} = \frac{{}^A d b^T}{dt} {}^b w, \tag{2.129}$$

note that the right hand side can be written as

$$\frac{{}^A d}{dt}((R_{ba} a)^T) {}^b w, \tag{2.130}$$

hence,

$$\frac{{}^A d\boldsymbol{w}}{dt} - \frac{{}^B d\boldsymbol{w}}{dt} = \frac{{}^A d}{dt}(a^T R_{ab}) {}^b w = (R_{ab} b)^T \frac{{}^A d}{dt}(R_{ab}) {}^b w = b^T R_{ba}(\frac{{}^A d}{dt} R_{ab}) {}^b w. \tag{2.131}$$

Therefore the result, which could also have been obtained from the dyadic representation, that:

$$^{b}\Omega_{AB} = R_{ba}\tfrac{^{A}d}{dt}R_{ab}. \tag{2.132}$$

It is left as an exercise for the reader to show from this definition and the fact that $R_{ba}R_{ab} = \mathbf{1}$ that $^{b}\Omega_{AB}$ is antisymmetric.

As stated above the antisymmetry is evident from the dyadic definition of angular velocity. Thus

$$^{A}\boldsymbol{\Omega}^{B} = (\tfrac{^{A}db}{dt})^{T}b \tag{2.133}$$

and the relations

$$b^{T}b = \boldsymbol{U} \tag{2.134}$$

$$\tfrac{^{A}db^{T}}{dt}b + b^{T}\tfrac{^{A}db}{dt} = 0 \tag{2.135}$$

show that for any vector $\boldsymbol{w}$

$$^{A}\boldsymbol{\Omega}^{B}\cdot\boldsymbol{w} = \tfrac{^{A}db^{T}}{dt}b\cdot\boldsymbol{w} = -b^{T}\tfrac{^{A}db}{dt}\cdot\boldsymbol{w} = -\boldsymbol{w}\cdot(\tfrac{^{A}db}{dt})^{T}b = -\boldsymbol{w}\cdot{}^{A}\boldsymbol{\Omega}^{B}, \tag{2.136}$$

that is $^{A}\boldsymbol{\Omega}^{B}$ is indeed antisymmetric. The advantage of this last demonstration is that it is independent of any particular standard triad, i.e. it directly shows the antisymmetry as an invariant property of the angular velocity dyad.

In summary the different rates of change of a vector as seen by observers attached to different bodies can be calculated using either the angular velocity dyad, matrix or vector. It should also be evident that the role played by the time as a differentiation variable in the above *can be replaced by other parameters, which describe the relative positions of the bodies in question.* Doing so leads to the idea of the Darboux vector, which is discussed in chapter 5, where it plays an important role in the elimination of rigid body constraint forces.

•Illustration

Returning to the motion of the pivoted rod, discussed in the illustration following the last section, we now calculate the two angular velocities $^{A}\boldsymbol{\omega}^{K}$ and $^{K}\boldsymbol{\omega}^{B}$ by means of equation 2.127. To apply this equation to these cases we need the derivatives of the unit vectors $\boldsymbol{k}_j$ and $\boldsymbol{b}_j$ with respect to the triads a and k respectively. Straightforward substitution gives the results that

$$\tfrac{^{A}d}{dt}\boldsymbol{k}_1 = \dot{\theta}\boldsymbol{k}_2, \tag{2.137}$$

$$\tfrac{^{A}d}{dt}\boldsymbol{k}_2 = -\dot{\theta}\boldsymbol{k}_1, \tag{2.138}$$

$$\tfrac{^{A}d}{dt}\boldsymbol{k}_3 = 0, \tag{2.139}$$

$$\tfrac{^{K}d}{dt}\boldsymbol{b}_1 = \dot{\psi}\boldsymbol{b}_3, \tag{2.140}$$

$$\tfrac{^{K}d}{dt}\boldsymbol{b}_2 = 0, \tag{2.141}$$

$$\tfrac{^{K}d}{dt}\boldsymbol{b}_3 = -\dot{\psi}\boldsymbol{b}_1. \tag{2.142}$$

Application of the formula then gives the results:

$$ {}^{A}\omega^{K} = \dot{\theta}\boldsymbol{a}_3, \tag{2.143} $$

$$ {}^{K}\omega^{B} = -\dot{\psi}\boldsymbol{k}_2. \tag{2.144} $$

It is interesting to compare these results with the angular velocity dyad between frames a and b derived in the last illustration. If we convert that result into its angular velocity vector we find that:

$$ \begin{aligned} {}^{A}\Omega^{B} &\rightarrow -\dot{\psi}\boldsymbol{b}_2 + \dot{\theta}(c_2\boldsymbol{b}_3 + s_2\boldsymbol{b}_1) \\ &= -\dot{\psi}\boldsymbol{k}_2 + \dot{\theta}\boldsymbol{a}_3. \end{aligned} \tag{2.145} $$

Thus, at least in this case the angular velocity vector ${}^{A}\omega^{B} = {}^{A}\omega^{K} + {}^{K}\omega^{B}$, where each term in this addition is quite simple, being the rate of change of the relevant angle times a unit vector along the axis of the rotation. This is no accident and the general conditions for this situation are discussed in the following section. •

2.7 Properties of Angular Velocity

No appeal has been made to any geometrical image connected to the quantity called the angular velocity. Its main role has been as an abstract operator which helps relate the derivative of vectors as calculated by different observers. The typical reader has most likely met up with angular velocity as defined by a geometric image relating the rates of turning of a vector about an axis. The advantage of the development above is that it lends a precise definition to the term, complete with a detailed specification on how to calculate it and its principle use in mechanics. The following material is intended to both provide some useful relations and to connect the formal definition with geometric intuition.

2.7.1 Antisymmetry of Angular Velocity

The geometric image does convey some useful further properties. If it describes the turning rate of one standard triad with respect to another then one would expect the angular velocity of a body B with respect to a body A to be the negative of the angular velocity of A with respect to B. That this is true is seen by a simple computation using the vectorial representation, thus for any vector $\boldsymbol{w}$:

$$ \frac{{}^{A}d\boldsymbol{w}}{dt} = \frac{{}^{B}d\boldsymbol{w}}{dt} + {}^{A}\omega^{B} \times \boldsymbol{w} \tag{2.146} $$

and

$$ \frac{{}^{B}d\boldsymbol{w}}{dt} = \frac{{}^{A}d\boldsymbol{w}}{dt} + {}^{B}\omega^{A} \times \boldsymbol{w}. \tag{2.147} $$

Addition of these two relations gives the result:

$$ ({}^{A}\omega^{B} + {}^{B}\omega^{A}) \times \boldsymbol{w} = 0. \tag{2.148} $$

Because the vector $\boldsymbol{w}$ is arbitrary this proves the important and expected property that:

$$^{A}\omega^{B} = -^{B}\omega^{A}. \tag{2.149}$$

2.7.2 Addition of Angular Velocities

Often it is both useful and necessary to deal with several bodies, each with its own standard triad. It can be an advantage to introduce intermediate or fictitious bodies simply for the purpose of computational convenience. In these cases effective use can be made of an additive property that exists between angular velocities relating different standard triads. It is simple to derive this property using the dyadic representation of angular velocity. Thus consider three observers, A, B and C, and any arbitrary vector $\boldsymbol{w}$. The straightforward application of our results for computing relative time derivatives shows that:

$$\frac{^{A}d\boldsymbol{w}}{dt} = \frac{^{B}d\boldsymbol{w}}{dt} + {^{A}\boldsymbol{\Omega}^{B}} \cdot \boldsymbol{w} \tag{2.150}$$

and

$$\frac{^{B}d\boldsymbol{w}}{dt} = \frac{^{C}d\boldsymbol{w}}{dt} + {^{B}\boldsymbol{\Omega}^{C}} \cdot \boldsymbol{w} \tag{2.151}$$

adding both sides of these equations to one another gives:

$$\frac{^{A}d\boldsymbol{w}}{dt} = \frac{^{C}d\boldsymbol{w}}{dt} + {^{A}\boldsymbol{\Omega}^{B}} \cdot \boldsymbol{w} + {^{B}\boldsymbol{\Omega}^{C}} \cdot \boldsymbol{w} = \frac{^{C}d\boldsymbol{w}}{dt} + ({^{A}\boldsymbol{\Omega}^{B}} + {^{B}\boldsymbol{\Omega}^{C}}) \cdot \boldsymbol{w}, \tag{2.152}$$

but

$$\frac{^{A}d\boldsymbol{w}}{dt} = \frac{^{C}d\boldsymbol{w}}{dt} + {^{A}\boldsymbol{\Omega}^{C}} \cdot \boldsymbol{w} \tag{2.153}$$

so that, as was to be demonstrated,

$${^{A}\boldsymbol{\Omega}^{C}} = {^{A}\boldsymbol{\Omega}^{B}} + {^{B}\boldsymbol{\Omega}^{C}}. \tag{2.154}$$

It should be clear that we can introduce further observers so that the following general result applies:

$${^{A_1}\boldsymbol{\Omega}^{A_n}} = {^{A_1}\boldsymbol{\Omega}^{A_2}} + \cdots + {^{A_{n-1}}\boldsymbol{\Omega}^{A_n}}.. \tag{2.155}$$

The relation between the angular velocity as a dyad and as a vector leads at once to the result that

$${^{A_1}\omega^{A_n}} = {^{A_1}\omega^{A_2}} + \cdots + {^{A_{n-1}}\omega^{A_n}}.. \tag{2.156}$$

Equations 2.155 and 2.156 provide a method for the convenient calculation of angular velocities. The technique is to find a sequence of intermediate observers which are simply related to one another and to calculate the intermediate angular velocities.

The above addition relations then provides the desired angular velocity. About the simplest relation one can expect between two observers is that they possess standard triads which have one vector in common. The remaining triad vectors are then related by a planar rotation which can be expressed in terms of one angle. The angular velocity between these two triads can be expressed in terms of one of the common vectors and the rate of change of the rotation angle. We shall follow Kane and call such a state of affairs *simple angular velocity.*

2.7.3 Simple Angular Velocity

To quantify the simple angular velocity relationship consider two observers, A and B, having respectively the standard triads a and b. Assume the vectors $\boldsymbol{a}_3$ and $\boldsymbol{b}_3$ are identical and are fixed in both bodies A and B. Let the angle going from the vector $\boldsymbol{a}_1$ to $\boldsymbol{b}_1$ be designated by θ, i.e. $Angle(\boldsymbol{a}_1, \boldsymbol{b}_1) = \theta$. Under these circumstances the reader should be readily able to verify that

$$\boldsymbol{b}_1 = \boldsymbol{a}_1 \cos\theta + \boldsymbol{a}_2 \sin\theta, \tag{2.157}$$

$$\boldsymbol{b}_1 = -\boldsymbol{a}_1 \sin\theta + \boldsymbol{a}_2 \cos\theta, \tag{2.158}$$

and

$$\boldsymbol{b}_3 = \boldsymbol{a}_3. \tag{2.159}$$

Use equation 2.127 to calculate ${}^{A}\boldsymbol{\omega}^{B}$. To accomplish this use the derivatives of the triad b with respect to observer A. From the above relations and considering that θ is a function of t it is seen that:

$$\frac{{}^{A}d\boldsymbol{b}_1}{dt} = (-\boldsymbol{a}_1 \sin\theta + \boldsymbol{a}_2 \cos\theta)\frac{d\theta}{dt} = \frac{d\theta}{dt}\boldsymbol{b}_2, \tag{2.160}$$

$$\frac{{}^{A}d\boldsymbol{b}_1}{dt} = (-\boldsymbol{a}_1 \cos\theta - \boldsymbol{a}_2 \sin\theta)\frac{d\theta}{dt} = -\frac{d\theta}{dt}\boldsymbol{b}_1, \tag{2.161}$$

and

$$\frac{{}^{A}d\boldsymbol{b}_3}{dt} = 0. \tag{2.162}$$

Inserting these results into 2.127 then gives the result that

$${}^{A}\boldsymbol{\omega}^{B} = \frac{d\theta}{dt}\boldsymbol{b}_3. \tag{2.163}$$

This result leads to the construction of a useful general rule. Before doing this it should be noted that if θ is defined as the angle going from $\boldsymbol{b}_1$ to $\boldsymbol{a}_1$, i.e. as $Angle(\boldsymbol{b}_1, \boldsymbol{a}_1)$ then the above calculation would give ${}^{A}\boldsymbol{\omega}^{B} = -\frac{d\theta}{dt}\boldsymbol{b}_3$. The *sign* obtained depends on the convention that we proceed from $\boldsymbol{a}_1$ to $\boldsymbol{b}_1$ and that the common vector about which the rotation takes place is in the direction of $\boldsymbol{a}_1 \times \boldsymbol{b}_1$. This later fact can be expressed by the relation that the sign of $\boldsymbol{a}_1 \times \boldsymbol{b}_1 \cdot \boldsymbol{a}_3$ is positive.

With the above in mind we can now formulate a general rule. Two observers A and B are in a relationship of *simple angular velocity dependence* if there exist three *unit vectors* $\boldsymbol{k}_A$, $\boldsymbol{k}_B$ and $\boldsymbol{e}$ such that $\boldsymbol{k}_A$ is *fixed* for observer A, $\boldsymbol{k}_B$ is *fixed* for observer B, $\boldsymbol{e}$ is *fixed* for *both* observers A and B, and $\boldsymbol{e}$ is orthogonal to $\boldsymbol{k}_A$ and $\boldsymbol{k}_B$. With

$$\epsilon = sign(\boldsymbol{k}_A \times \boldsymbol{k}_B \cdot \boldsymbol{e}), \tag{2.164}$$

the angular velocity can be expressed as

$${}^{A}\boldsymbol{\omega}^{B} = \epsilon\frac{d}{dt}Angle[\boldsymbol{k}_A, \boldsymbol{k}_B]\boldsymbol{e}. \tag{2.165}$$

The above requirements on the vectors $\boldsymbol{k}_A, \boldsymbol{k}_B$ and $\boldsymbol{e}$ can also be expressed by the relations:

$$\frac{{}^A d\boldsymbol{k}_A}{dt} = 0, \tag{2.166}$$

$$\frac{{}^B d\boldsymbol{k}_B}{dt} = 0, \tag{2.167}$$

$$\frac{{}^A d\boldsymbol{e}}{dt} = \frac{{}^B d\boldsymbol{e}}{dt} = 0 \tag{2.168}$$

and

$$\boldsymbol{e} \cdot \boldsymbol{k}_A = \boldsymbol{e} \cdot \boldsymbol{k}_B = 0. \tag{2.169}$$

The attainment of some skill in the application of the concept of simple angular velocity can increase ones problem solving ability in mechanics. Finally it should be noted that the concept of simple angular velocity in combination with the rule for the addition of angular velocity between different observers provides a connection with geometrical interpretations of angular velocity.

•Illustration

The direction cosine matrix relating one standard triad to another contains nine entries, however the orthogonality relations provide 6 relations among the entries, hence only 3 independent parameters are needed to specify the relation between the triads. It is convenient to have some standard means of specifying this 3 parameter group of orthogonal transformations, and in particular of calculating the angular velocity between the triads in terms of the same 3 independent quantities. In the determination of a motion obeying Newton's laws, what we shall call a 'Newtonian motion', it is necessary to refer quantities to a so called inertial triad. This is sometimes called the 'space axes' in contrast to a triad fixed to the moving body, sometimes called the 'body axes'. The concept of simple angular velocity and the additivity of angular velocities may be applied to the task of determining the angular velocity of the body axes with respect to the space axes.

The three independent parameters relating the two triads may be specified by specifying three rotation operations which takes the space axes to the body axes. Each rotation is simple in the sense of the discussion of simple angular velocity. As is easily verified, simple rotation operations lead to different results depending on the order in which they are carried out, in technical terminology, simple rotations do not commute. The angles associated with the three simple rotation operations needed to reach the body axes from the space axes are known as the *Euler angles.* There is no general agreement on how to pick the rotation operations or in which order they are to be applied. This discussion follows the choice given in the text of Goldstein.

Simple rotation operations can be labeled by specifying the axis about which the rotation takes place and the angle of the rotation, the latter being positive if it is in the counter-clockwise sense about the axis. It is also assumed that the starting and ending triads are right handed. The three possible positive simple rotations can then be specified as $R(+j, q)$ for a rotation in the positive sense about the jth triad vector of angular magnitude q. Figure 2.5 depicts these transformations, and the reader should be able to construct the direction cosine matrices for each one.

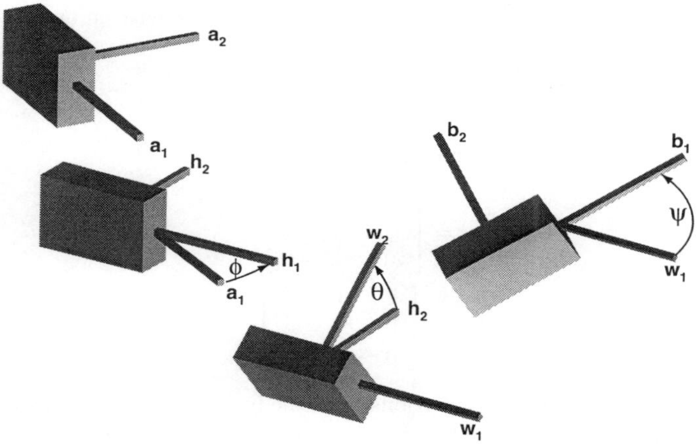

Figure 2.5: Euler Angles

The Euler angles are specified by three simple rotations, which can be taken as $(+3, \phi)$ followed by $(+1, \theta)$ followed in turn by $(+3, \psi)$. The names of the angles have also been chosen to agree with Goldstein. For our demonstration of simple angular velocity and the addition of angular velocities we take the first rotation from the triad a to h, the second from h to w and the last from w to b. In writing trigonometric functions the subscripts $1, 2$ and 3 for ϕ, θ and ψ respectively, e.g. $s_1 = \sin\phi$.

If the rotation angles are functions of time, each simple rotation corresponds to a simple angular velocity. Consider the rotation $(+3, \phi)$, which carries the triad a into the triad h. The unit vector $\boldsymbol{a}_1$ is fixed in frame A, the unit vector $\boldsymbol{h}_1$ is fixed in frame H, and both are orthogonal to $\boldsymbol{a}_3 = \boldsymbol{h}_3$, vectors which do not change in both frames A and H during the rotation. In addition the sign of $\boldsymbol{a}_1 \times \boldsymbol{h}_1 \cdot \boldsymbol{a}_3$ is positive. Therefore the result for simple angular velocity applies and

$$ {}^A\boldsymbol{\omega}^H = \dot{\phi}\boldsymbol{a}_3 = \dot{\phi}\boldsymbol{h}_3. \tag{2.170} $$

The reader new to these ideas is advised to verify, in the manner of the above, that for the rotations $(+1, \theta)$ from H to W and $(+3, \psi)$ from W to B the simple angular velocity results show that

$$ \begin{aligned} {}^H\boldsymbol{\omega}^W &= \dot{\theta}\boldsymbol{h}_1 && (2.171) \\ &= \dot{\theta}\boldsymbol{w}_1, \\ {}^W\boldsymbol{\omega}^B &= \dot{\psi}\boldsymbol{w}_3 && (2.172) \\ &= \dot{\psi}\boldsymbol{b}_3. \end{aligned} $$

The additive property of angular velocities now shows that:

$$ {}^{A}\omega^{B} = \dot{\phi}\boldsymbol{a}_3 + \dot{\theta}\boldsymbol{h}_1 + \dot{\psi}\boldsymbol{w}_3, \tag{2.173} $$

which, using the matrix containing vector notation, can be written as

$$ \begin{aligned} {}^{A}\omega^{B} &= (0,0,\dot{\phi})a + (\dot{\theta},0,0)h + (0,0,\dot{\psi})w \\ &= [(0,0,\dot{\phi}) + (\dot{\theta},0,0)R_{ha} + (0,0,\dot{\psi})R_{wh}R_{ha}]a. \end{aligned} \tag{2.174} $$

The direction cosine matrices being given by:

$$ R_{ha} = \begin{bmatrix} c_1 & s_1 & 0 \\ -s_1 & c_1 & 0 \\ 0 & 0 & 1 \end{bmatrix}, \tag{2.175} $$

$$ R_{wh} = \begin{bmatrix} 1 & 0 & 0 \\ 0 & c_2 & s_2 \\ 0 & -s_2 & c_2 \end{bmatrix}. \tag{2.176} $$

Carrying out all the indicated operations gives the final result for the angular velocity between the space axes and the body axes as referred to the space axes, thus

$$ \begin{aligned} {}^{A}\omega^{B} &= (\cos\phi\dot{\theta} + \sin\phi\sin\theta\dot{\psi})\boldsymbol{a}_1 \\ &+(\sin\phi\dot{\theta} - \cos\phi\sin\theta\dot{\psi})\boldsymbol{a}_2 \\ &+(\cos\theta\dot{\psi} + \dot{\phi})\boldsymbol{a}_3. \end{aligned} \tag{2.177} $$

It is informative to use the pattern of the rotations to predict what the result would look like when referred to the body axes and to verify the prediction by suitable calculation. •

2.8 Angular Acceleration

The principle of Newtonian Determinism indicates that in the development of mechanics we will need to consider second derivatives of displacements. Angular acceleration provides a convenient quantity for this task. It is simply defined as the appropriate time derivative of the angular velocity vector. Thus the angular acceleration is:

$$ {}^{A}\alpha^{B} = \frac{{}^{A}d\,{}^{A}\omega^{B}}{dt}. \tag{2.178} $$

A simple calculation, using the fact that the cross product of a vector with itself must vanish, shows that ${}^{A}\alpha^{B}$ could equally well be defined as

$$ {}^{A}\alpha^{B} = \frac{{}^{B}d\,{}^{A}\omega^{B}}{dt} \tag{2.179} $$

because

$$ \frac{{}^{A}d\,{}^{A}\omega^{B}}{dt} = \frac{{}^{B}d\,{}^{A}\omega^{B}}{dt} + {}^{A}\omega^{B} \times {}^{A}\omega^{B}. \tag{2.180} $$

Given two observers there are four different ways to compute second derivatives of a given vector. Thus for a vector $\boldsymbol{w}$ one can calculate the quantities:

$$\frac{{}^{A}d}{dt}\frac{{}^{A}d\boldsymbol{w}}{dt}, \frac{{}^{A}d}{dt}\frac{{}^{B}d\boldsymbol{w}}{dt}, \frac{{}^{B}d}{dt}\frac{{}^{A}d\boldsymbol{w}}{dt}, \frac{{}^{B}d}{dt}\frac{{}^{B}d\boldsymbol{w}}{dt}. \tag{2.181}$$

The effect of the rotational motion of the observers on the mixed second derivatives can be seen in the *commutation relation*

$$[A,B]_t\boldsymbol{w} = \frac{{}^{A}d}{dt}\frac{{}^{B}d\boldsymbol{w}}{dt} - \frac{{}^{B}d}{dt}\frac{{}^{A}d\boldsymbol{w}}{dt}. \tag{2.182}$$

Expressing this relation in terms of a triad in B, we find:

$$[A,B]_t\boldsymbol{w} = \left(\frac{{}^{B}d}{dt} + {}^{A}\boldsymbol{\omega}^{B}\times\right)\frac{{}^{B}d\boldsymbol{w}}{dt} - \frac{{}^{B}d}{dt}\left(\frac{{}^{B}d\boldsymbol{w}}{dt} + {}^{A}\boldsymbol{\omega}^{B}\times\boldsymbol{w}\right) \tag{2.183}$$

hence that:

$$[A,B]_t\boldsymbol{w} = -{}^{A}\boldsymbol{\alpha}^{B}\times\boldsymbol{w} = +{}^{B}\boldsymbol{\alpha}^{A}\times\boldsymbol{w}. \tag{2.184}$$

The above relation provides an elegant way to define the angular acceleration as an operator that measures the noncummutativity of rates of change of vectors for different observers, however we will stay with the simple idea of it being the time derivative of the angular velocity. It is interesting to note that the above commutator vanishes for observers that are related by constant angular velocities.

It is possible to verify a relation between unmixed derivatives given by:

$$\frac{{}^{A}d^2\boldsymbol{w}}{dt^2} - \frac{{}^{B}d^2\boldsymbol{w}}{dt^2} = {}^{A}\boldsymbol{\alpha}^{B}\times\boldsymbol{w} + {}^{A}\boldsymbol{\omega}^{B}\times({}^{A}\boldsymbol{\omega}^{B}\times\boldsymbol{w}) + 2\,{}^{A}\boldsymbol{\omega}^{B}\times\frac{{}^{B}d\boldsymbol{w}}{dt}. \tag{2.185}$$

Historical usage has given names to these various terms. Thus the term ${}^{A}\boldsymbol{\omega}^{B}\times({}^{A}\boldsymbol{\omega}^{B}\times\boldsymbol{w})$ is called the *centrifugal term*, and $2\,{}^{A}\boldsymbol{\omega}^{B}\times\frac{{}^{B}d\boldsymbol{w}}{dt}$ is called the *Coriolis term*. The right hand side of this relation, minus the Coriolis term is sometimes called the transport terms, for the reason that if $\boldsymbol{w}$ is constant for observer B the remaining differences are due to the relative motion or transport of the observers. The reader with a background in basic mechanics should find some comfort of familiarity in the above expression if the vector $\boldsymbol{w}$ represents a displacement between points in space. A precise discussion of this requires some care and will be given in the next chapter.

The property, similar to what holds for $\boldsymbol{\omega}$ that

$${}^{A}\boldsymbol{\alpha}^{B} = -{}^{B}\boldsymbol{\alpha}^{A} \tag{2.186}$$

should be obvious. The addition property, 2.156, for angular velocity is another matter, the relationship for angular acceleration being more complex. To examine this latter relation start with:

$${}^{A}\boldsymbol{\omega}^{C} = {}^{A}\boldsymbol{\omega}^{B} + {}^{B}\boldsymbol{\omega}^{C}. \tag{2.187}$$

Take the derivative of this with respect to observer A, so that

$${}^{A}\boldsymbol{\alpha}^{C} = \frac{{}^{A}d}{dt}({}^{A}\boldsymbol{\omega}^{B} + {}^{B}\boldsymbol{\omega}^{C}) = {}^{A}\boldsymbol{\alpha}^{B} + \frac{{}^{A}d\,{}^{B}\boldsymbol{\omega}^{C}}{dt} \tag{2.188}$$

hence:

$$ {}^A\boldsymbol{\alpha}^C = {}^A\boldsymbol{\alpha}^B + {}^B\boldsymbol{\alpha}^C + {}^A\boldsymbol{\omega}^B \times {}^B\boldsymbol{\omega}^C. \tag{2.189} $$

This relation clearly shows that simple additivity does not hold in the way it did for angular velocity.

•Illustration

The angular acceleration in terms of the Euler angles, introduced in the illustration following the last section, can be calculated by use of the formula for the composition of angular accelerations derived above. The reader is invited to compare this with a direct calculation. Instead of deriving the general formula, it is easiest to proceed with the case treated above, first composing the angular accelerations between A and H with H and W to obtain ${}^A\boldsymbol{\alpha}^W$. That is we first use the formula

$$ {}^A\boldsymbol{\alpha}^W = {}^A\boldsymbol{\alpha}^H + {}^H\boldsymbol{\alpha}^W + {}^A\boldsymbol{\omega}^H \times {}^H\boldsymbol{\omega}^W. \tag{2.190} $$

Using the previous results that

$$ {}^A\boldsymbol{\omega}^H = \dot{\phi}\boldsymbol{h}_3, \tag{2.191} $$
$$ {}^H\boldsymbol{\omega}^W = \dot{\theta}\boldsymbol{h}_1, \tag{2.192} $$
$$ {}^W\boldsymbol{\omega}^B = \dot{\psi}\boldsymbol{w}_3, \tag{2.193} $$

one can compute the angular accelerations:

$$ {}^A\boldsymbol{\alpha}^H = \ddot{\phi}\boldsymbol{h}_3, \tag{2.194} $$
$$ {}^H\boldsymbol{\alpha}^W = \ddot{\theta}\boldsymbol{h}_1, \tag{2.195} $$
$$ {}^W\boldsymbol{\alpha}^B = \ddot{\psi}\boldsymbol{w}_3. \tag{2.196} $$

The above formula then gives the result:

$$ {}^A\boldsymbol{\alpha}^W = \ddot{\phi}\boldsymbol{h}_3 + \ddot{\theta}\boldsymbol{h}_1 + \dot{\phi}\dot{\theta}\boldsymbol{h}_2. \tag{2.197} $$

A second application of the formula gives the desired result

$$ \begin{aligned} {}^A\boldsymbol{\alpha}^B &= {}^A\boldsymbol{\alpha}^W + {}^W\boldsymbol{\alpha}^B + {}^A\boldsymbol{\omega}^W \times {}^W\boldsymbol{\omega}^B \\ &= \ddot{\phi}\boldsymbol{h}_3 + \ddot{\theta}\boldsymbol{h}_1 + \dot{\phi}\dot{\theta}\boldsymbol{h}_2 + \ddot{\psi}\boldsymbol{w}_3 + (\dot{\phi}\boldsymbol{h}_3 + \dot{\theta}\boldsymbol{h}_1) \times (\dot{\psi}\boldsymbol{w}_3). \end{aligned} \tag{2.198} $$

Using the direction cosine matrix R_{wh}, or by inspection, it is seen that

$$ \boldsymbol{w}_3 = -s_2\boldsymbol{h}_2 + c_2\boldsymbol{h}_3, \tag{2.199} $$

so that in our notation

$$ {}^A\boldsymbol{\alpha}^B = (\ddot{\theta} + s_2\dot{\phi}\dot{\psi}, \dot{\theta}\dot{\phi} - s_2\ddot{\psi} - c_2\dot{\theta}\dot{\psi}, \ddot{\phi} + c_2\ddot{\psi} - \dot{\theta}\dot{\psi}s_2)h. \tag{2.200} $$

The direction cosine matrices can be used to refer this result to any one of the other triads a, w or b. Even though the result for composition of angular accelerations is not as simple as the angular velocity rule, it is still useful and worth keeping in mind.

•

2.9 Vector Representations and Computer Algebra

The material in this chapter has emphasized the importance of recognizing that the rate of change of vectors depends on the frame of the observer. This has been incorporated into the notation for a derivative and it has been seen that angular velocity dyads and vectors can be used to compare frame based vector derivatives of the same vector. A useful computer algebra system must also take account of these aspects of the transformation of vector derivatives between different standard frames.

The first order of business is to decide on a reasonable data structure to carry the information. We have already seen the use of lists for storing the components of vectors, however Maple as well as other computer algebra systems contain many alternative structures. While we will concentrate on a Maple implementation it should be clear how similar structures could be created in other systems.

Maple is provided with a very complete package for linear algebra, including data structures for matrices or arrays. Arrays may be used without loading the linear algebra package itself, but in general it will be useful to have the linear algebra package present in the workspace. This is accomplished by the command:

```
>with(linalg):
```

The array structure may in certain cases look very much like a list, however it is not treated internally at all like a list. The main difference is that an array has a fixed size. Matrices and vectors are special cases of arrays and if the linear algebra package is loaded, the commands 'vector' and 'matrix' can be used directly to form one and two dimensional arrays. To see how this works we examine a simple example in which all motions are parallel to a fixed plane. An articulated 'arm' consists of two links with a fixed rotary joint at point $\mathcal{O}$ a 'knuckle' joint at point $\mathcal{P}$ with the end 'effector' at point $\mathcal{Q}$. The configuration is described by the angles q_1 and q_2. The angle q_1 describes the positive rotation of the unit vector $\boldsymbol{a}_1$ fixed in the link $\mathcal{O} - \mathcal{P}$ with respect to the fixed unit vector $\boldsymbol{a}_1$. The angle q_2 describes the rotation angle of the link $\mathcal{P} - \mathcal{Q}$, the unit vector $\boldsymbol{b}_1$ being fixed in this link. The lengths of the links are taken as L_1 and L_2. Each of the vector sets $\boldsymbol{n}_j, \boldsymbol{a}_j$ and $\boldsymbol{b}_j$ are orthogonal unit vectors. We will now use symbolic computation to calculate the rate of change of the end point for an observer fixed in the $\boldsymbol{n}_j$ system with respect to the angles q_1 and q_2. This will also allow us to introduce the Maple array structure.

As a first attempt at a simple notation that distinguishes in which frame components are taken, indicate the frame name by an upper case letter attached to the name of the vector. Refer to the $\mathcal{O} - \mathcal{P}$ link as r1 and the $\mathcal{P} - \mathcal{Q}$ link as r2. The components of these vectors can then be represented as arrays using the Maple vector function (available only after loading the linear algebra package):

```
>r1A := vector([L1,0]);
```

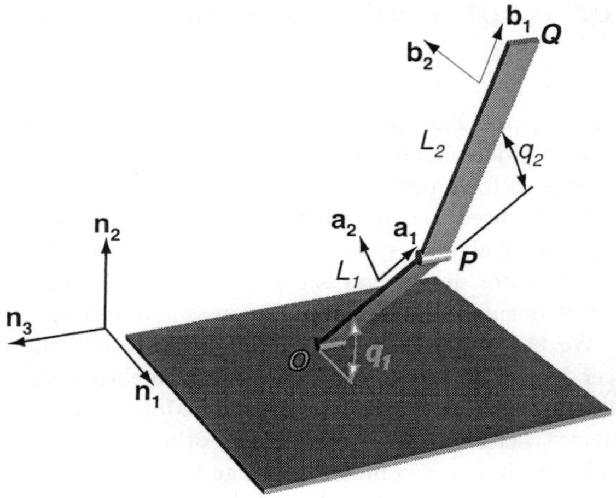

Figure 2.6: Simple Articulated Joint

$$[L1, 0]$$

```
>r2B := vector([L2,0]);
```

$$[L2, 0]$$

The Maple 'type' function returns true or false when given a data object and the name of one of the recognized data structures, thus:

```
>type(r1A,vector);
```

true

```
>type(r1A,list);
```

false

```
>type(r1A,matrix);
```

false

Some other recognized types are 'integer', 'float', 'odd', and 'even'. The Maple help facility provides a complete list of types. The basic step in calculating derivatives of vectors with respect to different frames was seen to be the transformation of the components into correct values for the respective frames. The most orderly technique for accomplishing this is to use the direction cosine matrices introduced earlier in the chapter. In the present two dimensional example this is done with the following commands using a notation that is close to that used in the text:

```
>rAN := matrix([[cos(q1),sin(q1)],[-sin(q1),cos(q1)]]);
```

$$\begin{bmatrix} \cos(q1) & \sin(q1) \\ -\sin(q1) & \cos(q1) \end{bmatrix}$$

```
>rBA := matrix([[cos(q2),sin(q2)],[-sin(q2),cos(q2)]]);
```

$$\begin{bmatrix} \cos(q2) & \sin(q2) \\ -\sin(q2) & \cos(q2) \end{bmatrix}$$

The related inverse transformations are simply obtained by transposition:

```
>rNA := transpose(rAN);
```

$$\begin{bmatrix} \cos(q1) & -\sin(q1) \\ \sin(q1) & \cos(q1) \end{bmatrix}$$

```
>rAB := transpose(rBA);
```

$$\begin{bmatrix} \cos(q2) & -\sin(q2) \\ \sin(q2) & \cos(q2) \end{bmatrix}$$

Finally the transformations between the N and B frames are obtained by matrix multiplication and transposition

```
>rNB := multiply(rNA,rAB);
```

$$\begin{bmatrix} \cos(q1)\cos(q2) - \sin(q1)\sin(q2) & -\cos(q1)\sin(q2) - \sin(q1)\cos(q2) \\ \sin(q1)\cos(q2) + \cos(q1)\sin(q2) & \cos(q1)\cos(q2) - \sin(q1)\sin(q2) \end{bmatrix}$$

```
>rBN := transpose(rNB);
```

$$\begin{bmatrix} \cos(q1)\cos(q2) - \sin(q1)\sin(q2) & \sin(q1)\cos(q2) + \cos(q1)\sin(q2) \\ -\cos(q1)\sin(q2) - \sin(q1)\cos(q2) & \cos(q1)\cos(q2) - \sin(q1)\sin(q2) \end{bmatrix}$$

At this point it is worthwhile to see how individual matrix entries can be retrieved. This should be clear from the following illustration:

```
>rAN[1,2];
```

$$\sin(q1)$$

The entries of the matrix rNB and rBN can clearly be simplified by use of trigonometric angular addition identities. Maple has a special operation called 'combine' which is useful for this task. It takes two arguments, an expression and the name of the type of required combinations, which in the present case is 'trig'. We illustrate this by applying the combine operation to the left upper diagonal component of the rNB direction cosine matrix:

```
>combine(rNB[1,1],trig);
```

$$\cos(q1 + q2)$$

The Maple 'map' function, introduced in the last chapter, also works on array structures as well as lists. Here is an example showing how the entire rNB direction cosine matrix can be simplified in one step:

```
>rNB := map(combine,rNB,trig);
```

$$\begin{bmatrix} \cos(q1 + q2) & -\sin(q1 + q2) \\ \sin(q1 + q2) & \cos(q1 + q2) \end{bmatrix}$$

The 'multiplication' function, which is part of the linear algebra package, takes matrices for arguments and, assuming that they are conformable, applies matrix multiplication to them. With this function and the above defined direction cosine matrices it is a simple matter to transform vector components from one standard frame to another. Use the notation r3N to indicate the vector representation in the N frame of the distance between $\mathcal{O}$ and $\mathcal{Q}$. To obtain this vector transform r1A and r2B to the N frame. In our notational convention these will be noted as r1N and r2N. Once this is done apply the Maple 'add' function, which is also contained in the linear algebra package and which simply adds arrays component by component. Thus:

```
>r1N := multiply(rNA,r1A);
```

$$[\cos(q1)L1, \sin(q1)L1]$$

```
>r2N := multiply(rNB,r2B);
```

$$[\cos(q1 + q2)L2, \sin(q1 + q2)L2]$$

```
>r3N := add(r1N,r2N);
```

$$[\cos(q1)L1 + \cos(q1 + q2)L2, \sin(q1)L1 + \sin(q1 + q2)L2]$$

The derivatives of this vector with respect to q_j can now be obtained using the standard 'diff' function, thus in an obvious notation:

```
>r3Nq1 := map(diff,r3N,q1);
```

$$[-\sin(q1)L1 - \sin(q1 + q2)L2, \cos(q1)L1 + \cos(q1 + q2)L2]$$

```
>r3Nq2 := map(diff,r3N,q2);
```

$$[-\sin(q1 + q2)L2, \cos(q1 + q2)L2]$$

The pattern of the above calculations can be used to solve most of the problems that arise when dealing with problems of frame based vectors. It is also clear that they involve a considerable amount of repetition. For this reason it is desirable to create a framework of Maple functions designed specifically to handle the situations that arise in technical mechanics. In the next section we begin the task of carrying out this program.

2.10 Sophia

The use of a standard symbolic manipulation system such as Maple can be of great assistance for formulating and solving problems in mechanics. A careful study of the example of the last paragraph also shows that a number of calculations follow standard patterns. It is therefore useful to create a specific package of tools for use with any given symbol manipulation system. A slightly more general point of view is to imagine a special system designed specifically for mechanics problems. Such a system might be implemented in its own right or it might be set on top of an existing package such as Maple. This point of view leads us to a very useful idea, an algorithmic language for describing and solving problems in mechanics. Therefore in what follows the reader can take the point of view that we are developing a technique for the general description of mechanics problems. The advantage of a means of specifying algorithmic descriptions for such problems is in the elimination of ambiguity. The prescriptions provided for describing and solving a problem must be understood by a concrete software system, hence they must be clear and orderly. We will call our language for this task 'Sophia', after the 19th century mathematician and contributor to mechanics, Sophia Kovalevsky. This immigrant to Sweden from the Russian nobility not only found a new class of solutions to the problem of the heavy top, but was the first woman to be appointed as Professor in a European University. There is no doubt that Professor Kovalevsky knew her mechanics very well indeed, so we may fantasize that in using the Sophia language we are describing our own problems to her. The implemented language running a particular problem might even be taken as a private consultation with her.

We will introduce 'Sophia' as we advance through the subject of this text. Eventually we will examine in depth the actual implementation of the Sophia statements in Maple. The set of programs which bring Sophia to life in Maple are provided in the appendix to this text. If you are using the text in a formal course the instructor will show you how to access the programs on your particular system. Sophia is useful even if you do not have access to an implemented version. Sophia's descriptive statements provide both a formal description of a mechanism and the steps to formulating useful equations of motion. At this stage we will not discuss how the Sophia statements are implemented but simply introduce them. Our first task will be to do this for the material in the present chapter. We will use the previous example of the simple articulated arm to illustrate the material.

2.10.1 Sophia Data Objects

Our algorithmic description of mechanisms requires data objects that correspond to vectors and dyads. In addition data objects must contain the relations to standard reference frames. Probably the most important of these is the vector. We will call a frame based vector in Sophia an 'Evector', the 'E' indicating that it pertains to the three dimensional Euclidian space familiar to applications of Newtonian mechanics. The Evector consists of two parts, a set of components and the identity of the standard triad to which the components refer. In the Maple implementation a function, Evector, outputs the data structure. It takes four arguments, the components of the vector in some frame and the name of the frame. The data object can be assigned to any desired symbol. It is useful to use symbols that have some meaning for the problem at hand. Thus a displacement Evector which is parallel to the unit vector $\boldsymbol{a}_1$ and has length L_1 is assigned to the symbol r1A by the statement:

```
>r1A := Evector(L1,0,0,A);
```

$$[[L1, 0, 0], A]$$

The semicolon causes Maple to show the output, which is the data object consisting of an array and a symbol, all gathered into a list. Another displacement, now parallel to the unit frame vector $\boldsymbol{b}_1$ is assigned to the symbol r2B:

```
>r2B := Evector(L2,0,0,B);
```

$$[[L2, 0, 0], B]$$

The first of the above corresponds in standard vector notation to the expression

$$\boldsymbol{r}_1 = L_1 \boldsymbol{a}_1.$$

For emphasis note again that in the Maple implementation the Evector is a list. The first element is a Maple vector which is an array containing the three components with respect to a given frame triad. The second component is the name of the frame triad. Thus using Maple's built in selector functions:

```
>r1A[1];
```

$$[L1, 0, 0]$$

while

```
>r1A[2];
```

$$A$$

This way of getting at the parts of an Evector depends on the particular implementation. A better way of doing the decomposition is to use specific operators for taking apart the data object. Sophia uses monadic operators for this purpose. Thus the above decomposition is accomplished by the statements:

```
>&vPart r1A:
>&fPart r1A:
```

The advantage of using such 'selector operators' is that we can change the basic way the Evector object is implemented and simply by defining new constructor and selector procedures retain previously written solutions. This process, called 'data abstraction', also allows the use of meaningful names and symbols for the operations. Individual components of the vector can be obtained with the command `Ec(eVector,axis)`. Thus

```
>Ec(r3N,2);
```

$$\sin(q1)L1 + L2\ \sin(q1)\cos(q2) + L2\ \cos(q1)\sin(q2)$$

Notice that the selector functions for picking out the component array and the frame name did not use parenthesis, though one could use them if so desired. This style of action in Maple is called an operator. Operators can be monadic as above, that is they act on the Maple object which follows. They can also be dyadic, an example being the common addition operation $a + b$. Operators that are added to the Maple system are denoted by the ampersand symbol '&', as seen in our selectors. In many cases we will provide an operator and a functional alternative for Sophia commands. Thus instead of the above way of defining an Evector you can use:

```
>r1A := A &ev [L1,0,0]:
```

While this latter style may take a little getting used to it can also be easier to understand due to the shorter and more concise form of the defined expressions. Another example is the operator form of the component selection command, thus to obtain the first component of r1A type

```
>r1A &c 1:
```

It is also convenient to be able to produce unit vectors in any frame. This is accomplished with the operator

```
&>
```

thus:

```
> A&>1;
```

$$[[1,0,0],A]$$

We have defined an object, the Evector, which contains frame information in the form of the frame name. It does not specify how frames are related to one another. In the articulated arm example we saw the need to define direction cosine arrays containing the information needed to transfer vector representations between frames. In most cases in mechanics the most direct way to specify the relationships between frames is by a sequence of simple rotations, i.e. rotations which take place about the axis of the involved frames. We will almost exclusively use this technique to set up frame relationships. To implement this technique introduce the rotation list, which is a list of lists. Each individual list contains four pieces of information, the name of the initial frame, the target frame, the axis of rotation (common to the two frames) and the angle of rotation. It is important to carefully note that this angle is taken as positive when the rotation proceeds from the initial to the target frame in the counter-clockwise sense about the rotation axis. That is rotations are positive or negative in accordance with the common 'right hand rule'. This of course follows the pattern discussed in the section on 'simple angular velocity'. The relationships between the N, A and B coordinate systems of the example are especially simple. A new data structure, the *simple rotation list* is now introduced.Thus the rotation from N to A is specified by the list [N, A, 3, q1], a rotation of q1 radians about the common 3 axis proceeding from N to A. A *rotation list* consists of a list of simple rotation lists. It should not be too difficult for the reader to see that the rotation list for the problem is [[N, A, 3, q1], [A, B, 3, q2]]. To tell the Sophia system that these are the required relations simply issue the command:

```
>chainSimpRot([[N,A,3,q1],[A,B,3,q2]]);
```

$$\begin{bmatrix} \cos(q1)\cos(q2) - \sin(q1)\sin(q2) & \sin(q1)\cos(q2) + \cos(q1)\sin(q2) & 0 \\ -\cos(q1)\sin(q2) - \sin(q1)\cos(q2) & \cos(q1)\cos(q2) - \sin(q1)\sin(q2) & 0 \\ 0 & 0 & 1 \end{bmatrix}$$

This command returns one of the direction cosine matrices among the frames. In fact all the needed matrices are stored in the system with names that follow the conventions of this chapter, i.e. RNA, RNB, RBN etc.. You are in for a surprise however if you expect to see the contents of these matrices by typing `RNA;` at your keyboard. In most cases Maple evaluates all symbols until it arrives at objects which only evaluate to themselves. Arrays however are an exception, their evaluation comes under the rule of the *last name*. Thus when you type the name of a symbol which has an array as its value it will only be evaluated up to the name of the array, the 'last name'. To get at the value of the symbol, i.e. the array itself you must use the Maple function 'op'. Thus:

```
>RNA;
```

$$RNA$$

```
>op(RNA);
```

$$\begin{bmatrix} \cos(q1) & -\sin(q1) & 0 \\ \sin(q1) & \cos(q1) & 0 \\ 0 & 0 & 1 \end{bmatrix}$$

To obtain the vector between the base and the end effector of the arm we need to add the displacements r1A and r2B. In the last section this required us to make use of the direction cosine matrices and to keep some careful book keeping so as to retain track of the needed objects. This is now done automatically, thus to add the vectors we can use the following:

```
>r3N := EaddList([r1A,r2B],N):
```

which gives the previously obtained result. Note that EaddList contains two arguments, a list of the Evectors which are to be added and the name of the frame in which the result is to be represented. An addition operator is also provided, however it presents its results in the frame of the second Evector. Thus

```
>r3B := r1A &++ r2B;
```

$$[[\cos(q2)L1 + L2, -\sin(q2)L1, 0], B]$$

It is of course useful to have a command which transfers Evector representations from one frame to another. Our name for this is 'sexpress' which stands for express and simplify and which has the operator form &to. Thus the following statements are equivalent:

```
>r3N := sexpress(r3B,N):
```

and

```
>r3N := N &to r3B:
```

Here is an example with output shown:

```
>r1N := N &to r1A;
```

$$[[\cos(q1)L1, \sin(q1)L1, 0], N]$$

The derivatives of this quantity with respect to the coordinate parameters can now be found as in the previous section. As frame based differentiation is such a common operation in mechanics it is convenient to introduce commands which carry it out and automatically take care of required transformations between representations. In addition we would like to have a convenient way of dealing with frame based differentiation when the coordinate parameters are functions of time. In chapter one we set up substitution sets that converted symbols such as q1 to quantities which indicated time dependences of the form q1(t). This is of course quite specific to Maple. In other systems time representation might take alternative forms. For the most part it is not necessary to indicate such a direct time representation and we will use the convention in Sophia that symbols of the form q1t and q1tt represent first and second derivatives of the quantity q1 with respect to time. Time itself will alway be represented by the symbol 't'. In Sophia the fact that a quantity is time dependent must be declared by the 'declare' statement. If more than one quantity is involved one can use the 'declareList' statement. The first takes a single quantity while the second takes a list of quantities as argument. Thus

```
>declare(q1);
```

q1 declared

```
>declare(q2);
```

q2 declared

The same result is obtained using

```
>declareList([q1,q2]):
```

Another useful form is given by `dependsTime(q1,q2,....)`, which can take an arbitrary number of arguments. At any point in a calculation additional parameters can be so declared. No problem occurs if a parameter is declared as time dependent more than once. It is important to note that a parameter will only be considered a time dependent function from the point in a calculation where it has been declared as such.

Once this step is taken we can use Sophia's frame based differentiation commands on functions with depend on time. Here is a list of ones we have defined:

```
diffTime(scaler)
diffVectorTime(Evector)
diffFrameTime(Evector,Frame)
Frame    &fdt  Evector
diffFrame(Evector,variable,Frame)
DiffAngVelOp(Evector,AngularVelocity,answerFrame)
```

The first operates on scalars, e.g. q1, and differentiates them with respect to time. No frame information is involved, thus

```
>diffTime(q1);
```

q1t

The second differentiates a vector with respect to time but does not take account of frame information, thus the expression of the representation is left to the user. The third and its operator syntax equivalent are the most important. It differentiates Evectors with respect to time and a specific frame. For example:

```
>diffFrameTime(r1A,N):
```

$$[[0, \sin(q1)^2 q1t\, L1 + \cos(q1)^2 q1t\, L1, 0], A]$$

Note that an obvious simplification has not occurred. Simplification is usually a time consuming operation. For this reason it is generally left to the user to decide when to simplify an expression. It will be evident that a number of simplification operations are available. Here is another example in operator form:

```
> B &fdt r1A;
```

$$[[0, -\sin(q2)^2 q2t\, L1 - \cos(q2)^2 q2t\, L1, 0], A]$$

The command diffFrame(Evector,variable,Frame) permits frame based differentiation with respect to variables which are explicitly present in expressions, for example q1. Thus:

```
>diffFrame(r1A,q1,N);
```

$$[[0, \sin(q1)^2 L1 + \cos(q1)^2 L1, 0], A]$$

Note again that automatic simplification is not included in the final result. Apply a simplification function defined for Evectors, thus:

```
>Esimplify(diffFrame(r1A,q1,N));
```

$$[[0, L1, 0], A]$$

The final differentiation command provides a means of using angular velocity information directly to compute frame based derivatives. This function is simply an implementation of the formula:

$$\frac{{}^{A}d}{dt}\boldsymbol{w} = \frac{{}^{B}d}{dt}\boldsymbol{w} + {}^{A}\boldsymbol{\omega}^{B} \times \boldsymbol{w}$$

Suppose that w_j are the components of angular velocity between frames N and B as represented in frame A. We can then write

```
>wNBA := Evector(w1,w2,w3,A):
r2BtN := DiffAngVelOp(r2B,wNBA,N);
```

$$[[-q2t\ L2\ \sin(q1 + q2) - w3\ L2$$

$$w3\ L2\ \cos(q1 + q2), w1\ L2\ \sin(q2) - w2\ L2\ \cos(q2)], N]$$

The angular velocity ${}^{A}\boldsymbol{\omega}^{B}$ between two frames can also be conveniently calculated using the Sophia command `A &av B`. In using the Sophia command we can think of the angular velocity of frame B in A. One reason for this inversion is to obtain some conformity with other computer algebra systems for mechanics. Taking as an example the calculation of ${}^{N}\boldsymbol{\omega}^{B}$ in the articulated arm problem, we find:

```
>wNB := N &aV B;
```

$$[[0, 0, q1t + q2t], B]$$

as expected from our previous discussion of simple angular velocity. Note that the answer is in given in the 'target' frame, which in this case is B. Of course the result would be the same in the frame N. Suppose we wished to determine the components of wNBA taken above as w1, w2 and w3. This is easily accomplished using the `sexpress` command. Thus

```
>wNBcoord := sexpress(wNB,A);
```

$$[[0, 0, q1t + q2t], A]$$

Because in this case the angular velocity is in the direction of the rotation axis the components are the same as in the B representation. The Evector is of course different as the frame name is now A. It is trivial to see that only the w3 component is non-zero, with value $q1t + q2t$.

Dyads, like vectors, must be considered in terms of components in each frame. Sophia includes a data object called an Edyad. An Edyad is formed with the command

```
Edyad(d11,d12,d13,d21,d22,d23,d31,d32,d33,FrameName)
```

The Edyad we will have most to do with is the representation of the inertia tensor. Later we will describe a special Sophia command for dealing with this symmetric tensor, which in general has only 6 independent components. The commands for determining the components in a different reference frame also work on the Edyad structure.

To complete this introduction to Sophia we introduce the command for carrying out the 'dot' product among Evectors and Dyads. This is

```
&o
```

where the arguments may be Evectors or Edyads and may be expressed in any frame as long as the proper frame relationships are known to the system. The vector cross product is given by the command

```
EcrossMixed(Ev1,Ev2,answerFrame)
```

This can also be expressed in terms of the infix operator &xx where the result is automatically presented in the frame of the Evector on the right side of the operator. The result can be expressed in the frame of the other Evector by taking the negative of the reversed ordered product, i.e.

```
(e1 &xx e2)
```

is in the same representation as e2, i.e. its frame is

```
&fpart e2
```

To multiply an Evector by a scalar use the command

```
Esm(scaler,Evector)
```

which has the operator counterpart

```
scaler &** Evector
```

The vector cross product of e1 into e2 expressed in the frame of e1 is thus

```
(-1) &** (e2 &xx e1)
```

This multiplication operator in combination with Evector addition gives us another way of inputing Evectors. Thus

```
(x&**(A&>1)) &++ (y&**(A&>2)) &++ (z&**(A&>3))
```

is equivalent to the statement

```
A &ev [x,y,z]
```

The advantage of the former is that it is easy to mix frames in the specification of a vector. Both methods will be used.

We now have almost enough Sophia to deal easily with the material presented earlier in this chapter. In the next section we will once again examine some of the material covered in this chapter from the Sophia viewpoint.

2.10.2 Sophia in Action

Sophia provides us with two major benefits. The most obvious is the reduction of tedious and error prone expression manipulation. Less obvious but of equal or more importance is that we can express the formulation of a problem as a series of Sophia statements. These statements not only lead to the solution by use of the Maple system but also provide a unique and clear prescription of how the problem is both specified and solved! To see this it is useful to examine some of the previous work of this chapter with the aid of Sophia.

In section 2.2 (figure 2.3) we considered the problem of expressing the vectorial distance between two points in a mechanism consisting of three hinged plates. This involved the use of frames in each plate as well as the specification of suitable direction cosine matrices. Using Sophia we can solve the given problem by the following sequence of statements:

```
>rotList:= [ [A,B,2,q1],[B,C,1,q2]]:
>chainSimpRot(rotList):
>r01 := A &ev [-L,0,0]:
>r12 := B &ev [-L,0,0]:
>r23 := C &ev [0,L,0]:
>r03 := ( r01 &++ r12) &++ r23;
```

$$[[-L - \cos(q1)L + \sin(q1)\sin(q2)L, \cos(q2\ )L, \sin(q1)L + \cos(q1)\sin(q2)L], A]$$

The reader should compare this with the discussion of the illustration in section 2.2. The rotList variable gives a definite and clear statement of the relations between the frames in the plates. The vector between the desired points is easily expressed in terms of simple Evectors in each frame. Finally the end result is given in the desired frame. Once one is familiar with the language this can be seen as a simplification of the more lengthy discussion of the problem solution given previously. A great advantage is that when run with the Sophia program we even obtain an explicit result which we can manipulate as we will. For example it is no problem at all to put the result into another frame, thus in C it would appear as:

```
>C &to r03;
```

$$[[-L - \cos(q1)L, L - \sin(q1)\sin(q2)L, -\sin(q1)\cos(q2)L], C]$$

For another example consider the pivoted rod used as an illustration in section 2.5. The task was to find the angular velocity dyad of the rod. A series of Sophia statements easily gives the angular velocity vector in frame B:

```
>rotList := [ [A,K,3,theta],[K,B,2,-psi]]:
>chainSimpRot(rotList):
>dependsTime(theta,psi):
>wAB:= angularVelocity(B,A);
```

$$[[\sin(\psi)\,thetat, -psit, thetat\,\cos(\psi)], B]$$

The angular velocity dyad is simply the skew symmetric matrix that corresponds to this vector. For convenience two Sophia functions have been defined as Maple monadic operators for this task. They are:

```
&VtoD     taking a vector to a skew symmetric dyad
&DtoV     taking the skew symmetric part of a dyad to a vector
```

Therefore the rotation dyad is computed by the Sophia statement:

```
>WAB := VtoD(wAB);
```

$$[\begin{bmatrix} 0 & -thetat\,\cos(\psi) & -psit \\ thetat\,\cos(\psi) & 0 & -\sin(\psi)\,thetat \\ psit & \sin(\psi)\,thetat & 0 \end{bmatrix}, B]$$

In section 2.8, as part of the discussion of simple angular velocity, it was shown how the angular velocity associated with the Euler Angles is calculated. This is in fact a somewhat tedious task, especially as there are many ways to define three angular rotations that relate two frames in space. For the case given in the text the angular velocity is found by the following Sophia statements:

```
>rotList := [[A,H,3,phi],[H,W,1,theta],[W,B,3,psi]]:
>chainSimpRot(rotList):
>dependsTime(theta,psi,phi):
>wABA := A &to (A &av B);
```

$$[[\cos(\phi)\,thetat + \sin(\phi)\sin(\theta)\,psit,$$

$$\sin(\phi)\,thetat - \cos(\phi)\sin(\theta)\,psit,$$

$$phit + \cos(\theta)\,psit], A]$$

In section 2.9 this result was used to obtain the angular acceleration. The Sophia frame based differentiation command is simply applied to the angular velocity, thus:

```
aABA := H &to (A &fdt wABA);
```

$$[[thetatt + phit\ \sin(\theta) psit,$$

$$phit\ thetat - \cos(\theta) thetat\ psit - \sin(\theta) psitt,$$

$$phitt - \sin(\theta) thetat\ psit + \cos(\theta) psitt], H]$$

This is expressed in frame H for comparison with the earlier results.

2.10.3 Four Bar Linkage

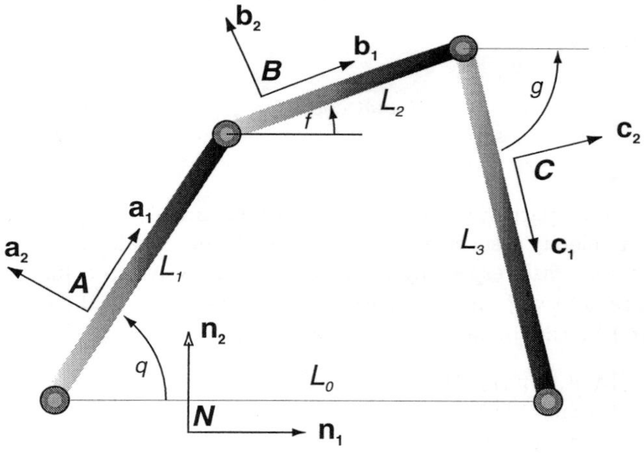

Figure 2.7: Four Bar Linkage

A somewhat more involved example for study is the so called four bar linkage. In fact this consists of three movable links two of which are fixed to 'ground' points as indicated in the figure. The length of the respective links are given as L_0 for the ground link and L_1 through L_3 for the moving links. Without loss of generality we can measure all lengths as ratios with the ground link, hence we take the ground link as having unit length, i.e. length 1. The orientation of each link is specified with respect to the ground link and is given by the angles q, f and g as indicated. If, as usual, the angles are measures with the counter-clockwise direction being positive the indicated angle for g in the figure would be negative for the shown orientation. Assign a standard frame to each link with the 1 base vector along the link as indicated by the arrow in the figure. Thus the $\boldsymbol{n}_1$ vector defines the orientation of the ground link and $\boldsymbol{n}_1 \cdot \boldsymbol{a}_1 = \cos(q)$. Clearly the angles q, f and g are not independent but are related by the constraint of the linkage. To express this constraint note that the sum of the vector distances along the linkage must vanish. This condition is easily implemented with Sophia's help:

```
>rotList := [[N,A,3,q],[A,B,3,f],[B,C,3,g]]:
>chainSimpRot(rotList):
>r0 := N &ev [1,0,0]:
>r1 := A &ev [L1,0,0]:
>r2 := B &ev [L2,0,0]:
>r3 := C &ev [L3,0,0]:
>r01230B := (r1 &++ r2) &++ (r3 &++ ((-1) &** r0)):
>eq1 := (r01230B &c 1) = 0:
>eq2 := (r01230B &c 2) = 0:
>eqs := combine({eq1,eq2},trig);
```

$$\{\sin(g)L3 - \sin(f)L1 + \sin(q+f) = 0,$$
$$\cos(f)L1 + L2 + \cos(g)L3 - \cos(q+f) = 0\}$$

From the above two relations we can see that the geometric state of the four bar linkage depends on only one angle, i.e. the mechanism has only one degree of freedom. One of the angles can be taken as independent and the other two can then be expressed for example as functions of q. The problem with this is that the problem has multiple solutions and one must exert care in tracking different branches. In kinematics we are interested in describing the motions that are possible for constrained mechanisms irrespective of the dynamics and forces. In fact all the kinematic information needed for the present problem is obtainable from the above equations. A useful way of extracting this information, especially with the availability of numerical routines and computers, is to obtain the rate of change of the dependent angles in terms of the independent angle. For the linkage problem this leads to two first order differential equations for the derivatives of f and g with respect to q. A possible initial state is first determined with the aid of the constraint relations. A kinematic motion from this state is then found by integrating the differential equations. Bifurcations to different branches are seen as special points in the space of dependent variables. All this is most easily seen from the study of particular examples. The first step is to use Maple to obtain the required differential relations. Thus:

```
>toFunction:={f=f(q),g=g(q),fq=diff(f(q),q),gq=diff(g(q),q) }:
>toExpression:={f(q)=f,g(q)=g,diff(f(q),q)=fq,diff(g(q),q)=gq}:
>eqs := subs(toExpression,diff(subs(toFunction,eqs),q)):
>kineEqs := combine(subs(toFunction,solve(eqs,{fq,gq})),trig):
>kineEqs := subs(toExpression,kineEqs);
```

$$\{fq = \frac{\sin(g+q+f)}{L1\ \sin(f+g) - \sin(g+q+f)},$$

$$gq = \frac{L1\ \sin(q)}{L3\ (L1\ \sin(f+g) - \sin(g+q+f))}\}$$

2.11 Problems

•Problem 2.1

The line $\mathcal{O} - \mathcal{Q}$ is fixed in the reference frame A. A plane mechanism consists of a collar, free to move along the line, a rod, of length l_1, between the collar and the center of a disc of radius l_2. The disc is free to rotate. The motion is in a plane orthogonal to the vector $\boldsymbol{a}_3$, with $\boldsymbol{a}_2$ along the line $\mathcal{OQ}$, the triad a and the point $\mathcal{O}$ forming a reference frame. The rod connecting the collar to the disc is at an angle ϕ to the vector $\boldsymbol{a}_1$, and a line fixed in the disc is at an angle ψ to the rod. Attach a triad b to the disc such that the vector $\boldsymbol{b}_1$ is parallel to this fixed line. The angles ϕ and ψ as well as the distance of the collar from the point $\mathcal{O}$ are to considered as functions of the time t. A vector $w = \boldsymbol{b}_1 + t\boldsymbol{b}_2$. Find $\frac{{}^A\partial w}{\partial \phi}$, $\frac{{}^A\partial w}{\partial \psi}$ and $\frac{{}^A dw}{dt}$ and express the results in terms both the triads a and b.

•Problem 2.2

(a) Find the projection operator dyad for the plane perpendicular to the vector $\sin t\boldsymbol{a}_1 + \cos t\boldsymbol{a}_2 + t\boldsymbol{a}_3$.

(b) Find the matrix representation of this dyad for a triad b in which $\boldsymbol{b}_3 = \boldsymbol{a}_3$ and $\boldsymbol{a}_1 \cdot \boldsymbol{b}_1 = \sqrt{3}/2$.

•Problem 2.3

Describe the geometric position of a disc rolling on a plane. In this description make use of 4 triads. The triad a is taken to be at rest with the disc rolling on a plane parallel to the plane orthogonal to $\boldsymbol{a}_3$. The triad b is such that $\boldsymbol{b}_3 = \boldsymbol{a}_3$ and the vector $\boldsymbol{b}_1$ is tangent to the edge of the disc touching the plane on which the rolling occurs. The triad c is defined so that $\boldsymbol{c}_1 = \boldsymbol{b}_1$ and $\boldsymbol{c}_3$ is perpendicular to the plane of the disc. Finally the triad d is fixed in the disc, with $\boldsymbol{d}_3 = \boldsymbol{c}_3$. Define all necessary angles and write expressions for all the direction cosine matrices that relate these various triads, e.g. $R_{ab}, R_{ac}, R_{ad}, R_{bc} \ldots$

•Problem 2.4

Show that any dyad can be expressed as the sum of a symmetric and antisymmetric dyad. Given an arbitrary dyad $\boldsymbol{D}$ define vect$(\boldsymbol{D})$ as vect of the antisymmetric part of $\boldsymbol{D}$.

•Problem 2.5

Show that the frame based derivative of the vectorial distance between two points in space is invariant for standard frames that are fixed in the same extended body.

Chapter 3

Configuration and Motion

The calculation of the behavior of a mechanical system requires a careful description of the system's geometrical form. The theory of systems composed of interconnected sets of rigid bodies must begin with such a geometrical description. Interconnection implies that the geometric form is constrained, hence a major task is to understand the implications of constraint. The motion of the system can be described in terms of the velocity and acceleration of all its components. This kinematical task will make use of the material concerning the relationships between observers discussed in the previous chapter. In the chapters that follow equations of motion are developed for the system in terms of its dynamic parameters and applied forces. This objective can be considerably simplified if we are able to eliminate direct consideration of the forces which constrain the mechanical systems geometry. The key idea for this objective is to determine certain directions at the points of application of forces. Consideration of multiple body systems is simplified by the introduction of an abstract configuration space. The required directions specify a tangent hyperplane to the instantaneous configuration hypersurface in such a space. Kane has singled out related geometric vectors, which he calls *partial velocities.* The ultimate purpose of this chapter is to derive this concept from a natural geometric viewpoint, that is to clearly answer the question as to what are partial velocities, why are they important and how they may be chosen and transformed!

3.1 Primary Observers

To describe a mechanical system we must have a *Viewpoint,* that is a place from which to make measurements. We also need a means for the measurement of distance and angles. This matter is formalized by defining a *primary observer* as a set of three objects, $\{A, a, \mathcal{O}\}$, that is a body, a standard triad prescribing the orientation, and a point fixed in the body. Note that the term body here is used in the sense of the last chapter. Thus a primary observer is an observer and a reference point. At times this reference point will be called the origin.

The primary observers main function is to view and describe a *mechanism.* The term mechanism is used to denote a restricted class of physical objects consisting of a set of *rigid* bodies. These bodies may or may not be connected, and some or all of them may simply consist of single points. In the following chapters various dynamic properties, such as mass, are assigned to these bodies. Some authors refer to a collection of particles and rigid bodies as a system, but it is felt that the term mechanism is more descriptive. The term is taken to encompass a range, from such objects as the solar system considered as a collection of interacting point masses, to a clockwork or robot arm composed of very stiff components. While the components may be stiff, interconnections using such idealizations as massless springs are allowed. Figure 3.1 shows a schematic view of a mechanism. The notation in the figure, which is used to denote bodies and orientations, is explained in the rest of this chapter.

The most important property of a mechanism at this stage is its *configuration.* A configuration is taken to be the totality of the mechanisms points and their spatial relationships. The points can be labeled in terms of numbers available to a primary observer. That is the primary observer can label or name all points of the mechanism, and given the label of any particular point can produce a vector, $\boldsymbol{r}^{\mathcal{OP}}$. Thus the primary observer can determine the distance and angular position from a viewpoint to any point in the mechanism. The requirement that the bodies which compose the mechanism are rigid considerably simplifies the task of labeling. For the purpose of

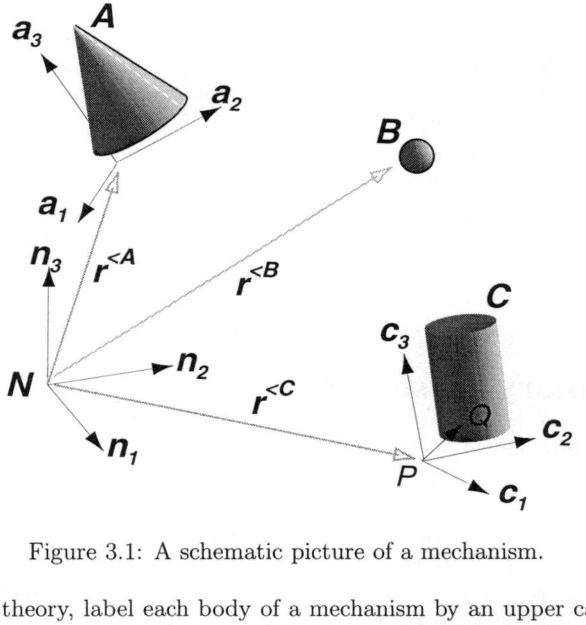

Figure 3.1: A schematic picture of a mechanism.

developing the theory, label each body of a mechanism by an upper case letter, e.g. N, A, B, C and assume that there are K bodies. For each body denote some point

$\mathcal{P}$ which is *fixed* to the body, an *index point.* In some cases the body will consist of only a single point and the index point will coincide with the entire body. It is also possible to choose an index point that is not within the confines of the physical body. The important property of the index point is that it is fixed for any observer attached to the body. A standard triad attached to the body is needed to specify the location of all its points, for example C . Points in each body can now be specified by referring them to the index point, that is by giving $\boldsymbol{r}^{\mathcal{PQ}}$ in terms of projections on the triad $c = (\boldsymbol{c}_1, \boldsymbol{c}_2, \boldsymbol{c}_3)^T$. The important fact here is that the body moves about as the configuration alters, but the internal labeling system of points in each body will not change! What will change is the position of the index points and the orientation of the standard triad associated with each body of the mechanism.

The results of the last chapter shows that each standard triad, a, may be specified by a relation of the form

$$a = R_{an} n, \tag{3.1}$$

relating it to the triad of the primary observer. The orthogonality of the matrix R_{an} implies that its elements are characterized by three independent quantities, i.e. that the equation $R_{an} R_{na} = 1$ specifies six independent relations between the nine components of the direction cosine matrix. The position vector $\boldsymbol{r}^{\mathcal{OP}}$ requires three quantities for its specification. Therefore each body of the mechanism can be characterized by six quantities. In the special case of point bodies only three quantities are required, as long as other circumstances do not call for an orientation being attached to the point in question. These considerations show that the configuration of a mechanism consisting of K bodies can usually be determined in terms of at most $6K$ parameters. Thus the configuration of a mechanism consisting of 4 unconnected rigid bodies moving about in space can be specified in terms of 24 coordinates giving the positions of reference points on the bodies and the attitude of each body. The quantities needed to specify a constrained configuration are called the *generalized coordinates* of the mechanism.

•Illustration

The mechanism sketched in Figure 3.2 consists of a rod, base fixed at the origin $\mathcal{O}$ of a prime observer and aligned with the $\boldsymbol{n}_3$ component of the observer's standard triad. The collar, A, which slides along the rod is taken as a 'point' object for this discussion. The rigid rod of length l connects the collar to the square plate, B, so as to allow the plate to rotate about the rod axis. The rod is pinned to the collar in the manner shown. We consider the bodies making up the mechanism to consist of the collar A and the plate B. The two rods provide the structure which maintains the configurational properties of the mechanism. Depending on the circumstances, one or both rods might also be considered part of the mechanism. Later when we discuss the dynamics of mechanisms we will assign mass to the involved bodies. In the context of such an assignment the rod connecting the collar to the plate would have negligible mass. The rod about which the collar moves is considered to be frictionless

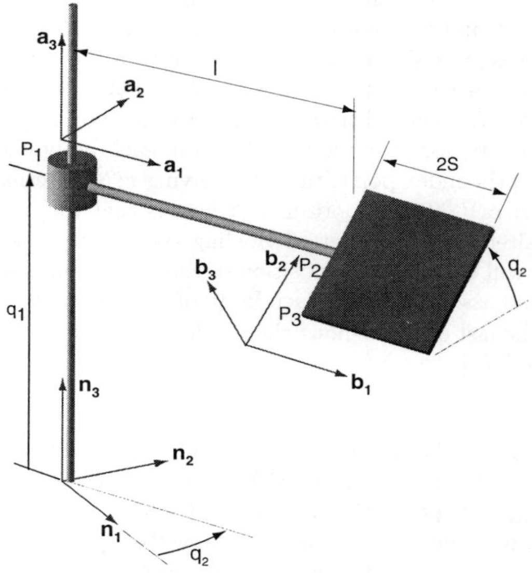

Figure 3.2: An example of a 'mechanism'.

and perfectly rigid. For our present purposes these matters are of no concern, it is only necessary to construct the mechanism in such a manner as to make it possible to specify all configurations which obey the given constraints!

To specify the configuration we provide a standard triad for each body, which we label as a and b for the collar and plate respectively. A set of generalized coordinates is given by the quantities q_1, q_2 and q_3; where q_1 is the distance of the collar from the prime observer's origin, q_2 is the angle between the vectors $\boldsymbol{n}_1$ and $\boldsymbol{a}_1$, and q_3 is the angle between $\boldsymbol{a}_2$ and $\boldsymbol{b}_2$. These are indicated in the figure. From the figure we can deduce the direction cosine matrices R_{an} and R_{ba} which relate the triads. Thus:

$$R_{an} = \begin{bmatrix} c_2 & s_2 & 0 \\ -s_2 & c_2 & 0 \\ 0 & 0 & 1 \end{bmatrix} \tag{3.2}$$

and

$$R_{ba} = \begin{bmatrix} 1 & 0 & 0 \\ 0 & c_3 & s_3 \\ 0 & -s_3 & c_3 \end{bmatrix}. \tag{3.3}$$

Using the results of chapter 2, we can now calculate:

$$R_{nb} = R_{na}R_{ab} = \begin{bmatrix} c_2 & -s_2c_3 & s_2s_3 \\ s_2 & c_2c_3 & -c_2s_3 \\ 0 & s_3 & c_3 \end{bmatrix}. \tag{3.4}$$

Using the direction cosine matrices it is possible to express the location of the index points $\mathcal{P}_1$ and $\mathcal{P}_2$ in terms of the primary observer's standard triad. Thus:

$$\boldsymbol{r}^{\mathcal{O}\mathcal{P}_1} = q_1\boldsymbol{n}_3, \tag{3.5}$$

and

$$\begin{aligned} \boldsymbol{r}^{\mathcal{O}\mathcal{P}_2} &= q_1\boldsymbol{n}_3 + l\boldsymbol{a}_1 \\ &= (0,0,q_1)n + (l,0,0)a \\ &= (0,0,q_1)n + (l,0,0)R_{an}n \\ &= (0,0,q_1)n + l(c_2,s_2,0)n \\ &= lc_2\boldsymbol{n}_1 + ls_2\boldsymbol{n}_2 + q_1\boldsymbol{n}_3. \end{aligned} \tag{3.6}$$

Take $\mathcal{P}_3$ on the indicated corner of the square plate as an example of an arbitrary point on the body B, so that:

$$\boldsymbol{r}^{\mathcal{O}\mathcal{P}_3} = \boldsymbol{r}^{\mathcal{O}\mathcal{P}_2} - s\boldsymbol{b}_2. \tag{3.7}$$

After suitable simplification this has the representation in frame N given by

$$(lc_2 + ss_2c_3)\boldsymbol{n}_1 + (ls_2 - sc_2c_3)\boldsymbol{n}_2 + (q_1 - ss_3)\boldsymbol{n}_3. \tag{3.8}$$

Using the Sophia routines the problem is treated by first stating the relationships among the frames. It is then a simple matter to find the position of the point $\mathcal{P}3$. The following statements do the job:

```
>rotList := [[N,A,3,q2],[A,B,1,q3]]:
>chainSimpRot(rotList):
>r01 := N &ev [0,0,q1]:
>r12 := A &ev [1,0,0]:
>r23 := B &ev [0,-s,0]:
>r03 := r01 &++ r12 &++ r23;
```

$$[[l, \sin(q3)\,q1 - s, \cos(q3)\,q1], B]$$

The result in the N frame is given by:

```
> N &to r03;
```

$$[[\cos(q2)l + \sin(q2)\cos(q3)s, \sin(q2)l - \cos(q2)\cos(q3)s,$$

$$q1 - \sin(q3)s], N]$$

•

In the illustration the orientation of body B was built up by using a series of single angle transformations. It then is relatively simple to specify the position of any arbitrary point on the body by using the frame vectors b which specify the body's orientation. In many of the illustrations that follow we will see this same technique adopted with suitable variations.

3.2 Constraint

The word mechanism is being used in a slightly non-conventional manner as it includes the case where no constraining geometrical relationships are imposed on the configuration of bodies. Thus the planets represented as point masses will interact dynamically, but we do not imagine geometric constraints such as strong cables attached between Mars and Jupiter. Many mechanisms are of interest precisely because of imposed geometric constraints on their configuration. A robot arm with its parts disconnected is an object for the repair shop, not the customer! When we impose constraints on a mechanism we depend on the strength properties of the materials composing the mechanism to maintain the desired class of configurations that are appropriate for the mechanisms function. Thus the material is expected to provide forces that will maintain the configuration. Therefore as soon as we start to specify *constraints* on the configuration of a mechanism a new task is introduced, the determination of the forces which will keep the family of configurations required during the motion of the mechanism. A primary task in formulating equations of motion is to deal with this fact.

In the case of a free mechanism consisting of K rigid bodies one needs $6K$ generalized coordinates to specify the configuration. These will also be referred to as configuration coordinates, a term that is somewhat more descriptive of the function of these quantities. Consider two unconnected bodies. Now connect them so that their index points are forced to coincide. It should be clear that we now need nine rather than twelve configuration coordinates to specify the geometric state of the bodies. The minimum number of configuration coordinates that uniquely specify the configuration of the mechanism will be called the mechanisms *geometric degree of freedom, n*. Denote these coordinates by the symbol q, standing for the set of quantities $\{q_1, q_2, \ldots, q_n\}$. Finding suitable generalized coordinates for a particular mechanism can be extremely difficult, or even impossible. In many cases more than one set of coordinates is required for a global description, leading to the concept of a manifold.

The present discussion requires only the possibility of having a local description. The specification of a set of generalized coordinates is closely analogous to the situation in a subject such as thermodynamics, where the theory assumes that there exist variables of state having certain properties, without giving the details of how to find such variables. The practical task of producing a set of generalized coordinates are mainly dealt with by means of examples. For the development of the theory simply assume that such a set is a part of the specification of the mechanism under study. It is also required that there exist certain kinematical relations between the rates of change of such coordinates and the coordinates themselves.

The generalized coordinates determine the geometric configuration. This means that *any* point in the mechanism can be specified by giving the values of the n generalized coordinates. Therefore if $\mathcal{P}$ is any point in the mechanism, the vector displacement $\boldsymbol{r}^{\mathcal{OP}}$ from the primary observers origin to the point will be a function of the q's, i.e.

$$\boldsymbol{r}^{\mathcal{OP}} = \boldsymbol{r}^{\mathcal{P}}(q_1, q_2, \ldots, q_n, t). \tag{3.9}$$

This is the essential *test* of a particular set of generalized coordinates, i.e. can *every point* in the mechanism be determined as a function of the generalized coordinates and the time! To simplify the discussion consider mechanisms composed of points. Then the location of $3K$ points determine the configuration. For a free mechanism, consisting of K points, any specific configuration requires $3K$ numbers. This means that any specific configuration can be thought of as the location of a point in a $3K$ dimensional cartesian space. This point is referred to as the *mechanism's configuration point.* In the example of a point confined to a surface, discussed in chapter one, we saw a case where constraint restricted the position to a two dimensional subspace of the space available to the free particle. In that case two generalized coordinates were sufficient to specify any allowed configuration. In general the $3K$ cartesian coordinates that specify the mechanism are functions of the n generalized coordinates. The constraint conditions confine the configuration point to an n dimensional surface embedded in the full $3K$ dimensional space of configurations allowed the free bodies composing the mechanism. This object is the mechanisms *configuration surface.*

It is possible to probe the configuration surface by changing the values of the generalized coordinates in a continuous manner. A systematic way of doing this is to take each generalized coordinate as a function of some parameter, say λ. Thus assume that $q_j = f_j(\lambda)$ for j running from 1 to n, and that the f_j are continuous and smooth functions. The configuration point will then traverse a curve that is embedded in the time fixed or frozen configuration surface. Such a curve is called a *test motion.* In general the configuration surface will have dimension greater than 2, however geometric intuition can be assisted using schematic pictures such as Figure 3.3.

In many cases the constraints are such as to make the configuration surface independent of time. One can also think of the situation in which they do depend on time as a sequence or family of time fixed surfaces. What is of interest here is the

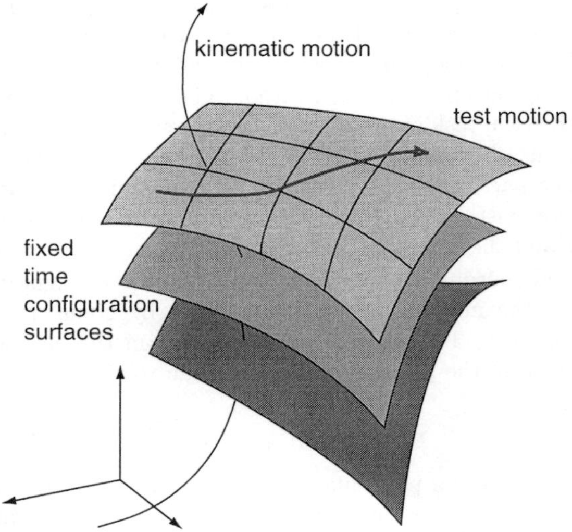

Figure 3.3: Configuration Surfaces

geometry of the configuration surface. If the constraints are independent of time this is determined by a single surface. If the constraints are time dependent it is necessary to consider the one parameter family of surfaces described by varying the time. The important geometric properties of the surface can be examined by following paths through the surface. In this sense kinematics is the study of curves embedded in the configuration hypersurface!

A *kinematic motion* of the mechanism will be determined by specifying all the generalized coordinates as functions of t, and then generating the curve for the configuration point. The curve generated in the configuration space will pass through different members of the family of configuration surfaces. If the constraints are not time dependent only one surface will be involved and it is not necessary for the path parameter to be the time. A path which is confined to one member of the family of time dependent configurations is a *test motion*. This geometric picture of the constraint condition is the key to understanding the mechanics of multibody mechanisms. Dynamics, which takes account of the body's inertial properties and the applied forces, acts in the arena of the configuration geometry. The motions observed are the result of the forces and the constraints invoked by the geometry.

The situation is more complicated for the case where the mechanism consists of rigid bodies rather than points. As above one can still consider a $3K$ dimensional space in which locations correspond to sets of coordinates for the index points of the involved bodies. As each body can vary its attitude, three attitude parameters are required for each body. Therefore the configuration space requires $6K$ dimensions and the configuration surface will be an n dimensional surface embedded in this space. This situation shall be treated in chapter 5.

•Illustration

The two body mechanism shown in Figure 3.4 provides a specific example that illustrates some of the above ideas. A rod is attached to a shaft at the origin of the prime observer. The prime observer's standard triad is designated by the column of orthogonal unit vectors a. The rod is forced to rotate about the shaft, i.e. about the $\boldsymbol{a}_3$ axis so that the projection of the rod onto the plane of $\boldsymbol{a}_1$ and $\boldsymbol{a}_2$ makes an angle $\theta(t)$ with the $\boldsymbol{a}_1$ axis. Note that this angle is a *given* function of time! Also the rod is at an angle μ to this plane. At a distance l from the prime observer's origin along the rod, another rod is attached so as to make an angle γ with the plane perpendicular to the first rod. Both rods serve as constraints for the motion of the particles located at points $\mathcal{P}_1$ and $\mathcal{P}_2$. In addition both rods are coplanar with the shaft. Thus the mechanism consists of 'beads' which are allowed to slide along the respective rods. There are a number of ways to choose generalized coordinates for this mechanism.

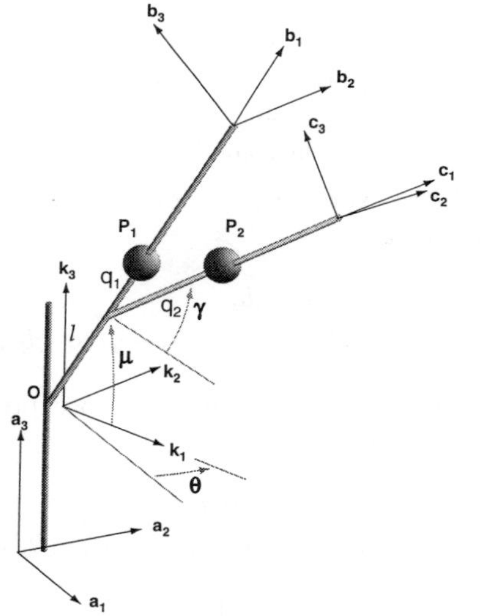

Figure 3.4: Two bead mechanism.

The constraints provide a fixed line for each of the two beads to move on. Therefore the mechanism has two degrees of freedom, one for each bead. Choose the quantities q_1 and q_2 to be the distances of the beads from the point where the two rods meet. It must now be shown that one may express the positions of the beads in the reference frame of the prime observer.

It is quite difficult to determine relations between sets of standard triads when more than one rotational angle is involved. Therefore it is usually helpful to introduce intermediate sets of triads. One can think of these as being fixed on fictitious bodies. In the present case use a triad $k = (\boldsymbol{k}_1, \boldsymbol{k}_2, \boldsymbol{k}_3)^T$, where $\boldsymbol{k}_3 = \boldsymbol{a}_3$ and $\boldsymbol{k}_1$ is at an angle $\theta(t)$ to $\boldsymbol{a}_1$. Therefore the direction cosine matrix relating the triads a and k is given by:

$$R_{ka} = \begin{bmatrix} \cos\theta & \sin\theta & 0 \\ -\sin\theta & \cos\theta & 0 \\ 0 & 0 & 1 \end{bmatrix} \tag{3.10}$$

The rod attached to the origin is now fixed with respect to the k triad in the plane of $\boldsymbol{k}_1$ and $\boldsymbol{k}_3$ and makes an angle μ with the vector $\boldsymbol{k}_1$. We also remove any ambiguity in the placement of the second rod by assuming that it is also in this plane. We introduce a triad b with $\boldsymbol{b}_1$ oriented along the first rod and with $\boldsymbol{b}_2 = \boldsymbol{k}_2$. Therefore we have the direction cosine matrix

$$R_{bk} = \begin{bmatrix} \cos\mu & 0 & \sin\mu \\ 0 & 1 & 0 \\ -\sin\mu & 0 & \cos\mu \end{bmatrix}, \tag{3.11}$$

hence

$$R_{ba} = R_{bk}R_{ka} = \begin{bmatrix} \cos\mu\cos\theta & \cos\mu\sin\theta & \sin\mu \\ -\sin\theta & \cos\theta & 0 \\ -\sin\mu\cos\theta & -\sin\mu\sin\theta & \cos\mu \end{bmatrix}. \tag{3.12}$$

We can now use the triad b to express the positions of both particles in terms of the generalized coordinates q. Thus

$$\boldsymbol{r}^{\mathcal{O}\mathcal{P}_1} = (l + q_1)\boldsymbol{b}_1 \tag{3.13}$$

$$\boldsymbol{r}^{\mathcal{O}\mathcal{P}_2} = l\boldsymbol{b}_1 + q_2\sin\gamma\boldsymbol{b}_2 - q_2\cos\gamma\boldsymbol{b}_3. \tag{3.14}$$

The direction cosine matrix R_{ba} can now be used to express these vectors in the reference frame of the prime observer, i.e.

$$\boldsymbol{r}^{\mathcal{O}\mathcal{P}_1} = (l + q_1, 0, 0)R_{ba}a \tag{3.15}$$

$$\boldsymbol{r}^{\mathcal{O}\mathcal{P}_2} = (l + q_2\sin\gamma, 0, -q_2\cos\gamma)R_{ba}a. \tag{3.16}$$

Carrying out the indicated operations we obtain

$$\boldsymbol{r}^{\mathcal{O}\mathcal{P}_1} = (l + q_1)\cos\mu\cos\theta\boldsymbol{a}_1 \tag{3.17}$$

$$+(l + q_1)\cos\mu\sin\theta\boldsymbol{a}_2 + (l + q_1)\sin\mu\boldsymbol{a}_3$$

$$\boldsymbol{r}^{\mathcal{O}\mathcal{P}_2} = ((l + q_2\sin\gamma)\cos\mu + q_2\sin\mu)\cos\theta\boldsymbol{a}_1 + \tag{3.18}$$

$$((l + q_2\sin\gamma)\cos\mu + q_2\sin\mu\cos\gamma)\sin\theta\boldsymbol{a}_2 +$$

$$((l + q_2\sin\gamma)\sin\mu - q_2\cos\gamma\cos\mu)\boldsymbol{a}_3.$$

In the above we expressed the position of the second bead in the frame b, however we could have introduced an additional frame c, as indicated in the figure, to assist in this task. In that case the displacement from the attachment point would have been expressed as $q_2\boldsymbol{c}_1$. It would also be required to introduce the direction cosine matrix relating frame c to frame b.

The positional equations describe a family of two dimensional surfaces embedded in the six dimensional space with coordinates

$$x_\beta^{(L)} = \boldsymbol{r}^{\mathcal{O}\mathcal{P}_\beta} \cdot \boldsymbol{a}_\beta, \tag{3.19}$$

where $L = 1, 2$ and $\beta = 1, 2, 3$.

A simple special case is given by the conditions that $\gamma = \mu = 0$, for which:

$$x_1^{(1)} = (l + q_1)\cos\theta \tag{3.20}$$
$$x_2^{(1)} = (l + q_1)\sin\theta \tag{3.21}$$
$$x_3^{(1)} = 0 \tag{3.22}$$
$$x_1^{(2)} = l\cos\theta \tag{3.23}$$
$$x_2^{(2)} = l\sin\theta \tag{3.24}$$
$$x_3^{(2)} = -q_2. \tag{3.25}$$

An example of a *test motion* would be $t = t_0$ fixed and $q_1 = \lambda$, $q_2 = \lambda$, with λ varying so as to form a curve in the surface appropriate to $t = t_0$. A kinematic motion would be given by $q_1 = Vt$, $q_2 = 2Vt$ and $\theta = \theta(t)$, with t varying to form a curve that passes through the family of constant t surfaces.

Because of the simplicity of the present example we could also imagine the constant t surfaces to be planes passing through the $\boldsymbol{a}_3$ axis. Each of these surface is one of the family of surfaces generated by the plane rotating about the axis, as shown in Figure 3.5. The configuration point would have the coordinates given by $(l + q_1)$ and $-q_2$. An example of both a kinematic and a test motion is shown in the figure for this special case.

When considering the constraint forces that will act on the particle we must assume that the rod material is 'ignorant' in the sense that it can only sense the behavior of the particles at the 'present' instant. Thus it has no way of knowing how a kinematic motion will evolve or has developed. Therefore it is a reasonable assumption that the constraint forces it produces will be orthogonal to the fixed time configuration surface, that is *orthogonal to any test motion.* •

3.3 Velocity and Acceleration

The displacement vector between an observer's origin and a point depends on the origin. An observer attached to the same body, but using a different point of the body for an origin, will record a different position vector. Observers attached to a

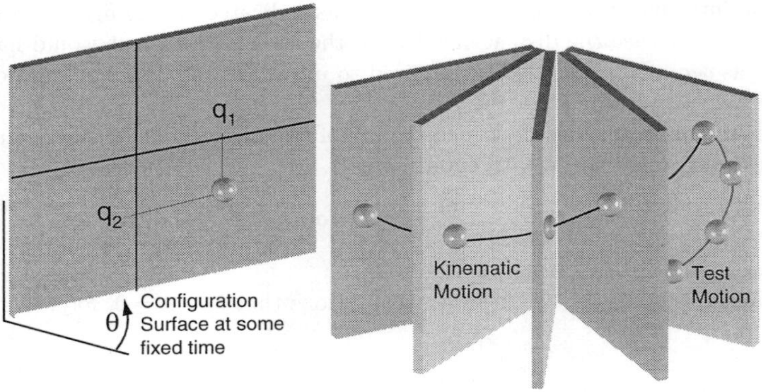

Figure 3.5: Example of test and kinematic motion

body B, at the points $\mathcal{Q}$ and $\mathcal{N}$ will assign the displacement vectors $\boldsymbol{r}^{\mathcal{QP}}$ and $\boldsymbol{r}^{\mathcal{NP}}$ for the location of the point $\mathcal{P}$. The relation between these displacements is given by

$$\boldsymbol{r}^{\mathcal{QP}} = \boldsymbol{r}^{\mathcal{QN}} + \boldsymbol{r}^{\mathcal{NP}}, \tag{3.26}$$

where $\boldsymbol{r}^{\mathcal{QN}}$ is a vector displacement *fixed* in B. If the point $\mathcal{P}$ is altering its position as a function of time one computes the vectorial change of position relative to the observer fixed at $\mathcal{Q}$ as

$$^{B}\boldsymbol{v}^{\mathcal{P}} = \frac{^{B}d}{dt}\boldsymbol{r}^{\mathcal{QP}}. \tag{3.27}$$

The fact that $\boldsymbol{r}^{\mathcal{QN}}$ is fixed in B implies that

$$^{B}\boldsymbol{v}^{\mathcal{P}} = \frac{^{B}d}{dt}\boldsymbol{r}^{\mathcal{NP}}. \tag{3.28}$$

and hence that the velocity of $\mathcal{P}$ relative to the body B is *independent* of the observers origin. This is also clearly true of higher derivatives, in particular the acceleration

$$^{B}\boldsymbol{a}^{\mathcal{P}} = \frac{^{B}d}{dt}{}^{B}\boldsymbol{v}^{\mathcal{P}} \tag{3.29}$$

is independent of the observers origin, thus the velocity and acceleration are the same for any observer fixed in the body B.

Given a mechanism, in the formal sense defined above, it is possible to obtain the velocity and acceleration of any point of the mechanism in terms of the velocity of the index points and the angular velocities of the individual bodies. Before showing how this is done we first introduce a notational convention that will allow us to use less cluttered symbols when discussing a mechanisms kinematics.

To define this notation let $\mathcal{P}$ serve as the index point for body B. Choose $\mathcal{Q}$ as any other point. In many cases it will be *fixed* in body B. Let the symbol $<$ be attached to the origin of the preferred observer, i.e. the point $\mathcal{O}$ and let $>$ be attached to the point $\mathcal{Q}$. The notational idea is now to indicate the sequence of the points $\mathcal{O}, \mathcal{P}$ by $<$, and B. The points $\mathcal{P}$, and $\mathcal{Q}$ are given by B, and $>$. Finally $\mathcal{O}, \mathcal{Q}$ is denoted by $<, B>$. With this convention we write $\boldsymbol{r}^{\mathcal{O}\mathcal{P}}$ as $\boldsymbol{r}^{<B}$, $\boldsymbol{r}^{\mathcal{P}\mathcal{Q}}$ as $\boldsymbol{r}^{B>}$ and $\boldsymbol{r}^{\mathcal{O}\mathcal{Q}}$ as $\boldsymbol{r}^{<B>}$. Figure 3.6 is a pictorial representation of these relationships. In terms of this

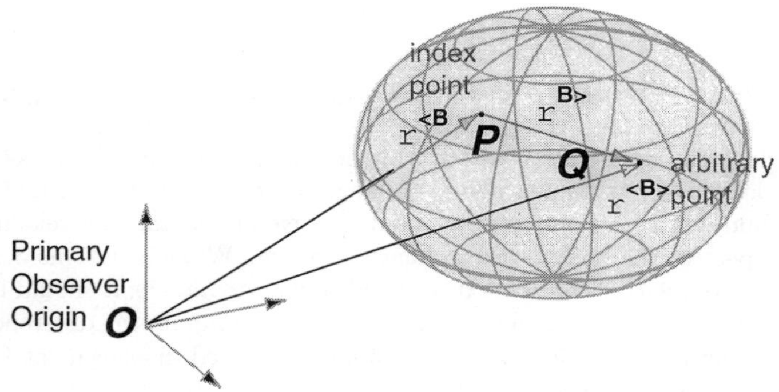

Figure 3.6: Position relations in a mechanism.

notation we have the relationship:

$$\boldsymbol{r}^{<B>} = \boldsymbol{r}^{<B} + \boldsymbol{r}^{B>}. \tag{3.30}$$

From the definition of velocity:

$${}^{A}\boldsymbol{v}^{\mathcal{Q}} = \frac{{}^{A}d\boldsymbol{r}^{<B>}}{dt} = \frac{{}^{A}d}{dt}(\boldsymbol{r}^{<B} + \boldsymbol{r}^{B>}) \tag{3.31}$$

Following our notational convention it is reasonable to write $\boldsymbol{v}^{<B>}$ in place of ${}^{A}\boldsymbol{v}^{\mathcal{Q}}$ and $\boldsymbol{v}^{<B}$ in place of ${}^{A}\boldsymbol{v}^{\mathcal{P}}$. So doing, and using the definition for ${}^{A}\boldsymbol{v}^{\mathcal{P}}$, we see that:

$$\boldsymbol{v}^{<B>} = \boldsymbol{v}^{<B} + \frac{{}^{A}d}{dt}\boldsymbol{r}^{B>}. \tag{3.32}$$

Restricting ourselves to the case where the point $\mathcal{Q}$ is fixed in body B and using the angular velocity ${}^{A}\boldsymbol{\omega}^{B}$, denoted by $\boldsymbol{\omega}^{(B)}$, we find that

$$\boldsymbol{v}^{<B>} = \boldsymbol{v}^{<B} + \boldsymbol{\omega}^{(B)} \times \boldsymbol{r}^{B>}. \tag{3.33}$$

Note carefully that 3.33 depends on the assumption that the point $\mathcal{Q}$ is fixed in body B. The importance of this result is that given the velocity of the index point and the angular velocity of the body we have sufficient information to find the velocity of any point in the body!

Equation 3.33 can be differentiated once more with respect to time to find the acceleration of the point $\mathcal{Q}$. Therefore:

$$\boldsymbol{a}^{<B>} = \boldsymbol{a}^{<B} + \tfrac{^A d}{dt}(\boldsymbol{\omega}^{(B)} \times \boldsymbol{r}^{B>}), \tag{3.34}$$

which can be expressed as

$$\boldsymbol{a}^{<B>} = \boldsymbol{a}^{<B} + \tfrac{^B d}{dt}(\boldsymbol{\omega}^{(B)} \times \boldsymbol{r}^{B>}) + \boldsymbol{\omega}^{(B)} \times (\boldsymbol{\omega}^{(B)} \times \boldsymbol{r}^{B>}) \tag{3.35}$$

finally giving the result that:

$$\boldsymbol{a}^{<B>} = \boldsymbol{a}^{<B} + \boldsymbol{\alpha}^{(B)} \times \boldsymbol{r}^{B>} + \boldsymbol{\omega}^{(B)} \times (\boldsymbol{\omega}^{(B)} \times \boldsymbol{r}^{B>}). \tag{3.36}$$

Equation 3.36 provides the acceleration of an arbitrary point fixed in a body as a function of the acceleration of the index point, the angular velocity and the angular acceleration of the body. The latter is simply the derivative of the angular velocity taken either with respect to the body or the primary observer. While equations 3.33 and 3.36 have been derived within the context of the idea of a *mechanism* it should be apparent that they apply to any situation where it is desired to compute the difference between the velocities and accelerations at two points with a fixed displacement for one observer as seen by another observer. This result proves to be of considerable practical value in kinematic analysis.

In the previous calculation it was assumed that the point $\mathcal{Q}$ was fixed in the body B and hence that the vector $\boldsymbol{r}^{B>}$ was fixed in the same body. To remove this restriction, consider the point in question to be moving with respect to both the preferred observer and an observer attached to B. The use of the angular velocity as a means of evaluating the difference in time derivatives for different observers should now be clear, hence the intermediate calculations are not displayed. Thus:

$$\boldsymbol{v}^{<B>} = \tfrac{^A d}{dt}(\boldsymbol{r}^{<B} + \boldsymbol{r}^{B>}) = \boldsymbol{v}^{<B} + (\tfrac{^B d}{dt} + \boldsymbol{\omega}^{(B)} \times)\boldsymbol{r}^{B>} \tag{3.37}$$

leading to the result that

$$\boldsymbol{v}^{<B>} = \boldsymbol{v}^{<B} + \boldsymbol{v}^{B>} + \boldsymbol{\omega}^{(B)} \times \boldsymbol{r}^{B>}. \tag{3.38}$$

The extra term, $\boldsymbol{v}^{B>}$ clearly vanishes under the assumption that $\boldsymbol{r}^{B>}$ is fixed in B. Again note that quantities referred to the body of the primary observer contain a $<$ as part of the superscript. When $\boldsymbol{\omega}^{(B)}$ vanishes this reduces to the result expected for non-rotating bodies.

Proceeding to take the derivative of the above leads to the well known result, known as the Coriolis theorem.

$$\boldsymbol{a}^{<B>} = \boldsymbol{a}^{<B} + \boldsymbol{a}^{B>} + \boldsymbol{\alpha}^{(B)} \times \boldsymbol{r}^{B>} + \boldsymbol{\omega}^{(B)} \times (\boldsymbol{\omega}^{(B)} \times \boldsymbol{r}^{B>}) + 2\boldsymbol{\omega}^{(B)} \times \boldsymbol{v}^{B>}. \tag{3.39}$$

This formula provides the general relation for the acceleration of a point as computed by two different observers. Again it reduces to the case of the displacement fixed between two points for one observer when $\boldsymbol{r}^{B>}$ becomes fixed for the observer attached to B.

There is another interesting way to look at equations 3.37 and 3.39 which has considerable practical use when carrying out a kinematic analysis. It was noted that the last result reduced to the case of the point $\mathcal{Q}$ fixed in the body . To appreciate this fully, leave the context of the general mechanism for the moment and assume that we have the typical situation where an observer is at a reference point $\mathcal{O}$, fixed to body A, and there is another body, B with a fixed point at $\mathcal{P}$. If a point $\mathcal{Q}$ is also fixed in B, equations 3.33 and 3.36 provide expressions for the differences between the velocities and accelerations of of the points $\mathcal{Q}$ and $\mathcal{P}$ as seen by the observer in A. The same equations expressed with this general viewpoint are:

$$^{A}\boldsymbol{v}^{\mathcal{Q}} = {}^{A}\boldsymbol{v}^{\mathcal{P}} + {}^{A}\boldsymbol{\omega}^{B} \times \boldsymbol{r}^{\mathcal{P}\mathcal{Q}} \tag{3.40}$$

and

$$^{A}\boldsymbol{a}^{\mathcal{Q}} = {}^{A}\boldsymbol{a}^{\mathcal{P}} + {}^{A}\boldsymbol{\alpha}^{B} \times \boldsymbol{r}^{\mathcal{P}\mathcal{Q}} + {}^{A}\boldsymbol{\omega}^{B} \times ({}^{A}\boldsymbol{\omega}^{B} \times \boldsymbol{r}^{\mathcal{P}\mathcal{Q}}). \tag{3.41}$$

The second viewpoint we can take is that the point $\mathcal{Q}$ is moving for observers in both A and B. The equations corresponding to 3.37 and 3.39 can be simplified if it is realized that they can be applied at any *fixed* instant of time. Suppose at such a fixed instant the point $\mathcal{P}$ is chosen so that it has the *same* position as the moving point $\mathcal{Q}$. To indicate this fact use the notation $B(\mathcal{Q})$ indicating that we pick a point that is fixed in B but is located at the instant of consideration at the position of the moving point $\mathcal{Q}$. Then apply equations 3.37 and 3.39 with $\mathcal{P} \to B(\mathcal{Q})$. Under these circumstances the term $\boldsymbol{r}^{\mathcal{P}\mathcal{Q}}$ vanishes, so that the equations relating the velocity and acceleration of a moving point as seen by observers in A and B become:

$$^{A}\boldsymbol{v}^{\mathcal{Q}} = {}^{A}\boldsymbol{v}^{B(\mathcal{Q})} + {}^{B}\boldsymbol{v}^{\mathcal{Q}} \tag{3.42}$$

and

$$^{A}\boldsymbol{a}^{\mathcal{Q}} = {}^{A}\boldsymbol{a}^{B(\mathcal{Q})} + {}^{B}\boldsymbol{a}^{\mathcal{Q}} + 2\,{}^{A}\boldsymbol{\omega}^{B} \times {}^{B}\boldsymbol{v}^{\mathcal{Q}}. \tag{3.43}$$

The practical application of these equations calls for expressing the kinematics of the fixed point $B(\mathcal{Q})$ in a *general form* in terms of the coordinates, i.e. it must be any point in B through which the point $\mathcal{Q}$ could pass! Comparison of these equations 3.37 and 3.39 show that they really contain nothing new and are simply a way of organizing the calculation of velocities and accelerations as seen by different observers. Even so they are well worth remembering as they provide a very convenient approach to the solution of specific problems.

The notation $B(\mathcal{Q})$ is very convenient for expressing kinematic relationships involving the motion of a point with some special properties. A general point in motion will occupy many points of the space of an arbitrary observer. This notation provides a means of labeling particular fixed points in a given observers frame that have been occupied by the general point $\mathcal{Q}$. The velocity ${}^{A}v^{B(\mathcal{Q})}$ will be referred to as *the frame fixed velocity of the point $\mathcal{Q}$ in the frame B*. Rolling motion without slip is a case of particular importance in which the frame fixed velocity concept is useful. The contact point between two bodies rolling on each other is not a material point of either body. It is simply the point which at some given time satisfies the condition of contact. The conditions of rolling require the specification of the velocity of the particular material points involved in the contact with respect to various fixed frames. We will return to this example later in this chapter and in our discussion of nonholonomic constraints.

•Illustration

In this illustration of the theory we will calculate the velocity relative to a prime observer of an arbitrary point on a mechanism in two ways. First we will use the relations involving angular velocity to relate velocities of different points. In general this indirect method involves less computational labor than the other technique we will use, direct differentiation of the position vector of the point and use of direction cosine matrices to express results in the frame of the primary observer.

The mechanism we will use for this example is shown in Figure 3.7. It consists of a square plate of side l, which is constrained to move in a plane which is rotating about the $\boldsymbol{n}_3$ axis of the prime observer. The rotation angle, θ, is considered to be a given function of the time. A rigid rod projects orthogonally from the center of the plate. A blade is fixed to the rod and its center is at a distance h from this base point. The distance h is fixed, but the attached 'blade' of length $2S$ is free to rotate about the rod as axis. The mechanism is taken as the square plate, body A and the blade, body B, the other parts providing the constraining framework. Generalized coordinates, which determine the geometrical configuration, are chosen as the position of the plate center relative to the prime observer's origin, the angle made by one edge of the plate with a line orthogonal to the axis of rotation, and the angle made by the blade with one edge of the plate. In addition to the standard triad of the prime observer we introduce a help triad, k, where $\boldsymbol{k}_3 = \boldsymbol{n}_3$ and the angle θ is between $\boldsymbol{n}_1$ and $\boldsymbol{k}_1$. The triads a and b are fixed in the plate and blade respectively. The vector $\boldsymbol{a}_1$ is at an angle q_3 to the vector $\boldsymbol{k}_1$. Also $\boldsymbol{b}_1$ makes an angle q_4 with $\boldsymbol{a}_1$. The position vector of the center of the plate is given by:

$$\boldsymbol{r}^{<A} = q_1\boldsymbol{k}_1 + q_2\boldsymbol{k}_3. \tag{3.44}$$

This is also the position vector of the index point for body A. The index point for body B will be taken as:

$$\boldsymbol{r}^{<B} = \boldsymbol{r}^{<A} + h\boldsymbol{k}_2, \tag{3.45}$$

which is the center of the blade. An arbitrary point in the plate, A, is given by

$$\boldsymbol{r}^{A>} = \xi_1\boldsymbol{a}_1 + \eta_2\boldsymbol{a}_3, \tag{3.46}$$

where $-l/2 \leq \xi_1, \eta_1 \leq l/2$. An arbitrary point on a line passing through the center and parallel to the long edge of the blade, B, has the position

$$\boldsymbol{r}^{B>} = \xi_2\boldsymbol{b}_1, \tag{3.47}$$

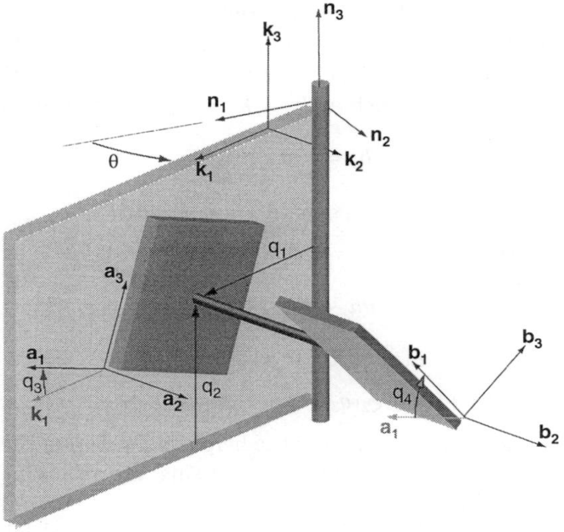

Figure 3.7: Rotating blade attached to rotating plate in rotating plane.

with $-S \leq \xi_2 \leq S$. The indirect calculation of velocities requires the use of the angular velocity vectors $\boldsymbol{\omega}^{(A)}$ and $\boldsymbol{\omega}^{(B)}$, between the bodies and the frame of the primary observer. These are most easily obtained by use of the concepts of 'simple angular velocities' and the addition law for angular velocities, which was discussed in chapter two. This gives the results that:

$$\boldsymbol{\omega}^{(A)} = \dot{\theta}\boldsymbol{k}_3 - \dot{q}_3\boldsymbol{k}_2 \tag{3.48}$$

$$\boldsymbol{\omega}^{(B)} = \boldsymbol{\omega}^{(A)} - \dot{q}_4\boldsymbol{k}_2. \tag{3.49}$$

The negative signs in these equations are a result of the fact that

$$\boldsymbol{k}_1 \times \boldsymbol{a}_1 \cdot \boldsymbol{k}_2 = -1 \tag{3.50}$$

$$\boldsymbol{a}_1 \times \boldsymbol{b}_1 \cdot \boldsymbol{k}_2 = -1. \tag{3.51}$$

Using these angular velocities we can compute $\boldsymbol{v}^{<B>}$, the velocity of an arbitrary point in body B relative to the frame of the primary observer, by the following sequence of 4 equations. The first two and the fourth equations use the result for the difference in velocities between two fixed points in a body and the third equation uses the result for the case where a point is moving relative to another point in the body:

$$\boldsymbol{v}^{<B>} = \boldsymbol{v}^{<B} + \boldsymbol{\omega}^{(B)} \times \boldsymbol{b}_1\xi_2 \tag{3.52}$$

$$\boldsymbol{v}^{<B} = \boldsymbol{v}^{<A} + \boldsymbol{\omega}^{(A)} \times \boldsymbol{k}_2 h \tag{3.53}$$

$$\boldsymbol{v}^{<A} = \boldsymbol{v}^{<A(A)} + \dot{q}_1\boldsymbol{k}_1 + \dot{q}_2\boldsymbol{k}_3 \tag{3.54}$$

$$\boldsymbol{v}^{<A(A)} = \dot{\theta}\boldsymbol{k}_3 \times (q_1\boldsymbol{k}_1 + q_2\boldsymbol{k}_3). \tag{3.55}$$

The notational convention that the body name is also used as the name of the index point, hence $\boldsymbol{v}^{<A(A)} = {}^{N}v^{A(\mathcal{P}_A)}$ is the frame fixed velocity of the center of the base plate relative to frame N. Carrying out the sequential substitution, using the properties of the cross product such as $\boldsymbol{k}_3 \times \boldsymbol{k}_1 = \boldsymbol{k}_2$, and the relations that

$$\begin{aligned} \boldsymbol{b}_1 &= c_4\boldsymbol{a}_1 + s_4\boldsymbol{a}_3 \\ &= (c_4c_3 - s_4s_3)\boldsymbol{k}_1 + (c_4s_3 + s_4c_3)\boldsymbol{k}_3, \end{aligned} \tag{3.56}$$

we obtain the result that

$$\begin{aligned} \boldsymbol{v}^{<B>} &= (\dot{q}_1 - \dot{\theta}h - \xi_2(\dot{q}_3 + \dot{q}_4)(s_3c_4 + c_3s_4))\boldsymbol{k}_1 \\ &+(q_1\dot{\theta} + \xi_2\dot{\theta}(c_3c_4 - s_3s_4))\boldsymbol{k}_2 \\ &+(\dot{q}_2 + \xi_2(\dot{q}_3 + \dot{q}_4)(c_3c_4 - s_3s_4)\boldsymbol{k}_3. \end{aligned} \tag{3.57}$$

This result has been expressed in terms of the help triad k. Use of the relation between this system and the triad n would allow expressing the result in the triad of the primary observer.

The same result can be obtained, with more computational labor, by the direct method of differentiating the vector $\boldsymbol{r}^{<B>}$ expressed in terms of the triad n. We have

$$\begin{aligned} \boldsymbol{r}^{<B>} &= q_1\boldsymbol{k}_1 + h\boldsymbol{k}_2 + q_2\boldsymbol{k}_3 + \xi_2\boldsymbol{b}_1 \\ &= (q_1, h, q_2)k + (\xi_2, 0, 0)b \\ &= [(q_1, h, q_2)R_{kn} + (\xi_2, 0, 0)R_{bn}]n. \end{aligned} \tag{3.58}$$

The direction cosine matrix R_{bn} satisfies the relation

$$R_{bn} = R_{ba}R_{ak}R_{kn}, \tag{3.59}$$

and examination of the relations between the standard triads provides the various direction cosine matrices needed for the calculation, thus

$$R_{kn} = \begin{bmatrix} c_\theta & s_\theta & 0 \\ -s_\theta & c_\theta & 0 \\ 0 & 0 & 1 \end{bmatrix}, \tag{3.60}$$

where we have used a subscript to indicate terms such as $\cos\theta$. The other matrices are:

$$R_{ak} = \begin{bmatrix} c_3 & 0 & s_3 \\ 0 & 1 & 0 \\ -s_3 & 0 & c_3 \end{bmatrix}, \tag{3.61}$$

$$R_{ba} = \begin{bmatrix} c_4 & 0 & s_4 \\ 0 & 1 & 0 \\ -s_4 & 0 & c_4 \end{bmatrix} \tag{3.62}$$

and

$$R_{bn} = \begin{bmatrix} c_\theta z_1 & s_\theta z_1 & z_2 \\ -s_\theta & c_\theta & 0 \\ -c_\theta z_3 & -s_\theta z_3 & z_1 \end{bmatrix}, \tag{3.63}$$

where

$$z_1 = c_3 c_4 - s_3 s_4 \tag{3.64}$$

$$z_2 = s_3 c_4 + c_3 s_4 \tag{3.65}$$

$$z_3 = c_3 s_4 + s_3 c_4. \tag{3.66}$$

Differentiating equation 3.58, the position vector of the arbitrary point in the blade expressed in terms of the primary observer's reference triad, with respect to time:

$$\begin{aligned} \boldsymbol{v}^{<B>} = & [(\dot{q}_1, 0, \dot{q}_2)R_{kn} + (q_1, h, q_2)\dot{R}_{kn} \\ & +(\xi_2, 0, 0)\dot{R}_{bn}]R_{nk}k. \end{aligned} \tag{3.67}$$

Carrying out all the indicated operations we again obtain the previous result for $\boldsymbol{v}^{<B>}$. The reader can judge the comparative computational labor! The second method does have the advantage of being straightforward and does not require much of the decision making needed in using the concept of angular velocity. Thus it might be preferable for computer based calculations using either numerical methods or computer algebra. It remains to carry out the calculations for acceleration of points located in the two bodies, a task which is left as an exercise for the student. •

3.4 Geometry of the Configuration Surface

The configuration of a constrained mechanism can be represented in terms of the geometry of an n dimensional surface. In typical geometrical formulations of mechanics this surface is considered intrinsic, that is its properties are studied independently of any embedding space. While this has some advantages there are also reasons for considering the configuration surface to be embedded in the space of all possible unconstrained configurations. The theory is less abstract and the calculation of constraint forces is direct in the embedding approach. The major disadvantages arise from problems involving the global topology of the surface. For the purposes of the present work these problems can be ignored. Assume that the dimension of the coordinate space of the unconstrained configuration is m. If the constrained system requires n coordinates it will be represented by an n-dimensional surface embedded in the m dimensional coordinate space of the unconstrained mechanism. Each point on the surface represents some possible configuration at some fixed time in terms of the generalized coordinates. The generalized coordinates, represented by q_j, $(j = 1 \ldots n)$, are surface coordinates for the n-surface. If all the generalized coordinates are taken as functions of some parameter, λ, and the time is held fixed, there is generated a curve or path in the surface, which is called a *test motion.* The path followed when the time, t varies, and the q_j are functions of t, is referred to as a *kinematic motion.* If the position of the configuration surface varies with time, so that there is a family of configuration surfaces parameterized by the time, a kinematic path will not be fixed in one member of the family for any fixed value of t. These facts must be expressed in appropriate mathematical form related to the geometry of the configuration surfaces. A tangent hyperplane is associated with each point of the configuration surface. This hyperplane is the space defined by all possible linear combinations of vectors *tangent to test motions passing through the point in question.* Any arbitrary path in the coordinate space that passes through a point on the configuration surface will also have a tangent vector at that point. It is important to note that the term tangent vector applies to any vector tangent to a possible motion, test or kinematic. The vector space of all tangent vectors to all paths passing through such a point can be decomposed into two non-intersecting subspaces. One is taken as the subspace tangent to the configuration surface. The other is the complement of this subspace with respect to the tangent space of all possible paths. An arbitrary tangent vector can be written as the sum of two vectors, one of which is tangent to the configuration surface. The *projection* of a tangent vector onto the configuration surface is simply this part of the vector.

In addition to motion we are concerned with forces. For the purposes of understanding the geometry of forces and motions it is necessary to introduce a slightly more abstract viewpoint than used in elementary mechanics. Thus instead of starting with the concept of force, the concept of *power* is taken as basic. This is based on the elementary definition of power as the scalar or dot product of force with velocity, which has the units of energy per unit time. Thus given a velocity and a number or scalar representing power we can find an infinite number of forces that will yield the power in question. The set of all possible forces can be abstractly thought of as the set of all *linear functionals* that act on the space of all tangent vectors to all possible motion paths in the coordinate space. In linear algebra this set is shown to form a vector space in its own right. This is the *dual space* to the space of tangent vectors. To express the matter another way, *velocities live in the space of tangent vectors, forces in the dual space.* The linear functional of forces acting on velocities yields power. In these abstract terms our discussion of D'Alembert's principle in chapter one generalizes to the statement that *the linear functional of constraint forces acting on vectors tangent to the configuration surface must vanish.* This is sometimes called *Jourdain's Principle* or the *the principle of virtual power.* Its physical interpretation is clear. It is simply the requirement that constraint forces, by definition, have no influence on motion in the configuration surface. The power of the principle is that it is also possible to split the applied forces into parts that go with the constraint forces and parts that actually influence the motion. While it is called a 'principle' it is no more than a statement as to the behavior of the type of ideal constraints assumed in rigid body mechanics. We now define the mathematical machinery to implement this principle.

First consider a mechanism that is composed of K points, labeled by $1 \dots K$, and interconnected in such a way as the configuration is determined by n generalized coordinates. Thus assume the existence of the position vectors $\boldsymbol{r}^{<1}(q,t), \dots \boldsymbol{r}^{<K}(q,t)$, where q stands for the set of quantities $q_1, q_2 \dots q_n$. The $3K$ component vector that represents a configuration point is indicated by the convention of dropping the label from $\boldsymbol{r}^{<L}$, that is by writing:

$$\boldsymbol{r}^{<} = \begin{bmatrix} \boldsymbol{r}^{<1} \\ \vdots \\ \boldsymbol{r}^{<K} \end{bmatrix} \tag{3.68}$$

with each component of the K column being a position vector to the Lth point.

With this notation the velocity of the configuration point relative to the primary observer is represented as

$$\frac{{}^A d}{dt}\boldsymbol{r}^{<} = \boldsymbol{v}^{<} = \begin{bmatrix} \boldsymbol{v}^{<1} \\ \vdots \\ \boldsymbol{v}^{<K} \end{bmatrix} \tag{3.69}$$

These 'K vectors' are thus ordered sets of K three dimensional vectors. From now on we will use the name *Kvectors.* They are vectors in their own right, and in fact are $3K$ dimensional vectors which can be considered tangent to any path of motion in the coordinate space. In the same manner we can define a force vector composed of the forces acting on each of the K particles. These vectors will be designated as K*vectors to indicate their relationship to Kvectors. This relationship is given by the sum of the scalar products of all the 'force vectors' with all the 'velocity vectors', i.e.

$$\boldsymbol{f}^{<} \bullet \boldsymbol{w}^{<} = \sum_{L=1}^{L=K} \boldsymbol{f}^{<L} \cdot \boldsymbol{w}^{<L}. \tag{3.70}$$

The fat dot is used to indicate this linear functional, which will sometimes be called the fat dot product. The physical interpretation of this is that this functional gives the *power* associated with the force-velocity system. The important point is that the physical meaning to a fat dot product of a tangent vector with another tangent vector, or of a force vector with another force vector is, in general, not meaningful. The dual or force vectors will also be called *cotangent vectors* following the conventions of differential geometry. For the most part no notation will be used to distinguish tangent from cotangent vectors. When a distinction is required an underbar will be used for the tangent vectors and an overbar for the cotangent vectors, thus

$$\overline{\boldsymbol{f}}^{<} \bullet \underline{\boldsymbol{w}}^{<} = \sum_{L=1}^{L=K} \overline{\boldsymbol{f}}^{<L} \cdot \underline{\boldsymbol{w}}^{L}. \tag{3.71}$$

With these definitions we can think of our Kvectors just as we think about vectors in Euclidian three space. Simple diagrams, based on our notions of ordinary vectors can be used as a guide to obtaining useful relationships.

Now consider a *test motion* in which each q_j is a function of λ. Using the chain rule we find that:

$$\frac{{}^A d}{d\lambda}\boldsymbol{r}^< = \sum_{j=1}^{j=n}(\frac{{}^A\partial}{\partial q_j}\boldsymbol{r}^<)\frac{dq_j}{d\lambda}. \tag{3.72}$$

The n quantities,

$$\boldsymbol{\tau}_j^< = (\frac{{}^A\partial}{\partial q_j}\boldsymbol{r}^<) \tag{3.73}$$

are tangent to the configuration surface at the point $\boldsymbol{r}^<(q(\lambda), t)$, and in fact are tangent to the *coordinate lines*. The latter are the curves traced out by $\boldsymbol{r}^<$ for q_j varying with $q_1, \ldots, q_{j-1}, q_{j+1}, \ldots, q_n$, and t fixed.

For a kinematic motion, the generalized coordinates are given functions of time, t. The total time derivative with respect to the primary frame now represents the velocity of each point. Evaluation by the chain rule thus gives:

$$\frac{{}^A d}{dt}\boldsymbol{r}^< = \sum_{j=1}^{j=n}(\frac{{}^A\partial}{\partial q_j}\boldsymbol{r}^<)\frac{dq_j}{dt} + (\frac{{}^A\partial}{\partial t}\boldsymbol{r}^<) = \boldsymbol{v}^<. \tag{3.74}$$

The kinematic motion will in general have a component pointing out from the fixed t configuration surface. This of course reflects the fact that the configuration constraints can in general change with time. Instead of a single configuration surface there is a family of surfaces parameterized by the time. It is important to appreciate the fact that if the constraints vary with time the tangent vector to a kinematic path may have a component that is *not tangent* to the time fixed configuration surface. As the vectors $\boldsymbol{\tau}_j^<$ are tangent to the surface, the vector

$$\boldsymbol{\tau}_t^< = (\frac{{}^A\partial}{\partial t}\boldsymbol{r}^<) \tag{3.75}$$

can have an out of surface component. It should also be noted that in general, $\boldsymbol{\tau}_t^<$ will also have a component that is tangent to the current configuration surface, i.e. a projection onto the plane tangent to the surface! A simple illustration of this is provided at the end of this section.

The expected properties of constraint forces have been noted in the introduction and also in the proceeding part of this chapter. The main assumption made about such constraint forces is that they have *a vanishing projection in the instantaneous configuration surface.* The instantaneous configuration surface is the member of the family of configuration surfaces which is appropriate to the current instant of time, thus this is also referred to as the current configuration surface.

By projecting all forces onto the directions of vectors tangent to the configuration surface it is possible to eliminate constraint forces from the equations of motion. One can determine tangent directions by considering test motions, and in fact a set of n tangent vectors $\boldsymbol{\tau}_j^<$ can be found by differentiating $\boldsymbol{r}^<$ with respect to each of the n generalized coordinates. This particular set of tangent vectors, obtained by differentiation, are called *coordinate tangent vectors.*

The set of n coordinate tangent vectors form the basis for a local tangent surface to the current configuration surface. The velocity of a kinematic motion can broken up into two parts, the part that can be projected onto the tangent surface and the remainder. This tangent surface is equally well determined from a test motion. The tangent hyper- plane is sometimes denoted as $T_n(q,t)$, showing that it depends both on the value of the time and the point on the configuration hypersurface. It is possible to take linear combinations of the coordinate tangent vectors to form new vectors. Any set of n linear combinations of the n coordinate tangent vectors will also form a basis for the tangent hyperplane. In general a new set of basis tangent vectors will not be obtainable as derivatives of the position vector in terms of n scalar variables. That is an arbitrary basis will not in general consist of vectors which are tangent to coordinate lines for *any* possible choice of coordinates. Even so such general sets of tangent vectors can be very useful for formulating equations of motion. It is one of Kane's contributions to demonstrate the usefulness of doing this. While something of this can be found in the works of Gibbs and Appell, Kane deserves the credit for showing how these arbitrary tangent vectors can be exploited directly for the effective solution of mechanics problems. The vector components of an arbitrary set of linear combinations of coordinate tangent vectors, are called *partial velocities* and the equations of motion obtained from their direct use are called *Kane's equations.* For example the ordinary vector $\boldsymbol{\tau}_L^{<k}$ is the partial velocity associated with the jth tangent vector and the Lth particle.

The derivatives of the coordinates with respect to time, dq_j/dt, are sometimes called generalized velocities, even though they are not vectorial in character. The convention is adopted of representing total time differentiation by an over dot, i.e. $\dot{q}_j$. The linear form that results from the chain rule showed that in a kinematic motion, the motion component in the surface is given by a sum of products of coordinate tangent vectors with the generalized velocities. If a possible non-coordinate basis for $T_n(q,t)$ is used, the in surface motion can be represented by a sum of products of quantities linearly related to the generalized velocities with each of the non-coordinate basis vectors. These new quantities which are linear combinations of the generalized velocities are, again following Kane, called generalized speeds. The details follow in the next section.

•Illustration

The main purpose of this illustration is to explicitly exhibit the set of tangent vectors for a time dependent configuration constraint and to show that the vector $\boldsymbol{\tau}_t^<$ has components that are not tangent to the local configuration tangent space. To accomplish this task we examine the problem of a point mass confined to move in an infinite wave like sheet which is itself moving relative to the frame of the prime observer. This situation is shown in Figure 3.8. If $\boldsymbol{r}$ is taken as the position of the particle, the particle position in the prime observers frame can be expressed by the 'coordinates'

$$x_\beta = \boldsymbol{r} \cdot \boldsymbol{a}_\beta, \tag{3.76}$$

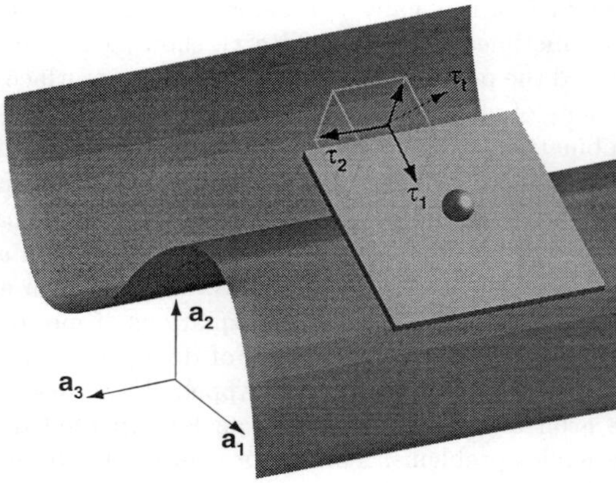

Figure 3.8: Particle moving in a wave like surface

with $\beta = 1..3$. We choose the generalized coordinates so that:

$$x_1 = q_1 \tag{3.77}$$

$$x_3 = q_2. \tag{3.78}$$

The constraint is expressed by the fact that

$$x_2 = h \sin \frac{q_1 - Ut}{l}, \tag{3.79}$$

where h measures the height of the surface, l the 'wavelength', and U is a constant with the dimensions of velocity, hence a measure of the wave speed in the prime observer's frame. For convenience we define the 'phase' as

$$\phi = \frac{q_1 - Ut}{l}, \tag{3.80}$$

and 'aspect ratio' as

$$\epsilon = \frac{h}{l}. \tag{3.81}$$

Then we have

$$\boldsymbol{r} = q_1 \boldsymbol{a}_1 + h \sin \phi \boldsymbol{a}_2 + q_2 \boldsymbol{a}_3. \tag{3.82}$$

In this case, where we have only one particle, we do not need the full apparatus of K vectors, which of course reduce to ordinary vectors when $K = 1$. The illustration following this one will give an example of the full use of the K vector concept. The tangent vectors $\boldsymbol{\tau}_\beta$ are now easily calculated from their definition above, hence:

$$\boldsymbol{\tau}_1 = \boldsymbol{a}_1 + \epsilon \cos\phi \boldsymbol{a}_2 \tag{3.83}$$

$$\boldsymbol{\tau}_2 = \boldsymbol{a}_3 \tag{3.84}$$

$$\boldsymbol{\tau}_t = -\epsilon U \cos\phi \boldsymbol{a}_2. \tag{3.85}$$

The velocity of the particle is given in terms of the tangent vectors by the expression:

$$\boldsymbol{v}^< = u_1\boldsymbol{\tau}_1 + u_2\boldsymbol{\tau}_2 + \boldsymbol{\tau}_t, \tag{3.86}$$

where

$$u_1 = \dot{q}_1 \tag{3.87}$$

$$u_2 = \dot{q}_2. \tag{3.88}$$

From the above equations it is possible to write

$$\boldsymbol{\tau}_t = -U\boldsymbol{\tau}_1 + U\boldsymbol{a}_1. \tag{3.89}$$

The vector $\boldsymbol{a}_1$ can be considered a tangent vector for motion which breaks the constraint. Therefore the three vectors $\boldsymbol{\tau}_1$, $\boldsymbol{\tau}_2$ and $\boldsymbol{a}_1$ span the coordinate space of the unconstrained particle. In terms of this decomposition the projection of the vector $\boldsymbol{\tau}_t$ onto the tangent space to the configuration is $-U\boldsymbol{\tau}_1$.

We do not have to use the coordinate basis tangent vectors $\boldsymbol{\tau}_j$. Any linear combinations of these vectors are still tangent to the configuration surface. As an example of another set consider:

$$\boldsymbol{\beta}_1 = \boldsymbol{\tau}_1 + \boldsymbol{\tau}_2 \tag{3.90}$$

$$\boldsymbol{\beta}_2 = \boldsymbol{\tau}_1 - \boldsymbol{\tau}_2. \tag{3.91}$$

Solving for the $\boldsymbol{\tau}_j$ in terms of the $\boldsymbol{\beta}_i$ and substituting into the velocity relation gives:

$$\boldsymbol{v}^< = \frac{1}{2}(u_1 + u_2)\boldsymbol{\beta}_1 + \frac{1}{2}(u_1 - u_2)\boldsymbol{\beta}_2 + \boldsymbol{\tau}_t. \tag{3.92}$$

The term $\boldsymbol{\tau}_t$ can also be written as:

$$\boldsymbol{\tau}_t = U\boldsymbol{a}_1 - \frac{1}{2}U(\boldsymbol{\beta}_1 + \boldsymbol{\beta}_2). \tag{3.93}$$

Therefore the velocity expansion can be put into the form

$$\boldsymbol{v}^< = \frac{1}{2}(u_1 + u_2 - U)\boldsymbol{\beta}_1 + \frac{1}{2}(u_1 - u_2 - U)\boldsymbol{\beta}_2 + U\boldsymbol{a}_1. \tag{3.94}$$

This suggests the introduction of the generalized speeds w_j such that

$$w_1 = \frac{1}{2}(u_1 + u_2 - U) \tag{3.95}$$

$$w_2 = \frac{1}{2}(u_1 - u_2 - U). \tag{3.96}$$

With the definition $\boldsymbol{\beta}_t = U\boldsymbol{a}_1$ the velocity expansion takes the form

$$\boldsymbol{v}^< = w_1\boldsymbol{\beta}_1 + w_2\boldsymbol{\beta}_2 + \boldsymbol{\beta}_t. \tag{3.97}$$

The relations between the time derivatives of the generalized coordinates and the new generalized speeds, w_j are:

$$\dot{q}_1 = w_1 + w_2 + U \tag{3.98}$$

$$\dot{q}_2 = w_1 - w_2. \tag{3.99}$$

In this illustration of the theory we have seen that it is possible to extend the set of tangent vectors to the configuration to a set of tangent vectors for the entire coordinate space. Given such a complete set of vectors the $\boldsymbol{\tau}_t$ term could be projected onto the tangent surface. It was also shown that by using the velocity expansion equation and by taking linear combinations of the coordinate tangent vectors we could obtain new sets of tangent vectors. The velocity expansion equation could then be used to define new generalized speeds. Finally it was seen that the projection terms could be absorbed into the new generalized speeds. Thus a linear relation between sets of basis tangent vectors for the tangent hyperplane imply an affine relation between sets of generalized speeds. All of these transformations are based on keeping the same linear form for the velocity in terms of an expansion in tangent vectors and generalized speeds.•

•Illustration

This example is intended to demonstrate the way one works with Kvectors. It is relatively simple so that the reader can concentrate on the mechanics of dealing with configuration surfaces that are embedded in spaces with dimension exceeding three. Figure 3.9 shows the mechanism we use for this purpose. The mechanism is restricted in its motion to a plane passing through the prime observers origin and parallel to the vectors $\boldsymbol{a}_1$ and $\boldsymbol{a}_2$. It consists of two particles which can slide along a 'wire' which is oriented at an angle $\theta(t)$ to the direction of the triad vector $\boldsymbol{a}_1$, where the angle $\theta(t)$ is assumed to be a given function of the time. The position of the first particle is given by the generalized coordinates, q_1 and q_2. The second particle is at a distance q_3 along the wire away from the first particle. As there are two particles the unconstrained system would require six parameters for its geometric determination. Therefore we have a situation in which the constrained system is represented by a family of three dimensional hypersurfaces embedded in a six dimensional configuration space. Each

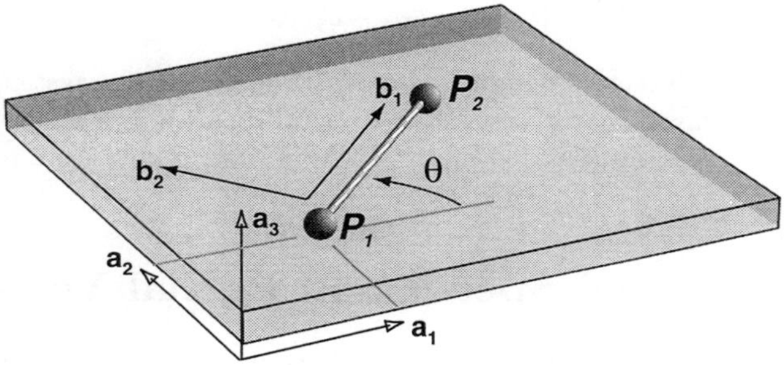

Figure 3.9: Particles on a moving rod.

value of the time determines a new angle $\theta(t)$, hence a different member of the family of configuration hypersurfaces.

The mechanism configuration is seen from the above description to be given by the equations:

$$\boldsymbol{r}^{<1} = q_1 \boldsymbol{a}_1 + q_2 \boldsymbol{a}_2, \tag{3.100}$$

$$\boldsymbol{r}^{<2} = (q_1 + q_3 \cos\theta)\boldsymbol{a}_1 + (q_2 + q_3 \sin\theta)\boldsymbol{a}_2. \tag{3.101}$$

In terms of our Kvector notation, here $K = 2$, we have:

$$\boldsymbol{r}^{<} = \begin{bmatrix} \boldsymbol{r}^{<1} \\ \boldsymbol{r}^{<2} \end{bmatrix}. \tag{3.102}$$

The coordinate tangent vectors are given by

$$\boldsymbol{\tau}_j^{<} = \begin{bmatrix} \frac{^A\partial}{\partial q_j}\boldsymbol{r}^{<1} \\ \frac{^A\partial}{\partial q_j}\boldsymbol{r}^{<2} \end{bmatrix} \tag{3.103}$$

Therefore the tangent Kvectors that define the instantaneous configuration surface are given by:

$$\boldsymbol{\tau}_1^{<} = \begin{bmatrix} \boldsymbol{a}_1 \\ \boldsymbol{a}_1 \end{bmatrix} \tag{3.104}$$

$$\boldsymbol{\tau}_2^{<} = \begin{bmatrix} \boldsymbol{a}_2 \\ \boldsymbol{a}_2 \end{bmatrix} \tag{3.105}$$

$$\boldsymbol{\tau}_3^{<} = \begin{bmatrix} 0 \\ \cos\theta \boldsymbol{a}_1 + \sin\theta \boldsymbol{a}_2 \end{bmatrix}. \tag{3.106}$$

To define a kinematic path, which in this case will not be parallel to the instantaneous configuration hypersurface, we need the vector $\boldsymbol{\tau}_t^<$, that is

$$\boldsymbol{\tau}_t^< = \begin{bmatrix} 0 \\ -q_3 \sin\theta \boldsymbol{a}_1 + q_3 \cos\theta \boldsymbol{a}_2 \end{bmatrix} \dot{\theta}. \tag{3.107}$$

Each of the tangent Kvectors describes a mode of motion of the system. The reader should convince himself that $\boldsymbol{\tau}_1^<$ corresponds to a translation of the system parallel to the $\boldsymbol{a}_1$ axis, $\boldsymbol{\tau}_2^<$ to a translation parallel to the $\boldsymbol{a}_2$ axis and $\boldsymbol{\tau}_3^<$ to a rotation of the rod. •

3.5 Generalized Speeds and Partial Velocities

For a mechanism with n geometric degrees of freedom, a set of n independent Kvectors, such as $\boldsymbol{\tau}_j^<$, provide a basis in terms of which we can represent *any* Kvector belonging to the tangent hyperplane to the configuration manifold at the point (q,t). The special feature of the set $\boldsymbol{\tau}_j^<$, is that these vectors are obtained by differentiation of the position Kvector with respect to the generalized coordinates, i.e. that

$$\boldsymbol{\tau}_j^< = \frac{{}^A\partial}{\partial q_j} \boldsymbol{r}^<. \tag{3.108}$$

By taking n suitable linear combinations of these vectors we can form a new basis for the tangent hyperplane. In general such a new set of K vectors will not be obtainable as derivatives of position vectors with respect to some set of coordinates. Even so they will serve just as well as the $\boldsymbol{\tau}_j^<$ for representing the local geometry of the configuration hypersurface.

The theory of transformations between local configuration hyperplane basis sets provides us with a flexible tool for the optimal formulation of equations of motion. To develop this theory we start with a known basis that is derivable from coordinates, hence we use $\boldsymbol{\tau}_j^<$ to denote members of this basis. Using such a basis related to a particular set of generalized coordinates q, we have seen that the velocity Kvector, tangent at the point (q,t) to a kinematic motion, is given by the relation:

$$\boldsymbol{v}^< = \sum_{j=1}^{j=n} \dot{q}_j \boldsymbol{\tau}_j^< + \boldsymbol{\tau}_t^<, \tag{3.109}$$

where

$$\boldsymbol{\tau}_t^< = \frac{{}^A\partial}{\partial t} \boldsymbol{r}^<. \tag{3.110}$$

A basic theorem of linear algebra tells us that any set of linearly independent vectors can be expanded into a basis by adding suitable vectors to the set. In the present case the vector space consists of all tangent vectors to all possible kinematic motions at a point on the configuration surface. The coordinate tangent vectors provide a basis to a subset of this vector space. This set describes the local tangent plane to the configuration surface. Note that the word 'tangent vector' applies to any Kvector that is tangent to a possible kinematic motion at the point in question. The coordinate tangent vectors are special! They provide a basis for the vector subspace that is tangent to all possible test motions, i.e. the hyperplane that is tangent to the configuration surface. A Kvector which can be expressed only in terms of tangent vectors that are tangent to the configuration hypersurface is termed *parallel to the hypersurface.* The term $\boldsymbol{\tau}_t^<$ in the above expansion of the Kvector velocity may or may not be parallel. This was seen in the examples of the last section.

Any linear combination of coordinate tangent Kvectors that span the tangent hyperplane will also serve as a basis for the hypersurface's tangent space, which will be denoted as $T(q,t)$. This notation emphasizes that we are focusing attention at a point on the hypersurface for time t which is determined by the set of generalized coordinates $(q_1 \cdots q_n)$.

With the above facts in mind it is natural to examine what happens to the structure of the Kvector velocity for a kinematic motion when we switch to another set of basis vectors for $T(q,t)$. The generalized velocities, $\dot{q}_j$ act as multipliers of each of the tangent Kvectors $\boldsymbol{\tau}_j^<$. Also, depending on how we expand the basis of coordinate tangent Kvectors to a basis for all kinematic motions, we can split the term $\boldsymbol{\tau}_t^<$ into two parts, one of which is parallel to $T(q,t)$. Note that such a split is totally arbitrary and not unique. Even if $\boldsymbol{\tau}_t^<$ vanished, the natural situation when the constraints are time independent, we could add and subtract an arbitrary linear sum of coordinate tangent Kvectors to the velocity Kvector. Thus in general we can write:

$$\boldsymbol{\tau}_t^< = \sum_{j=1}^{n} h_j \boldsymbol{\tau}_j^< + \boldsymbol{\tau}_t'^<, \tag{3.111}$$

and thus

$$\boldsymbol{v}^< = \sum_{j=1}^{n} (\dot{q}_j + h_j)\boldsymbol{\tau}_j^< + \boldsymbol{\tau}_t'^<. \tag{3.112}$$

This shows that even if we retain the given coordinate tangent vectors as a basis for $T(q,t)$ we can preserve the structure of the velocity expansion by defining new parameters to replace the generalized velocities $\dot{q}$. Thus write

$$u_j = \dot{q}_j + h_j \tag{3.113}$$

and hence

$$\boldsymbol{v}^< = \sum_{j=1}^{n} u_j \boldsymbol{\tau}_j^< + \boldsymbol{\tau}_t'^<. \tag{3.114}$$

The new, more general parameters u are the generalized speeds discussed above. The generalized velocities are now given by the differential equations

$$\dot{q}_j = u_j - h_j. \tag{3.115}$$

Equations for the generalized velocities in terms of generalized speeds will be called *kinematic differential equations.* They are extremely important role in our formulation of mechanics! Note that the relation $\dot{q}_j = u_j$ is a simple but important special case of a kinematic differential equation.

The above considerations show that a change in basis for $T(q,t)$ that preserves the structure of the Kvector velocity expansion involves a change in the velocity parameters. While the change in basis is purely linear, the change in velocity parameters is in general *affine.* This is simply a formal way of saying that it also involves a possible additive term. From another point of view the additive term, independent of $\dot{q}_j$, corresponds to the physical fact that there is no distinguished origin for velocities in the tangent hyperplane $T(q,t)$. From its definition the only requirement is one of parallelism. The addition of an arbitrary parallel velocity Kvector has no effect on this structure. In summary it has been seen that the form of the velocity expansion for a kinematic motion is preserved as a sum of hyperplane tangent vectors and an arbitrary term independent of the generalized velocities by a linear transformation of the hyperplanes basis vectors and an affine transformation of the velocity parameters. The latter, generalized speeds, are the coefficients of the hyperplane tangent vector expansion.

The structure of the transformation theory of basis Kvectors can be better appreciated if we adopt some of the notational ideas of chapter two. The symbol τ by itself will indicate a column matrix of n basis Kvectors $\boldsymbol{\tau}_j^<$. Again the superscript T will indicate the transpose matrix, hence τ^T will correspond to a row matrix of $\boldsymbol{\tau}_j^<$. Note that now the individual entries are not ordinary vectors but Kvectors! Thus

$$\tau = \begin{bmatrix} \boldsymbol{\tau}_1^< \\ \vdots \\ \boldsymbol{\tau}_n^< \end{bmatrix} = \begin{bmatrix} \begin{bmatrix} \boldsymbol{\tau}_1^{<1} \\ \vdots \\ \boldsymbol{\tau}_1^{<K} \end{bmatrix} \\ \vdots \\ \begin{bmatrix} \boldsymbol{\tau}_n^{<1} \\ \vdots \\ \boldsymbol{\tau}_n^{<K} \end{bmatrix} \end{bmatrix}. \tag{3.116}$$

which we might call 'super' Kvectors. Then if we let

$${}^{\tau}h = \begin{bmatrix} {}^{\tau}h_1 \\ \vdots \\ {}^{\tau}h_n \end{bmatrix}, \tag{3.117}$$

one can replace summations with matrix multiplications as in chapter two. For example in this notation we have

$$\boldsymbol{\tau}_t^< = \tau^T\,{}^{\tau}h + \boldsymbol{\tau}_t'^<. \tag{3.118}$$

The reader should also note that in the above expansion, ${}^{\tau}h$ is a function of (q, t) but *not* of $\dot{q}$.

Now use this notation to describe the technique of transforming to a new set of basis vectors. Let such a new basis be indicated by β, where of course

$$\beta = \begin{bmatrix} \boldsymbol{\beta}_1^{<} \\ \vdots \\ \boldsymbol{\beta}_n^{<} \end{bmatrix}. \tag{3.119}$$

As each $\boldsymbol{\beta}_j^{<}$ can be expressed as a linear sum over the vectors $\boldsymbol{\tau}_i^{<}$, we can find an array $W_{ij}^{\beta\tau}$, such that in the new notation

$$\beta^T = W^{\beta\tau}\tau, \tag{3.120}$$

or more explicitly

$$\boldsymbol{\beta}_j^{<} = \sum_{i=1}^{i=n} W_{ij}^{\beta\tau}\boldsymbol{\tau}_i^{<}. \tag{3.121}$$

It will be assumed that $W^{\tau\beta}$, the inverse transformation matrix exists so that

$$W^{\beta\tau}W^{\tau\beta} = U. \tag{3.122}$$

The essential observation in what follows is that for the time dependent constraint situation the velocity Kvector is composed of two parts, one linearly dependent on the generalized velocities $\dot{q}$ and another that depends on q but not $\dot{q}$. This second part will in general have a part parallel to $T(q, t)$, and in fact we can always introduce such a parallel component artificially as discussed above. Thus from the above considerations

$$\boldsymbol{v}^{<} = \dot{q}^T\tau + \boldsymbol{\tau}_t^{<}, \tag{3.123}$$

which by substitution of the new set of basis vectors is transformed to

$$\boldsymbol{v}^{<} = \dot{q}^T W^{\tau\beta}\beta + \boldsymbol{\tau}_t^{<}. \tag{3.124}$$

Now, following the above remark, the last term can be split into a component parallel to the tangent plane and the remaining part. The split is arbitrary and the remainder may still have parallel components. We can expand some of the parallel part in terms of the new tangent vectors. If ${}^{\beta}f$ denotes the matrix column of expansion coefficients, this takes the form

$${}^{\beta}f^T\beta. \tag{3.125}$$

In terms of this we can write

$$\boldsymbol{v}^{<} = (\dot{q}^T W^{\tau\beta} + {}^{\beta}f^T)\beta + \boldsymbol{\beta}_t^{<}. \tag{3.126}$$

We now define the *generalized speeds* as

$$u^T = \dot{q}W^{\tau\beta} + {}^{\beta}f^T. \tag{3.127}$$

Note the arbitrary nature of the splitting off of parallel components, hence the arbitrary nature of $\boldsymbol{\beta}_t^{<}$. The Kvector velocity expansion is now given by

$$\boldsymbol{v}^{<} = u^T \beta + \boldsymbol{\beta}_t^{<}. \tag{3.128}$$

The splitting of $\boldsymbol{\tau}_t^{<}$ is thus seen to provide flexibility in the form of the transformation.

Use the basis transformation and the fact that $W^{\tau\beta} W^{\beta\tau} = 1$, to solve for the time derivatives of the generalized coordinates, the generalized velocities, thus:

$$\dot{q}^T = (u^T - {}^{\beta}f^T) W^{\beta\tau} \tag{3.129}$$

which can also be written as

$$\dot{q} = (W^{\beta\tau})^T u - {}^{\tau}f. \tag{3.130}$$

where

$${}^{\tau}f^T = {}^{\beta}f^T W^{\beta\tau} \tag{3.131}$$

These relations are the general form for the *kinematic differential equations.*

The kinematic differential equations could be used to define u, the *generalized speeds.* Thus we have the choice of either choosing a new basis for the tangent hyperplane or a new set of generalized speeds. As one change implies the other the choice is not independent.

The component vectors of the set of basis K vectors for the tangent hyperplane are the *partial velocities.* In other words the ordinary vectors which make up the entries in the Kvector are partial velocities. Basis vectors which are derivable from differentiation of position vectors with respect to configuration surface coordinates (coordinate basis vectors) can be considered as a special case. Kane has shown that by proper use of the flexibility one has in introducing generalized speeds and partial velocities, it is possible to considerably simplify the process of deriving equations of motion that are free of unknown constraint forces. It has been claimed by some authors that they are implicit in the work of classic investigators such as Appell and Gibbs. This seems somewhat irrelevant, as it is certainly Kane who has developed the conceptual structure to arrive at equations of motion by the direct use of generalized speeds and partial velocities. These quantities can be exceptionally helpful when dealing with so called non-holonomic constraints. At this point it can simply be said that the reason for this is that the partial velocities need not be coordinate basis vectors! The key to the practical use of these relations is that one can start the analysis of a mechanism by determining its possible kinematic motions. If the mechanism has n geometric degrees of freedom it is possible to choose a suitable set of n generalized speeds to express any kinematic motion. Moreover we can then determine the tangent vectors, hence partial velocities by direct inspection, that is once we carry out a kinematical analysis in terms of n generalized speeds we observe that the coefficients of each u_j provide the corresponding tangent vector. The process of choosing generalized coordinates is separate from this task. Once such coordinates are determined, a parallel kinematic analysis determines the relations between the $\dot{q}_j$ and the u_j. These relations are the kinematic differential equations! Generalized velocities are in the present terminology a special case of generalized speeds. In such a case the tangent vectors will reduce to the coordinate basis tangent vectors associated with the particular set of generalized coordinates!

The components of the array, $(W^{\tau\beta})^T$ are called Y_{rs} in Kane and Levinson's equation 2.12.1, while the components of our ${}^{\beta}f$ are named Z_r. Their W_{sr} and X_s as used in equation 2.14.5 of their book corresponds to our $(W^{\beta\tau})^T$ and $-{}^{\tau}f$. The notation used here has the advantage that it conforms with the treatment of ordinary basis vectors in chapter two and it is easy to see the meaning and recall the various relations by the balanced placing of superscripts, though care must be taken due, in contrast to chapter two, to the general non-orthogonality of the transformation. Thus *the transpose of the coefficient matrix in the kinematic differential equations is the linear transformation from the coordinate basis tangent vectors to the new tangent vectors*, while the additive term is a component of the velocity Kvector generally arising from the imposition of time dependent constraints. The term ${}^{\tau}f$. can be chosen arbitrarily, different choices giving different forms to the velocity expansion in terms of tangent vectors, i.e. to the term ${}^{\beta}\mathcal{K}_t$. Because of this it is felt that it deserves a name and symbol of its own. Kane and Levenson call its components $-X_s$ which can be confused with coordinate names or forces as used in much of the mechanics literature. Since it appears in the kinematic differential equations as an additive term which is part of the sum of terms giving the generalized velocities $\dot{q}$ it will be called *the generalized selected velocity* which will be denoted by the symbol $\dot{Q}$ in this text, i.e.

$$\dot{Q} = {}^{\tau}f. \tag{3.132}$$

The kinematic differential equations are then put in a form with this term transposed to the left side.This allows a meaningful interpretation of this term, thus

$$\dot{q} + \dot{Q}(q,t) = (W^{\beta\tau}(q,t))^T u. \tag{3.133}$$

This shows explicitly that the term $\dot{Q}$ represents an arbitrary augmentation of the generalized velocities. It is emphasized that the transformation coefficients and the generalized selected velocities can be explicit functions of the generalized coordinates and time.

•Illustration

This example will illustrate how generalized speeds may be chosen without making a commitment to a set of generalized coordinates. Partial velocity K vectors can then be obtained by inspection of the expression for a kinematic motion. Once a choice of generalized coordinates is made it is possible to construct the appropriate set of kinematic differential equations. These equations in turn provide the transformation to the set of basis K vectors that relate to the generalized coordinates. The example, shown in Figure 3.10, is relatively simple, so that we can concentrate on the techniques

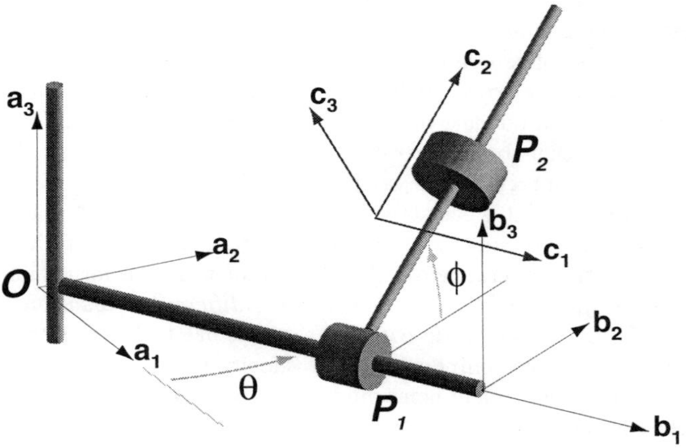

Figure 3.10: Example to illustrate a basis transformation

and formalism. The collar located at $\mathcal{P}_1$ moves along the line $\mathcal{O}\mathcal{P}_1$, which is rotating about the axis aligned in the $\boldsymbol{a}_3$ direction through the point $\mathcal{O}$. The rotation angle, $\theta(t)$, is assumed given as a function of time. A second collar, is constrained to move on the line $\mathcal{P}_1\mathcal{P}_2$, and this line rotates about the axis of the first collar. The prime observers standard triad is the set of unit vectors a, The standard triads b and c are aligned with the collars at $\mathcal{P}_1$ and $\mathcal{P}_2$. As indicated in the figure, $\boldsymbol{a}_1 \cdot \boldsymbol{b}_1 = \cos\theta$, and $\boldsymbol{b}_2 \cdot \boldsymbol{c}_2 = \cos\phi$. The distance $\mathcal{O}\mathcal{P}_1$ is l_1 and the distance $\mathcal{P}_1\mathcal{P}_2$ is l_2. While l_1, l_2 and ϕ could be taken as generalized coordinates, we start by simply considering them as convenient labels for the geometry of the system, i.e. they are easier to write than something like distance$(\mathcal{P}_1\mathcal{P}_2)$. Note that in general these quantities will vary with time, hence a particular set of time functions assigned to l_1, l_2 and ϕ defines what we have called a kinematic motion. If θ is fixed we have what we have defined as a test motion. Note also that the mechanism is so configured that $\boldsymbol{a}_3 = \boldsymbol{b}_3$ and $\boldsymbol{b}_1 = \boldsymbol{c}_1$.

If the mechanism consists of the two point bodies located at $\mathcal{P}_1$ and $\mathcal{P}_2$, the configuration hypersurface is defined by the K position vector:

$$\boldsymbol{r}^{<} = \begin{bmatrix} l_1\boldsymbol{b}_1 \\ l_1\boldsymbol{b}_1 + l_2\boldsymbol{c}_2 \end{bmatrix}. \tag{3.134}$$

In this case the direct approach of taking the derivative of this in the frame of the prime observer is the simplest route to calculating the velocity vector associated with a kinematic motion. The direct calculation is carried out using the angular velocities $\boldsymbol{\omega}^{(B)}$ and $\boldsymbol{\omega}^{(C)}$. Representations for these are easily obtained by using the concept of simple angular velocities and the rules for the addition of angular velocities. Carrying out this program we find that:

$$\boldsymbol{\omega}^{(B)} = \dot{\theta}\boldsymbol{b}_3, \tag{3.135}$$

$$\boldsymbol{\omega}^{(C)} = \dot{\theta}\boldsymbol{b}_3 + \dot{\phi}\boldsymbol{b}_1, \tag{3.136}$$

hence that

$$\boldsymbol{v}^{<} = \frac{^A d}{dt}\boldsymbol{r}^{<} = \begin{bmatrix} \boldsymbol{v}^{<1} \\ \boldsymbol{v}^{<2} \end{bmatrix}, \tag{3.137}$$

where

$$\boldsymbol{v}^{<1} = \dot{l}_1\boldsymbol{b}_1 + l_1\boldsymbol{\omega}^{(B)} \times \boldsymbol{b}_1, \tag{3.138}$$

$$\boldsymbol{v}^{<2} = \boldsymbol{v}^{<1} + \dot{l}_2\boldsymbol{c}_2 + l_2\boldsymbol{\omega}^{(C)} \times \boldsymbol{c}_2. \tag{3.139}$$

Carrying out the indicated operations and using the definitions of the angles θ and ϕ, the components of the Kvector velocity that is tangent to a kinematic motion are found to be:

$$\boldsymbol{v}^{<1} = \dot{l}_1\boldsymbol{b}_1 + l_1\dot{\theta}\boldsymbol{b}_2, \tag{3.140}$$

$$\boldsymbol{v}^{<2} = \boldsymbol{v}^{<1} + \dot{l}_2\boldsymbol{c}_2 - \dot{\theta}\cos\phi l_2\boldsymbol{b}_1 + \dot{\phi}l_2\boldsymbol{c}_3. \tag{3.141}$$

In many problems it is useful to think in terms of a velocity or kinematic analysis before making a commitment to a particular choice of generalized coordinates. We have seen that the Kvector $\boldsymbol{v}^{<}$ is tangent to a kinematic motion and that it can be expressed in terms of products of generalized speeds with so called tangent Kvectors plus a Kvector independent of the generalized speeds, i.e. an expression of the form $\beta^T u + \beta_t^{<}$. In the present case we have a mechanism with three geometric degrees of freedom. Hence the general motion state can be expressed in terms of three parameters, i.e. three generalized speeds. We now choose a particular set. If our choice is incorrect we will find inconsistencies in our results. As the point $\mathcal{P}_1$ is constrained to move along the direction of $\boldsymbol{b}_1$ it is clear that the projection on this vector of its velocity relative to a prime observer determines that velocity. We take this as the generalized speed u_1. The velocity of the other collar will have a projection onto the $\boldsymbol{a}_3$ direction, but to uniquely determine it we must also know some other datum. We take the projection of the angular velocity onto the direction of $\boldsymbol{b}_1$ for this additional information. Therefore our choice of generalized speeds becomes:

$$u_1 = \boldsymbol{v}^{<1} \cdot \boldsymbol{b}_1 \tag{3.142}$$

$$u_2 = \boldsymbol{v}^{<2} \cdot \boldsymbol{b}_3 \tag{3.143}$$

$$u_3 = \boldsymbol{\omega}^{(C)} \cdot \boldsymbol{b}_1. \tag{3.144}$$

This is not a unique choice, and it is certainly not the best possible choice. What is notable is the freedom we have in choosing generalized speeds. Now we can see if our choice is consistent and how we can obtain the partial velocities that go along with it. To do this we use the expression we obtained for $\boldsymbol{v}^{<}$ and apply the conditions determining the generalized speeds in terms of the geometry. This gives:

$$u_1 = \boldsymbol{v}^{<1} \cdot \boldsymbol{b}_1 = \dot{l}_1 \tag{3.145}$$

$$u_2 = \boldsymbol{v}^{<2} \cdot \boldsymbol{b}_3 = \dot{l}_2 \sin\phi + \dot{\phi} l_2 \cos\phi \tag{3.146}$$

$$u_3 = \boldsymbol{\omega}^{(C)} \cdot \boldsymbol{b}_1 = \dot{\phi}, \tag{3.147}$$

which we can solve for $\dot{l}_1, \dot{l}_2$ and $\dot{\phi}$. This gives:

$$\dot{l}_1 = u_1, \tag{3.148}$$

$$\dot{l}_2 = \frac{1}{\sin\phi} u_2 - l_2 \cot\phi u_3, \tag{3.149}$$

$$\dot{\phi} = u_3. \tag{3.150}$$

Inserting this last result into the expression for $\boldsymbol{v}^{<}$, we obtain

$$\boldsymbol{v}^{<1} = u_1 \boldsymbol{b}_1 + l_1 \dot{\theta} \boldsymbol{b}_2, \tag{3.151}$$

$$\boldsymbol{v}^{<2} = \boldsymbol{v}^{<1} - \dot{\theta} \cos\phi l_2 \boldsymbol{b}_1 + (\frac{1}{\sin\phi} u_2 - l_2 \cot\phi u_3) \boldsymbol{c}_2 + l_2 u_3 \boldsymbol{c}_3. \tag{3.152}$$

The coefficients of u_j are the partial velocity vectors $\boldsymbol{\beta}_j^{<}$, hence

$$\boldsymbol{\beta}_1^{<} = \begin{bmatrix} \boldsymbol{b}_1 \\ \boldsymbol{b}_1 \end{bmatrix}, \tag{3.153}$$

$$\boldsymbol{\beta}_2^{<} = \begin{bmatrix} 0 \\ \frac{1}{\sin\phi} \boldsymbol{c}_2 \end{bmatrix}, \tag{3.154}$$

$$\boldsymbol{\beta}_3^{<} = \begin{bmatrix} 0 \\ -l_2 \cot\phi \boldsymbol{c}_2 + l_2 \boldsymbol{c}_3 \end{bmatrix}. \tag{3.155}$$

The remaining part, which is independent of u, contains the information about components of the tangent Kvector which are not parallel to the local tangent hyperplane, thus

$$\boldsymbol{\beta}_t^{<} = \begin{bmatrix} l_1 \dot{\theta} \boldsymbol{b}_2 \\ -\dot{\theta} \cos\phi l_2 \boldsymbol{b}_1 + l_1 \dot{\theta} \boldsymbol{b}_2 \end{bmatrix}. \tag{3.156}$$

We are free to make any choice of generalized coordinates we wish, as long as they fully describe the geometric configuration of the mechanism. The simplest choice in the present context is to use the parameters l_1, l_2 and ϕ that we used to label the basic geometric features of our example. Then formally we have:

$$\dot{q}_1 = u_1, \tag{3.157}$$

$$\dot{q}_2 = \frac{1}{\sin q_3} u_2 - q_2 \cot q_3 u_3, \tag{3.158}$$

$$\dot{q}_3 = u_3. \tag{3.159}$$

Comparison with the equation relating generalized speeds under a change of basis, i.e. with $\dot{q} = (W^{\beta\tau})^T u - {}^{\tau}f$, gives us the result that:

$$(W^{\beta\tau})^T = \begin{bmatrix} 1 & 0 & 0 \\ 0 & \csc q_3 & -q_2 \cot q_3 \\ 0 & 0 & 1 \end{bmatrix}, \tag{3.160}$$

and that

$${}^{\tau}f = 0. \tag{3.161}$$

For this transformation we have

$$\det(W^{\beta\tau}) = \csc q_3, \tag{3.162}$$

and the inverse, when it is not singular, is

$$(W^{\tau\beta})^T = \begin{bmatrix} 1 & 0 & 0 \\ 0 & \sin q_3 & q_2 \cos q_3 \\ 0 & 0 & 1 \end{bmatrix}. \tag{3.163}$$

We can now check our work by computing the coordinate basis vectors for the tangent hyperplane by use of the transformation and by direct calculation from the expression for $\boldsymbol{v}^{<}$ expressed in terms of the generalized velocities, the latter of course being simply special cases of generalized speeds. The transformation is given by $\tau = W^{\tau\beta}\beta$, which gives

$$\boldsymbol{\tau}_1^{<} = \boldsymbol{\beta}_1^{<}, \tag{3.164}$$

$$\boldsymbol{\tau}_2^{<} = \sin q_3 \boldsymbol{\beta}_2^{<}, \tag{3.165}$$

$$\boldsymbol{\tau}_3^{<} = q_2 \cos q_3 \boldsymbol{\beta}_2^{<} + \boldsymbol{\beta}_3^{<}. \tag{3.166}$$

The expression for $\boldsymbol{v}^{<}$ in terms of the generalized coordinates q has the two components:

$$\boldsymbol{v}^{<1} = \dot{q}_1 \boldsymbol{b}_1 + q_1 \dot{\theta} \boldsymbol{b}_2, \tag{3.167}$$

$$-\dot{\theta} \cos q_3 q_2 \boldsymbol{b}_1 + \dot{q}_2 \boldsymbol{c}_2 + \dot{q}_3 q_2 \boldsymbol{c}_3. \tag{3.168}$$

Taking either route indeed gives the same result, i.e.

$$\boldsymbol{\tau}_1^{<} = \begin{bmatrix} \boldsymbol{b}_1 \\ \boldsymbol{b}_1 \end{bmatrix}, \tag{3.169}$$

$$\boldsymbol{\tau}_2^{<} = \begin{bmatrix} 0 \\ \boldsymbol{c}_2 \end{bmatrix}, \tag{3.170}$$

$$\boldsymbol{\tau}_3^{<} = \begin{bmatrix} 0 \\ q_2 \boldsymbol{c}_3 \end{bmatrix}. \tag{3.171}$$

In this particular case where $X^{\tau} = 0$ we have $\boldsymbol{\tau}_t^{<} = \boldsymbol{\beta}_t^{<}$. •

The practical application of the results of this section require the ability to deal effectively with the complexity of the algebraic manipulations. This is addressed in the discussion of the use of computer algebra that follows.

3.6 Orthogonality

In most of the remainder of this text the power functional will be treated as if it provided a true dot product between tangent Kvectors. The following is intended to justify this to readers that may be upset to see tangent vectors 'dotted' into tangent vectors. If this does not bother you please ignore the rest of this section.

Tangent vectors were defined as Kvectors tangent to a possible kinematic motion. The special class of tangent vectors, tangent to a test motion, define the local tangent hyperplane to a fixed time member of the class of configuration hypersurfaces. This led to the concept of Kvectors being parallel to the hyperplane. The power function, indicated by the fat dot, allowed the introduction of cotangent vectors. In this model Kvector velocities are tangent vectors and Kvector forces are cotangent vectors. As will be emphasized in chapter four, D'Alembert's principle is equivalent to the power functional of the constraint force Kvector of a system with any tangent Kvector to a test motion vanishing. The entire theory can be set up and discussed purely in terms involving these constructs without the introduction of any concept of orthogonality. For both computational and conceptual reasons it can be useful to introduce a concept of orthogonality of Kvectors. In the discussion of dynamics a natural way of doing this will involve the *mass operator*. The mass operator acting on a tangent Kvector produces a cotangent vector and the time derivative of this vector will be a force. The power functional thus produces objects with the correct physical dimensions. Once a mass operator is specified it allows a unique identification of any tangent vector with a cotangent vector. If $\boldsymbol{M}_K$ is the mass operator for an K body system we can write:

$$\boldsymbol{M}_K \cdot \underline{\boldsymbol{v}^<} = \overline{\boldsymbol{v}^<}. \tag{3.172}$$

Therefore we can define a *dot product* by the operation

$$\underline{\boldsymbol{v}^<} \bullet \underline{\boldsymbol{v}^<} = (\boldsymbol{M}_K \cdot \underline{\boldsymbol{v}^<}) \bullet \underline{\boldsymbol{v}^<}. \tag{3.173}$$

This notation allows the $\bullet$ symbol to be interpreted as a dot product and allows us to talk about tangent Kvectors being orthogonal when the dot product vanishes. If the operator $\boldsymbol{M}_K$ is the mass operator for a system of point masses the system's kinetic energy could be written as

$$\frac{1}{2}\boldsymbol{v}^< \bullet \boldsymbol{v}^< \tag{3.174}$$

it being understood that for the fat dot product to make sense we must convert the tangent vector to a cotangent vector using the mass operator. This can lead to some confusion in interpreting equations and we will not in general use the notation in this manner. Instead we introduce another operator, $\boldsymbol{M}_D$, which identifies tangent vectors with cotangent vectors. It is defined as a diagonal matrix of unit dyads, each multiplied by a unit constant mass. For example for a two point mass system:

$$\boldsymbol{M}_D = \begin{bmatrix} 1\boldsymbol{U} & 0 \\ 0 & 1\boldsymbol{U} \end{bmatrix} \tag{3.175}$$

If this operates on a tangent vector we have:

$$\begin{bmatrix} 1\boldsymbol{U} & 0 \\ 0 & 1\boldsymbol{U} \end{bmatrix} \cdot \begin{bmatrix} \boldsymbol{v}^{<1} \\ \boldsymbol{v}^{<2} \end{bmatrix} = \begin{bmatrix} 1\boldsymbol{v}^{<1} \\ 1\boldsymbol{v}^{<2} \end{bmatrix}. \tag{3.176}$$

The inverse of $\boldsymbol{M}_D$ simply replaces the unit masses by the inverse unit dimensional constants and provides a means of identifying cotangent vectors with tangent Kvectors. Thus simply by making the appropriate dimensional adjustments we can identify tangent and cotangent vectors. It is then unnecessary to use the apparatus of the operator $\boldsymbol{M}_D$. This identification will be called the *dimensional identification convention.* If we in fact use nondimensional units in our calculations the dimensional identification convention is totally automatic! From now on when we talk about the orthogonality of Kvectors we will simply assume the dimensional identification. Chapter five introduces the concepts of rigid bodies and we must extend our Kvector formalism to account for the attitude of such objects. It will also be seen that with a slight modification the dimensional identification convention is easily applied to this case.

The use of $\boldsymbol{M}_D$ to identify tangent and cotangent Kvectors is not unique. It essentially extracts the dimensional part of the mass operator and we could write $\boldsymbol{M}_K$ as $\boldsymbol{M}_D \cdot \boldsymbol{M}_S$ where the latter would contain the appropriate numerical magnitudes. Thus despite its nonuniqueness, $\boldsymbol{M}_D$ allows the geometric or kinematic aspects of the system to be discussed in terms of a concept of orthogonality. In numerical and even algebraic calculations the assumption of such an orthogonality might lead to less efficient algorithms than otherwise possible. Despite this it has been used, though not in the explicit form presented here. Constraint elimination by the use of orthogonal complementary matrices and so called pseudo inverses are equivalent to the our use of $\boldsymbol{M}_D$. As stated in the prelude to this section these matters need not concern a reader who is simply interested in producing correct equations of motion for multi-body systems.

3.7 Reciprocal Basis and Projection

In what follows it will be assumed, following the discussion of the previous section, that the fat dot product is indeed a dot product and that we can talk about the orthogonality of Kvectors. Once we have a concept of orthogonality we can use it to produce projection operators by means similar to our discussion of Euclidian three space in the previous chapter. Thus we could define an operator $\boldsymbol{\Pi}_*^<$ which projects Kvectors into the tangent hyperplane. The purpose of this section is to provide an algorithm for producing such operators when the tangent Kvectors may not be orthonormal.

In general the tangent K-vectors will not be orthonormal. This can be somewhat inconvenient when it is desired to form a projection operation, for example when it is desired to resolve a term such as $\boldsymbol{\Pi}^< \bullet \boldsymbol{\tau}_t^<$ into components with respect to $\boldsymbol{\beta}_j^<$.

Reciprocal base systems provide a formal algorithm for this task. They are also of some interest in their own right as they provide a useful tool for working with non-orthonormal systems of base vectors. To keep the notation simple the theory is first put in terms of vectors in $n-$dimensional Euclidian space with a well defined inner product. Once this is done it is trivial to express the main results in K-vector notation.

Let $\underline{\tau}$ be an $n-$ column of vectors which form a basis for n dimensional space. It is not assumed that the vectors are normal to each other or that they are unit vectors. The underline is part of the notation and will also be applied to each of the vectors in the column, thus $\underline{\boldsymbol{\tau}_j}$. In the case of our K-vectors these would consist of the tangent vectors such as $\underline{\boldsymbol{\tau}_j^{<}}$, the underline now being added to the notation. The use of an under and overline on vectors here is not the same as its use to distinguish tangent and cotangent Kvectors and only applies to this section.

The advantage of orthonormal base vectors is that it is very easy to find the components of an arbitrary vector. In chapter two this led us to the unit dyad $a^T a = \sum_i \boldsymbol{a}_i \boldsymbol{a}_i$, which applied to any vector gave the component expansion of that vector. Thus

$$\boldsymbol{w} = \sum_i \boldsymbol{a}_i \boldsymbol{a}_i \cdot \boldsymbol{w} = \sum_i {}^a\!w_i \boldsymbol{a}_i = {}^a\!w^T a, \tag{3.177}$$

with the last form using the matrix containing vectors notation. With this background consider the expansion of a vector in terms of given base vectors. This will take the form:

$$\boldsymbol{w} = \sum \overline{{}^a\!w_i} \underline{\boldsymbol{a}_i}. \tag{3.178}$$

The overline on the components matches the underline on the base vectors. As the base is not assumed to be orthonormal we can not obtain these quantities by the simple dot product. Taking dot products with each of the base vectors would lead to solving a system of linear equations for each case. Reciprocal base vectors provide a way of getting around this problem. They are defined by the relation

$$\overline{\boldsymbol{a}_i} \cdot \underline{\boldsymbol{a}_j} = \delta_{ij}. \tag{3.179}$$

In matrix containing vector notation this can be expressed as

$$\overline{a} \cdot \underline{a}^T = \underline{a} \cdot \overline{a}^T = \mathbf{1}. \tag{3.180}$$

. It can easily be seen by doting each of these vectors into $\sum \overline{{}^a\!w_i} \underline{\boldsymbol{a}_i}$ that

$$\boldsymbol{w} = \sum \underline{\boldsymbol{a}_i} \overline{\boldsymbol{a}_i} \cdot \boldsymbol{w}, \tag{3.181}$$

and that

$$\overline{{}^a\!w_j} = \overline{\boldsymbol{a}_j} \cdot \boldsymbol{w}. \tag{3.182}$$

Therefore the unit dyad can be expressed as:

$$\boldsymbol{U} = \sum \underline{\boldsymbol{a}_i} \overline{\boldsymbol{a}_i} = \sum \overline{\boldsymbol{a}_i} \underline{\boldsymbol{a}_i}. \tag{3.183}$$

It should be clear that using one of these forms of the unit dyad provides the above expansion in terms of the given or direct basis vectors. The expansion using reciprocal vectors is given by the other form, or by operating from the other side of the dyad. The components in the latter case are indicated by an underline, i.e. ${}^{a}\underline{w}_j$ or in column matrix form as ${}^{a}\underline{w}$.

To provide a convenient algorithm for the determination of the reciprocal base, start with the given or direct base as known. Each of the vectors in the reciprocal base can be expressed in terms of a sum over the direct base, thus

$$\overline{a} = \overline{G}\underline{a}, \tag{3.184}$$

where G is a matrix whose rows are the components of each of the reciprocal base vectors in terms of the direct base vectors. Now use the reciprocal base property to see that

$$\overline{a} \cdot \overline{a}^T = \overline{G}\underline{a} \cdot \overline{a}^T = \overline{G}\mathbf{1} = \overline{G}. \tag{3.185}$$

If

$$\underline{a} = \underline{G}\overline{a}, \tag{3.186}$$

similar reasoning to the above shows that

$$\underline{G} = \underline{a} \cdot \underline{a}^T. \tag{3.187}$$

Also the definition of $\overline{G}$ and $\underline{G}$ implies that

$$\overline{G}\underline{G} = \mathbf{1} \tag{3.188}$$

or

$$\overline{G} = \underline{G}^{-1}. \tag{3.189}$$

The components of the matrix $\underline{G}$ are $\underline{\boldsymbol{a}_i} \cdot \underline{\boldsymbol{a}_j}$, therefore it is clear how to calculate $\underline{G}$ which can then be inverted to obtain $\overline{G}$ and hence the reciprocal base $\overline{a}$. The matrix $\underline{G}$ is called the direct form of the *Euclidian metric* with respect to the a frame. The overline gives the reciprocal form.

Now suppose that the vectors $\underline{a}$ span a linear subspace S of a higher dimensional vector space N. From the above discussion it should be evident that the dyad $\underline{a}^T\overline{a}$ is an operator that projects vectors in N into S in the form of a linear sum over the direct base $\underline{a}$. The metric can be used to put this in a form containing only the direct basis vectors, thus

$$\boldsymbol{\Pi}_S = \underline{a}^T(\underline{a} \cdot \underline{a}^T)^{-1}\underline{a}, \tag{3.190}$$

where $\boldsymbol{\Pi}_S$ is the projection dyad from N onto S.

These results can easily be applied to the problem of representing $\boldsymbol{\tau}_t^<$ as a linear sum over the tangent vector base to the configuration hypersurface $\boldsymbol{\beta}_j^<$, i.e.

$$\boldsymbol{\Pi}^< = \beta^T(\beta \bullet \beta^T)^{-1}\beta. \tag{3.191}$$

Since the operator can be expressed only in terms of the direct base it is not necessary to use the over-underbar notation. A linear vector space with a dot product can be decomposed into two *orthogonal* subspaces in that any vector can be written as the sum of two vectors, one in each subspace, and the two vectors will have a vanishing dot product. The two subspaces are orthogonal complements. We can construct the projection operator $\boldsymbol{\Pi}_*^<$ onto this space in an entirely similar manner to way this was done for ordinary vectors. These projection operators now have an algorithmic definition in terms of previously found quantities. A formula for the projection onto the complementary space in terms of the unit dyad for the full coordinate space is easily derived. Thus extend the given set of base vectors to one that spans the full coordinate space and form the unit dyad for this space. We then have the result that

$$\boldsymbol{\Pi}_*^< = \boldsymbol{U}^< - \boldsymbol{\Pi}^<. \tag{3.192}$$

A final word about notation is in order. In general tensor analysis the convention is to use superscripts and subscripts for the direct and reciprocal basis. This avoids the introduction of over and under lines, however it still leaves a problem with the matrix representations used here. The final result is entirely in terms of the direct base, hence for our purposes the difference can be ignored, as in equation 3.191 above.

3.8 Symbolic Manipulation of K-vectors

The Kvector formalism introduced in this chapter is easily expressed in terms of computer algebra. In this section we will examine the implementation of Kvectors in Maple as part of the Sophia system. A Maple representation of a Kvector is called a *KMvector*. A KMvector is defined as a list of Evectors and a number giving the number of Evectors involved. We first define constructor and selection functions for this structure. To illustrate how KMvectors are formed and used we consider the classic double pendulum problem, shown in figure 3.11. Units are chosen such that the upper rod has length 1 and the lower rod length L, that is L is the ratio of rod lengths. The following Sophia statements sets up the frames for the problem, describe the configuration, define generalized speeds and calculate the velocities of the pendulum's mass points:

```
>chainSimpRot([[N,A,3,q1],[A,B,3,q2-q1]]):
>dependsTime(q1,q2,u1,u2):
>kde := {q1t=u1,q2t=u2}:
>r1:= A &ev [0,-1,0]:
>r2:= (B &ev [0,-L,0]) &++ r1:
>v1:= &simp subs(kde, (N &fdt r1)):
>v2:= &simp subs(kde, (N &fdt r2)):
```

The set of equations that define the generalized speeds u1 and u2 are presented as a Maple set with the name *kde, standing for kinematic differential equations.* This

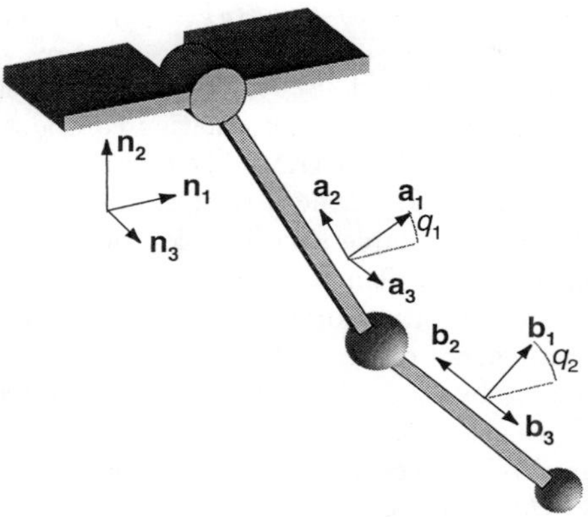

Figure 3.11: Frames for the Double Pendulum

will be our standard practice in using Sophia routines. This provides a convenient substitution set which is used to replace the explicit appearance of generalized coordinate time derivatives, i.e. generalized velocities, by generalized speed parameters. It can not be over emphasized that the choice of kinematic differential equations is a result of the choice of generalized speeds and that this is entirely up to the user. The velocity Evectors obtained above are:

```
>v1;
```

$$[[u1, 0, 0], A]$$

and

```
>v2;
```

$$[[\cos(q1 - q2)Lu2 + u1, -\sin(q1 - q2)Lu2, 0], A]$$

The KMvector is formed by presenting a list of Evectors to the operator &KM, thus the velocity KMvector is obtained by the statement:

```
&KM [v1,v2];
```

$$[[[u1, 0, 0], A],$$

$$[[\cos(q1 - q2)Lu2 + u1, -\sin(q1 - q2)Lu2, 0], A], 2]$$

Note that this consists of a three part list, the Evectors v1 and v2 and the number of involved Evectors, in this case 2. This information can be retrieved by use of the selector functions &vPartKM and &sPartKM. The first outputs a list of the Evectors, the second the number of Evectors in the list. It is also useful to have an operator which can be used to carry out Maple simplification operations on the components of the KMvector. This is called &Ksimp.

The main reason for defining Kvectors is that they provide a means for obtaining partial velocities as components of tangent Kvectors and that the fat dot product with the tangent vectors can be used to eliminate constraint forces from the dynamical equations of motion. Once generalized speeds are defined it was seen that the tangent vectors could be obtained by inspection of the velocity expression. This is automated by a Sophia command, KMtangents, which takes as arguments a velocity Kvector, the base name used for the generalized speeds and the number of expected independent tangent directions to the configuration manifold. The output is a list of the tangent KMvectors. In the present example:

```
>tau :=  KMtangents(vK,u,2):
```

$$[[[0, 0, 0], A], [[0, 0, 0], A], 2]$$

The alert reader will note that despite the appearance of the colon something is printed to the output device. In fact the tangent vectors have been stored in the list `tau`, however as an aid to see if results are reasonable the procedure prints out the KMvector part of the velocity expansion due to time dependent constraints or deliberate insertion of explicit time dependence into the representation. In the velocity expansion formula of previous sections this term was written as $\boldsymbol{\tau}_t^<$. As expected for the present problem this leads to a null KMvector! This part of the velocity can also be obtained by the operator &VKtime(Vk,u,n), which outputs the above KMvector. The individual KMvectors for each tangent KMvector are obtainable by use of the standard Maple selection operations from the list tau, thus

```
>tau[1];
```

$$[[[1, 0, 0], A], [[1, 0, 0], A], 2]$$

and, using the Maple function `op()` for variety,

```
>op(tau,2);
```

$$[[[0,0,0],A],[[\cos(q_1-q_2)L,-\sin(q_1-q_2)L,0],A],2]$$

For clarity and consistency with possible changes in Sophia it is recommended that one should use the defined Sophia selector functions rather than `op` as was done here! Also the reader is advised to verify these results by hand calculation, if for no other purpose then to gain confidence in the system.

The next step is to introduce algebraic operations on KMvectors. The most important for our work is the fat dot product, which we provide the Sophia operator &O. Note this is composed of the ampersand character and the uppercase letter O, while recall that the normal dot product between Evectors was represented by &o. The user need not worry about which frames are used in representing the component Evectors as this is automatically taken care of by the operator. To see an example consider the fat dot products of the tangent Kvectors found above. Thus

```
>tau[1] &O tau[1];
```

$$2$$

```
>tau[2] &O tau[2];
```

$$L^2$$

```
>tau[1] &O tau[2];
```

$$\cos(q_1-q_2)L$$

The *geometric identification* of tangent and cotangent vectors discussed above has been assumed. To complete the algebra of KMvectors operators are defined for scalar multiplication, addition and subtraction. These are:

```
s &*** vK
vK1 &+++ vK2
vK1 &--- vK2
```

We will see many applications of these operators in the work that follows. It is also convenient to have an operator which takes frame based time derivatives of all the component Evectors with respect to a particular frame. This is done with the operator &Kfdt. For example we can obtain the acceleration Kvector by the operation:

```
>aK := &Ksimp subs(kde,( N &Kfdt vK));
```

$$[[[u1t,u1^2,0],A],$$

$$[[\sin(q_1-q_2)Lu2^2+\cos(q_1-q_2)Lu2t+u1t,$$

$$u1^2+\cos(q_1-q_2)Lu2^2-\sin(q_1-q_2)Lu2t,0],A],2]$$

Note that the operations of substitution of the kinematic differential equations and simplification have all been combined with the differentiation.

3.9 Super KMvectors

The geometry of the configuration manifold as an embedded 'surface' in the space of all possible configurations is revealed in the tangent vectors. We have seen that the tangent vectors are represented by Sophia KMvectors. Collections of KMvectors describe subspaces and it is useful to also represent them as a Sophia data structure. In fact the object 'tau' derived by use of the KMtangent function is a typical case and we will consider lists of KMvectors as a new structure.Thus lists of KMvectors will be called *SKvectors* (for Super KMvectors). Like our other structures we define constructors and selectors, even though they are cases of already existing Maple structures we want to be able to redefine them if necessary, hence the computer science principle of data abstraction is best served by using only our special constructors and selectors in our calculations. The basic constructor of an SKvector is:

```
tau &SK [tau1,tau2]:
```

while the selector has the form:

```
>tau1 := tau &SKc 1:
>tau2 := tau &SKc 2:
```

The number of KMvectors in an SKvector is given by

```
> &SKn tau:
```

which in the present case would be 2. Sophia contains operators for an algebra of SKvectors, thus

```
>s &**** tau:
>beta &++++ tau:
>beta &---- tau:
```

implement the operations of multiplication by a scalar, addition and subtraction. In fact the most interesting operation on SKvectors is one which corresponds to linear transformations on the space of KMvectors. If such a linear transformation is defined by a matrix of coefficients, say `Wij`, then

```
>Wij &++** tau:
```

outputs a new SKvector or list of KMvectors which are obtained by matrix multiplication of the column of KMvectors obtained from the list tau, i.e.

$$\boldsymbol{\beta}_1^{<} = W_{11}\boldsymbol{\tau}_1^{<} + W_{12}\boldsymbol{\tau}_2^{<}$$

and

$$\boldsymbol{\beta}_2^{<} = W_{21}\boldsymbol{\tau}_1^{<} + W_{22}\boldsymbol{\tau}_2^{<}$$

would comprise the component KMvectors of the output list for the above operation in the case of our double pendulum example. Therefore linear transformation of Kvectors as discussed in this chapter is implemented in Sophia by the above operator.

In the discussion of reciprocal base Kvectors we saw that an important role was played by the matrix G, consisting of an array of the inner product of the base Kvectors. Sophia provides an operator which given any collection of independent KMvectors, i.e. an SKvector, produces the matrix that would apply to the subspace spanned by those vectors. For the case of our example we simply write:

```
>GKtau := &Kmetric tau;
```

$$\begin{bmatrix} 2 & \cos(q_1 - q_2)L \\ \cos(q_1 - q_2)L & L^2 \end{bmatrix}$$

which clearly corresponds to our above calculations. The reciprocal base is obtained by use of the inverse of the matrix, i.e.

```
GKtauC := inverse(GKtau):
```

where 'inverse' is a standard function of Maple's linear algebra package which is loaded into the system along with the Sophia routines. In many cases the only interest is in the reciprocal vectors and not in the matrix G, therefore Sophia contains an operator which directly computes it:

```
>tauC := &cv tau:
```

Any KMvector can be projected onto the subspace spanned by the members of an SKvector composed of linearly independent KM vectors. From the discussion of reciprocal base systems we see that the components of the projection can be obtained as our 'fat' inner product with the reciprocal base vectors. Sophia contains an operator

```
wSK &<> vK
```

to carry out this calculation. Given an SKvector and a KMvector it calculates an array of 'fat' inner products of each member of the SKvector with the KMvector. If the SKvector is composed of the reciprocal base vectors the members of the array will be the components of the expansion of the KMvector in terms of the corresponding base vectors. If the KMvector is not in the subspace this will correspond to its projection onto the subspace. In our pendulum example the velocity KMvector is in the tangent space (this is because the constraints are time independent) hence the components of vK with respect to tau should simply be the generalized speeds defined above. The calculation is:

```
>cvKtau := tauC &<> vK;
```

$$[u1, u2]$$

This result is in the form of a Maple array, which while it looks like a list is not one, so for example op(cvKtau) is required to output the array as it is subject to Maple's last name evaluation rule.

If we want to combine the expansion in terms of the coefficient array and the base vectors we use the Sophia operator:

```
coefficientArray &SKsum baseSKvector:
```

Which for our example is

```
>cvKtau &SKsum tau;
```

$$[[[u1, 0, 0], A], [[\cos(q_1 - q_2)Lu2 + u1,$$
$$- \sin(q_1 - q_2)Lu2, 0], A], 2]$$

We now have the tools needed to return to the problem of determining the kinematic differential equation and the implications of a specific determination.

3.10 Kinematic Differential Equations and Computer Algebra

The problem treated in this section is that given a set of kinematic differential equations, how does one in general determine the corresponding basis tangent Kvectors and expansion of the velocity in terms of such vectors. In the development of a formalism aimed at hand calculation this is usually done by suitable substitutions and inspection. With the computer algebra tools, gathered under the name of Sophia, more direct methods will have occasional practical as well as theoretical interest. The following discussion will use both the standard mathematical and the Sophia notation. Not all calculational results will be shown and the style of presenting the computer algebra steps as if engaged in an actual terminal session will not be followed. In other words from now on we will treat the Sophia commands as a prescriptive language which just happens to be understood by a Maple interpreter which incorporates the Sophia routines.

3.10.1 The Direct Kinematic Problem

The Direct Kinematic Problem is that given the kinematic differential equations find the the tangent Kvectors and the expansion of the velocity Kvector in terms of these vectors. That is given

$$\dot{q} = Yu + X, \tag{3.193}$$

or

$$\dot{q}_i = \sum_{j=1}^{n} Y_{ij} u_j + X_i, \tag{3.194}$$

obtain the array of tangent vectors β and the term $\boldsymbol{\beta}_t^<$ needed to form the expansion

$$\boldsymbol{v}^< = u^T \beta + \boldsymbol{\beta}_t^<. \tag{3.195}$$

From the discussion of Kvector transformations we see that the solution to the problem of obtaining the tangent vectors is that:

$$W^{\beta\tau} = Y^T, \tag{3.196}$$
$$\beta = W^{\beta\tau}\tau, \tag{3.197}$$

where the set of tangent vectors τ are obtained by using generalized velocities as generalized speeds, i.e. by taking $\dot{q} = u$. Both τ and β can be obtained by the inspection method, but since the set τ are coordinate based tangent vectors they can also be found by use of the relation

$$\boldsymbol{\tau}_j^< = \frac{{}^N\partial}{\partial q_j}\boldsymbol{r}^<. \tag{3.198}$$

In addition for such coordinate based vectors we have

$$\boldsymbol{\tau}_t^< = \frac{{}^N\partial}{\partial t}\boldsymbol{r}^<, \tag{3.199}$$

where the term $\boldsymbol{r}^<$ is the Kvector collection of configuration positions. We will see the natural extensions required for rigid body systems in chapter 5. The term $\boldsymbol{\beta}_t^<$ is computable both by the inspection method and by direct calculation. The latter follows by application of the projection operators defined previously. Thus

$$\begin{aligned} \boldsymbol{\beta}_t^< &= \boldsymbol{\tau}_t^< - {}^\beta f^T \beta \\ &= \boldsymbol{\tau}_t^< - \dot{Q}^T \beta. \end{aligned}$$

or in terms of the posed form of the kinematic differential equations:

$$\boldsymbol{\beta}_t^< = \boldsymbol{\tau}_t^< + X^T \beta. \tag{3.200}$$

It may also be useful to obtain the projection of an arbitrary Kvector on to the tangent hyperplane. To accomplish this we need the reciprocal basis of the set β. This is obtained by use of the β matrix G_β. Thus with $\overline{\beta}$ representing the array of reciprocal base vectors

$$\begin{aligned} G_\beta &= \beta \bullet \beta^T, \\ \overline{\beta} &= G_\beta^{-1} \beta. \end{aligned}$$

Then if $w^<$ is an arbitrary Kvector we have:

$$\boldsymbol{\Pi} \bullet w^< = (\overline{\beta}^T \bullet w^<)^T \beta, \tag{3.201}$$

and

$$\boldsymbol{\Pi}_* \bullet w^< = w^< - \boldsymbol{\Pi} \bullet w^<. \tag{3.202}$$

The generalization of this to any subspace is simple. Form the basis of the subspace from a set of independent vectors and obtain G and the reciprocal vectors for that set.

•Illustration

In this illustration it will be seen that the two methods of obtaining the tangent Kvectors appropriate to a particular choice of kinematic differential equations do indeed give the same result. This will also give an opportunity to show how the various Sophia routines work in practice. The twin pendula shown in the figure are

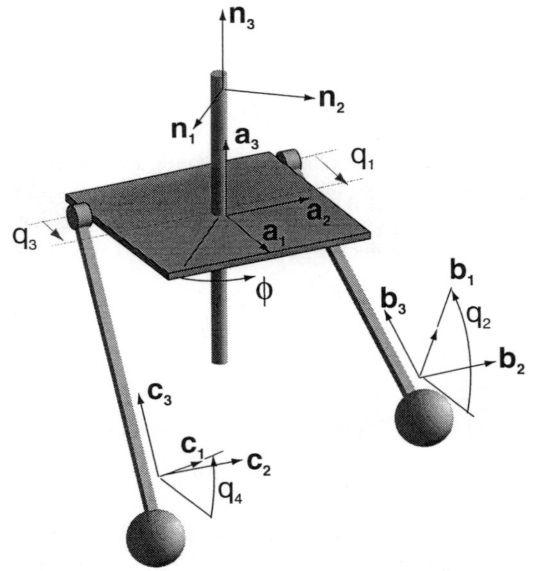

Figure 3.12: Twin Sliding and Rotating Pendula

attached to collars which can slide along the edges of the rotating square frame. In the illustration it will be assumed that the frame rotates about a vertical axis so that the angle made with the reference direction $\boldsymbol{n}_1$ is a given function $\phi(t)$. The pendula are assumed to rotate in planes orthogonal to the edges of the frame. The frame is designated as body A, the pendula as bodies B and C. The configuration of the system is described by the angle ϕ and the four parameters q_1 through q_4 where q_1 and q_3 describe the distance from the frame centerline along the edges and q_2 and q_4 are the angles made by the pendula arms with the vertical. The edges of the frame are taken as $2L$, the length of the pendula as L. The complete specification is in fact given by the Sophia statements that describe the configuration. These are:

```
> chainSimpRot([ [N,A,3,phi],[A,B,2,-q2]]):
> chainSimpRot([ [N,A,3,phi],[A,C,2,-q4]]):
```

which describe the relationships of the frames N,A,B and C, where N is the fixed reference frame. The position of the sliding collars and the pendulum ends are then given as:

```
> r1 := A &ev [L,q1,0]:
> r2 := r1 &++ (B &ev [0,0,-L]):
> r3 := A &ev [-L,q3,0]:
> r4 := r3 &++ (C &ev [0,0,-L]):
```

The generalized coordinates, q_j as well as the angle ϕ are functions of time, hence must be declared as such for the Sophia frame based time differentiation commands to work. This is accomplished with the statement:

```
> dependsTime(q1,q2,q3,q4,phi):
```

The velocities of each of the points of the mechanism with respect to the reference frame N is now easily calculated by differentiating the position vectors r_j with respect to the frame N. It is useful to carry out this somewhat repetitive action with the help of the Maple 'for' statement. This is a very general iterative statement with many optional features. The form that is most familiar to users of traditional computer languages is used in what follows, thus in conjunction with Maple's concatenation operator (indicated by the period or full stop) we can write:

```
> for j from 1 to 4
>   do
>     v.j := &simp (N &fdt r.j):
>   od:
> j:='j':
```

Depending on the system used this may have to be entered as a single line, however the above form has been used to make the statement easier to read. As in languages such as Fortran or Basic the Maple 'for' statement is given the first and last value of the running index j. The step of 1 is implicit unless the keyword 'by' is used to indicate another choice of step. The sequence of statements to be evaluated is set off by the key words 'do' and 'od'. This backward spelling convention is also used in the Maple 'if' statement which is terminated by the key word 'fi'. The concatenation operator does not evaluate the symbol to its left, in this case v, which in any event would evaluate to itself since it has not been assigned. On the other hand the index 'j' is evaluated each time the loop is iterated. This causes the assignment to be made to v1, v2 etc. as the loop is repeated. The same applies to the term r.j which is evaluated as r1,r2 etc.. Finally the index 'j' is 'unassigned'. This prevents problems from occurring due to it being left with the value of 5 at the end of the iteration. After the loop is finished we can examine the resulting values of the velocity, for example one can see that v3 has the value:

$$[[-phit\ q3,\ q3t - phit\ L, 0], A]$$

The 'standard' choice for the form of the kinematic differential equations is to take the generalized speeds to simply be the time derivatives of the generalized coordinates, i.e. to be what are normally called the generalized velocities in the literature of mechanics. We will designate these generalized speeds with the symbol u1 etc. so that the set of kinematic differential equations are given by:

```
> kde := {q1t=u1,q2t=u2,q3t=u3,q4t=u4}:
```

The configuration under study has four generalized coordinates and hence at least in this case four independent generalized speeds. Just as we were able to use our freedom of choice in picking the generalized coordinate we also have the option of choice in choosing generalized speeds, hence we need not adhere to the standard choice indicated in the above kinematic differential equation set. Here follows a description of how to make another of an infinity of possible choices. If we examine the velocities of the pendulum bobs as represented in frame N, easily done by using the Sophia commands

```
> N &to v2:
> N &to v4:
```

it is seen that the vertical or 3 components do not depend on the generalized coordinates time derivatives, while as expected from the general velocity expansion formula, the other components are linearly dependent on the the generalized velocities. Therefore we can define *four* new generalized speed so that these velocities will take the simple forms:

$$\boldsymbol{v}_1 = w_1\boldsymbol{n}_1 + w_2\boldsymbol{n}_2 + s_1\boldsymbol{n}_3$$

and

$$\boldsymbol{v}_2 = w_3\boldsymbol{n}_1 + w_4\boldsymbol{n}_2 + s_2\boldsymbol{n}_3.$$

The quantities w_j will now be taken as the new generalized speeds with the functions s_j to be determined by simple comparison. From the above comments it should be clear that they will not depend on the derivatives of generalized speeds. The new form of the kinematic differential equations are now obtained by using Sophia's selector functions together with Maple's ability to solve equation systems and to present the results as sets, suitable for the substitution command. The equations for the w are first formed, thus:

```
> v2q := &simp (N &to v2):
> v4q := &simp (N &to v4):
> eqks := { (v2q &c 1) - w1, (v2q &c 2) - w2,
                (v4q &c 1) - w3,(v4q &c 2) - w4}:
```

where some simplification has also been carried out using the Maple 'combine' function. In the next step the solve function is applied and the result further simplified to take account of trigonometric identities:

```
> kdew := combine(solve(eqks,{q1t,q2t,q3t,q4t}),trig):
```

The resulting set is assigned to the name 'kdew' to indicate that the new kinematic differential equations use the term 'w' for generalized speeds. The Maple set does not preserve order and we have no way of knowing what order the kinematic differential equations will be in as this depends on the Maple session. Therefore it is useful to convert the above information to a list form in which the right hand side of the equations appear as expressions in the order of the generalized speed designations. The following Maple code accomplishes this:

```
> KDEW := subs(kdew,[q1t,q2t,q3t,q4t]):
```

substituting the set kdew into the given list. Now for example

```
> KDEW[2];
```

gives the equation for $\dot{q}_2$ i.e.

$$\frac{w1\ \cos(\phi)}{L\cos(q2)} + \frac{phit\ q1}{L\cos(q2)} + \frac{\sin(\phi)w2}{L\cos(q2)}$$

Maple has the routine 'genmatrix' for abstracting coefficient matrices from list forms as given above. They can be used as part of the linear algebra package or you can use slightly modified forms that have been defined to work with the Sophia commands. These modified forms are now used to obtain the arrays needed for the transformation of tangent vectors and other terms, thus:

```
> Y := CoefficientArray(KDEW,[w1,w2,w3,w4]):
> X := SourceList(KDEW,[w1,w2,w3,w4]):
```

generates a matrix and a list respectively following the notation used in our discussion of SKvector transformations and generic forms for the kinematic differential equations. Thus the Maple code

```
> add(multiply(Y,[w1,w2,w3,w4]),X)
```

generates the rightside of the kinematic differential equations between $\dot{q}$ and w in the form $Yw + X$.

As the reader is certainly aware, even quite simple mechanical systems generate complex and unwieldy expressions. Maple provides a useful facility for dealing with this problem in the form of two functions, 'freeze' and 'thaw'. These are not present in a Maple session until the command `readlib(freeze)` is given. When applied to an expression freeze will replace it by a single system generated symbol of the form _Rj, where j is an integer. The expression can be recovered by the use of the command 'thaw'. If an expression is made up of other subexpressions the freeze command can be applied separately to the components using the 'map' command. The presentation of the above coefficient matrix provides a useful example, thus:

```
> Yf := map(freeze,op(Y));
```

produces the output:

$$\begin{bmatrix} \_R7 & \_R2 & 0 & 0 \\ \_R6 & \_R5 & 0 & 0 \\ 0 & 0 & \_R1 & \_R0 \\ 0 & 0 & \_R4 & \_R3 \end{bmatrix}$$

The various terms can be expressed in terms so the basic variable by use of the 'thaw' command:

```
> thaw(_R7)
```

produces the first term in the coefficient matrix

$$-\frac{\sin(\phi + q2)}{2\cos(q2)} - \frac{\sin(\phi - q2)}{2\cos(q2)}$$

From the general theory of tangent base transformations we can obtain the matrix relating the new tangent vectors, designated by the list or SKvector β as the transpose of the coefficient matrix Y, or in Maple:

```
> BetaTau := transpose(Y):
```

The coordinate basis of tangent vectors can be obtained either by differentiation of the position Kvector, rK defined below, or by the inspection method. The latter is implemented using the following set of steps in Sophia using the `KMtangents` and `vKtime` commands

```
> rK := &KM [r1,r2,r3,r4]:
> vK := N &Kfdt rK:
```

The velocity vectors for each of the four particles are contained in the Kvector vK which is obtained by frame based differentiation of the entire position Kvector with respect to the reference frame N. The first component, vK[1] is the Evector velocity of the particle at position r1 relative to frame N, which is vK[1] or more explicitly:

$$[[v11, v12, 0], A]$$

where

$$v11 = -\cos(\phi)^2 \tfrac{d\phi}{dt}\, q1 - \sin(\phi)^2 \tfrac{d\phi}{dt}\, q1$$

and

$$v12 = \sin(\phi)^2 \tfrac{d\phi}{dt}\, L + \sin(\phi)^2 q1t + \cos(\phi)^2 \tfrac{d\phi}{dt}\, L + \cos(\phi)^2 q1t$$

The velocity Kvector is now expressed in terms of the generalized velocities, $\dot{q}$ using the basic kinematic differential equations. At the same time some trigonometric simplification is carried out,

```
> vKtau := combine(subs(kde,vK),trig):
```

The coordinate basis tangent vectors are now stored in the list tau as what we are calling an SKvector. The term independent of the generalized velocities in the velocity expansion, $\tau_t^{<}$ is stored as the Kvector `taut`:

```
> tau := KMtangents(vKtau,u,4):
> taut := vKtime(vKtau,u,4):
```

Note that the command which produces the tangent vectors takes as arguments the Kvector velocity, the letter 'u' which is the common base symbol for the generalized velocities and 4, the number of independent tangent vectors expected. The result is that tau is given by the expression:

$$[[[[0,1,0],A],[[0,1,0],B],[[0,0,0],A],[[0,0,0],C],4],$$

$$[[[0,0,0],A],[[L,0,0],B],[[0,0,0],A],[[0,0,0],C],4],$$

$$[[[0,0,0],A],[[0,0,0],B],[[0,1,0],A],[[0,1,0],C],4],$$

$$[[[0,0,0],A],[[0,0,0],B],[[0,0,0],A],[[L,0,0],C],4]]$$

which the reader can verify is a list of four Kvectors. It is an excellent exercise to verify that these are indeed the correct tangent Kvectors. The new set of basis tangent vectors can also be obtained by the inspection method. To check the theory first use Sophia to obtain them directly from the transformation matrix, i.e.

```
> betaA := BetaTau &++** tau:
```

The notation betaA has been used to indicate that the direct method was used. The remaining part of the velocity expansion is given by

```
> betatA := taut &+++ (X &SKsum tau):
```

where again the symbol betatA indicates that the direct method was used and betatA corresponds to the term β_t^{κ}. The indirect method requires that the new kinematic differential equations are substituted into the velocity Kvector. The steps are:

```
> vKbeta := combine(subs(kdew,vK),trig):
> betaB := KMtangents(vKbeta,w,4):
> betatB := vKtime(vKbeta,w,4):
```

If all is correct the terms with the appended A should equal the terms with the appended B. This can be checked using operations on Kvectors and on SKvectors as follows:

```
> betatA &--- betatB;
> betaA &---- betaB;
```

which does result in a null Kvector and a null SKvector. As a final set of checks we can use the reciprocal basis theory and projection both on the tau and the beta set. This is done as follows for the tau set

```
> tauC &<> vKtau;
> TestTau:=tauC &<> (vKtau &--- taut);
```

The result is simply a list of the generalized velocities. A list of the new generalized speeds results from operations

```
> betaC := &cv betaA:
> TestBeta := betaC &<> (vKbeta &--- betatA);
```

The operations on beta involve computation of the 4 by 4 inverse matrix for the beta tangent vectors which is a somewhat costly operation and takes several minutes on a laptop machine. One must expect practical limits in applying some of these complex operations but it is encouraging that Maple even handles this case so well.
•

It should now be evident that with the aid of computer algebra we can explicitly apply the theory of this chapter to relatively complex cases, though we must expect limits due to the increasing size of expressions. In fact it is hard to imagine that we would have use for explicit algebraic results for very large systems and it is quit satisfying that the manipulations work so well for systems of intermediate complexity. Much more needs to be said about this but further discussion is postponed until we get to the determination of the dynamic equations of motion. It also appears that even for the simple cases treated so far the presentation of the results is a problem. Thus Maple is easily able to produce results that are fairly difficult to present if we do not wish to fill the text with endless pages of equations. The policy adopted is that only sample or important results will be shown explicitly. This does mean that for the reader to see full results it may be necessary to actually have a computer algebra system at hand!

3.11 Frame Fixed Velocity

The concept of 'frame fixed velocity' is very useful in the determination of conditions of constrained rolling. The most familiar and elementary example of this is shown in figure 3.13, where a cylinder rolls along a plane in a pure two dimensional motion. Two frames are involved in this problem, one fixed in the plane surface, the other in

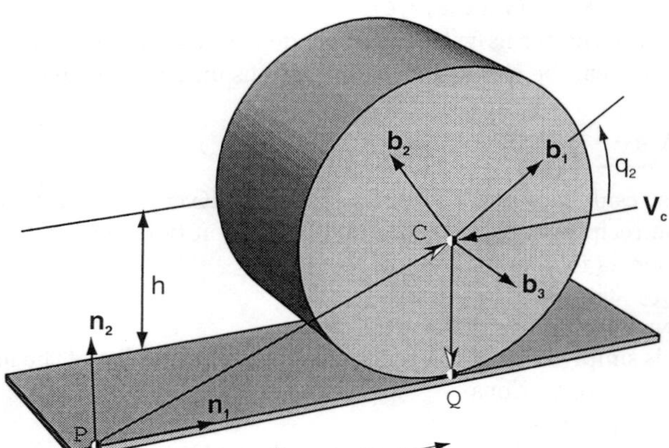

Figure 3.13: Simple Rotating Cylinder

the rolling cylinder. These are designated as N and B. Each frame is also assigned a fixed reference point. The plane surface reference point, P, is taken arbitrarily as any point which is on the surface, while the center C is a point chosen at the center of the cylinder. It is emphasized that each of these points are *fixed in their respective frames*, that is they have zero velocity relative to their frames at all times! The point Q is the *contact point* at which the cylinder touches the plane. This in fact is the defining characteristic of the point. *The contact point is not a point which is necessarily fixed in any body.* At each instant of time the contact point may be a different actual point on the surface of the cylinder or the plane. None of the points of the fixed plane are moving relative to the its frames reference point P. Whatever one of the surface points is the contact point at any time will have zero velocity relative to the plane. The idealized condition of *rolling with out slipping* or smooth rolling is the requirement that whatever point of body B, the cylinder, is the contact point at any instant it has the same velocity as the contact point in body N. The idea of a frame fixed velocity has in fact already been introduced in the discussion of section 3.3 where it was designated by the notation ${}^N\boldsymbol{v}^{B(Q)}$. A point Q is in general in motion with respect to frame B. At any instant the point Q coincides with one of the fixed points of B. This notation stands for the velocity of that latter point with respect to frame N. *It is very important to realize that it is a point that is a fixed part of frame B and need only have the same position as point Q at the instant in question.* This is exactly the situation we have in the rolling contact problem. The non-slip condition is thus stated as

$$ {}^N\boldsymbol{v}^{N(Q)} = {}^N\boldsymbol{v}^{B(Q)}. \tag{3.203}$$

In fact this should apply for any arbitrary frame filling the roll of N in the above, thus an equivalent statement would be:

$$ {}^A\boldsymbol{v}^{N(Q)} = {}^A\boldsymbol{v}^{B(Q)}. \tag{3.204}$$

By its very definition it should also be evident that in general

$$ {}^A\boldsymbol{v}^{A(Q)} = 0 \tag{3.205}$$

which simply states that points making up a frame or body do not move with respect to one another.

A quantitative picture is obtained by realizing that we must have a specific definition of the point Q, which can now be any point which may be in motion with respect to any frame. We need a general reference frame A with an origin O, in which case we take the position vector of Q as given in the form

$$ \boldsymbol{r}_(Q) = \boldsymbol{r}^{\mathcal{OQ}}. \tag{3.206}$$

We also need the velocity of the origin P of the frame B, ${}^A\boldsymbol{v}^P$ and the angular velocity ${}^A\boldsymbol{\omega}^B$ so that from our previous work

$$ {}^A\boldsymbol{v}^{B(Q)} = {}^A\boldsymbol{v}^P + {}^A\boldsymbol{\omega}^B \times \boldsymbol{r}^{\mathcal{PQ}}. \tag{3.207}$$

Equation 3.207 may now be taken as the definition of ${}^{A}\boldsymbol{v}^{B(Q)}$. For computational purpose this becomes most useful in the form

$$ {}^{A}\boldsymbol{v}^{B(Q)} = \frac{{}^{A}d}{dt}\boldsymbol{r}_P + {}^{A}\boldsymbol{\omega}^{B} \times (\boldsymbol{r}_Q - \boldsymbol{r}_P). \tag{3.208} $$

This is easily implemented in Sophia-Maple as

```
> (A &fdt rP) &++ ( wAB &xx (rQ &-- rP))
```

and because of its importance this has been added to Sophia in two forms. The first is

```
> FFV(A,rP,wAB,rQ)
```

where the angular velocity must be computed and inserted as wAB. The other form is as an infix operator. A typical example would be

```
> [A, B] &ffv [rP, rQ]
```

The angular velocity is then calculated internally. In some problems this may be time consuming and one may have other needs for the angular velocity, hence the first form may be more suitable.

Now that we have an implementation of the fixed point velocity concept it is interesting to examine some typical examples. First consider the simple rolling example discussed above. The term ${}^{N}\boldsymbol{v}^{B(Q)}$ is easily calculated by the following steps:

```
> dependsTime(q1,q2):
> chainSimpRot([[N,B,3,q2]]):
> rP := N &ev [q1,L,0]:
> rQ := N &ev [q1,0,0]:
> vNBQ:=[N,B] &ffv [rP,rQ]:
```

where q1 is the distance from the N frame origin to the contact point and q2 is the counterclockwise positive angle of the wheel. The result is:

$$ [[q1t + q2t\, L, 0, 0], N] $$

which in standard notation is

$$ {}^{N}\boldsymbol{v}^{B(Q)} = \frac{dq_1}{dt} + L\frac{dq_2}{dt}. $$

Setting this to zero gives the non-slip condition, familiar from elementary mechanics courses.

A slightly more challenging example is indicated in figure 3.14. The disk of radius L is taken as body B. It spins about the moving arm of length L taken as body A. The arm in turn rotates about an axis fixed in the frame of the plane taken as N. The problem is to determine the relative contact velocity of the spinning disk relative to the plane fixed in frame N, i.e. ${}^{N}\boldsymbol{v}^{B(Q)}$. For convenience we also introduce a new Sophia command **&kde n** which sets up the kinematic differential equations in generalized velocity form, and declares the time dependence of 2 n standard variables, q and u. In the present problem it is only the latter effect of the declaration that is of interest.

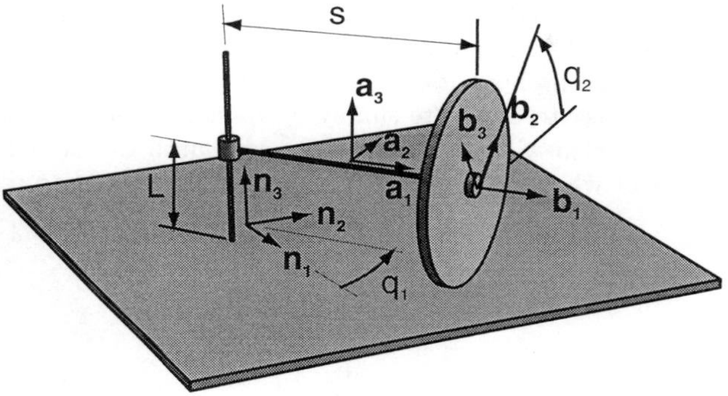

Figure 3.14: Spinning Disk on Rotating Arm

The first step is to define the configuration and to declare which variables are to be considered functions of time. This is done with the Sophia statements:

```
> &kde 2:
> chainSimpRot([ [N,A,3,q1],[A,B,1,q2] ]):
```

which will result in a message telling you that the variables q1, q2, u1 and u2 have been declared time dependent and that the simplest form of the kinematic differential equations, in which $\dot{q} = u$, has been assigned to the name kde. To complete the problem we will use the form of the frame fixed velocity function that takes angular velocity as an argument, therefore we first compute and store the angular velocity:

```
> wNB := N &aV B:
```

The rest of the calculation of ${}^{N}\boldsymbol{v}^{B(Q)}$ requires the position of the location of the origin of the rotating wheel frame B and the definition of the contact point, i.e. the Sophia statements:

```
> rB := s &** (A&>1):
> rQ := rB &-- (l &**  (A&>3)):
```

Finally the second form of the fixed frame velocity function is called to calculate the velocity of the contact point in the wheel relative to the fixed frame N:

```
> vNBQ := FFV(N,rB,wNB,rQ);
```

$$[[0, q1t\, s + q2t\, l, 0], A].$$

This last result is relatively obvious as it is the sum of the velocity of the end of the rotating arm and the velocity of the circumference of the wheel relative to that position and is clearly in the direction of the unit vector $\boldsymbol{b}_2$ at any fixed instant. The velocity of the contact point as a fixed point in frame N, ${}^{N}\boldsymbol{v}^{N(Q)}$, vanishes. Therefore the condition of non-slip contact gives the result that

$$sq_{1t} + lq_{2t} = 0.$$

The next example, shown in figure 3.15, is a classic problem for the illustration of so called nonholonomic constraints. For the moment the reader may consider these kind of constraints to be conditions relating the generalized speeds rather than the generalized coordinates. This is not always the case but further discussion is reserved for a later chapter. The problem itself is sometimes called the rolling coin or disk problem. It is quite easy to duplicate simply by tossing a coin onto a flat surface. With a little care the coin will roll about on its edge until the motion is overcome by frictional effects at which point the coin will most likely come to rest in the flat position.

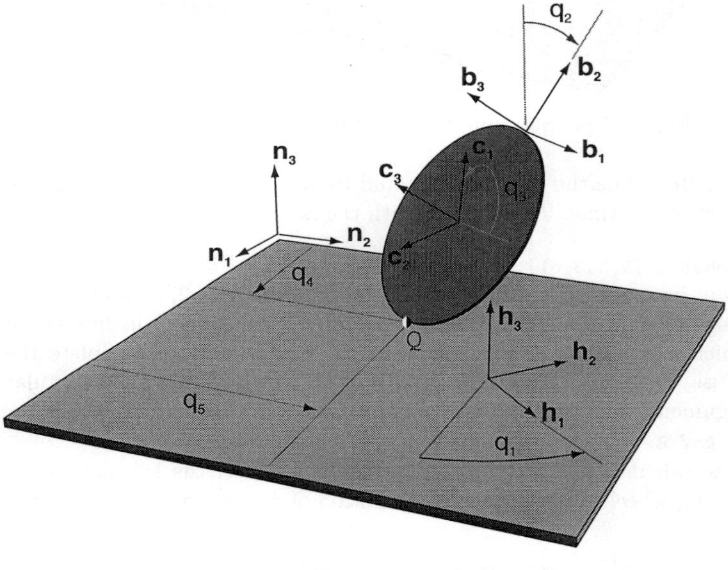

Figure 3.15: Rolling Coin

The configuration of the disk is specified by five generalized coordinates. To see this imagine a sequence of transformations that bring a disk from a reference position to an arbitrary final position. As reference take the disk to be parallel to the $\boldsymbol{n}_1 - \boldsymbol{n}_3$ plane with the contact point located at the origin of the N reference frame, i.e. the center will have coordinates $(0, 0, R)$ for a disk of radius R. The arbitrary configuration can be achieved by first moving the contact point to a position in the N frame with coordinates (q_4, q_5). The axis of the disk is now aligned with the $\boldsymbol{n}_2$ frame vector. Now turn the plane of the disk through an angle q_1. This gives us a new reference frame, H obtained by a rotation of q_1 about the $\boldsymbol{n}_3$ axis. Now the disk can be put into a slanted position by rotation of its axis, i.e. the frame B is obtained from H by rotation of $\boldsymbol{h}_2$ through a an angle of $\frac{\pi}{2} - q_2$. The final step is to rotate the disk about its own axis into the frame C by an angle q_3. The disk is at rest in this final frame! The relative velocity of the contact point is now obtained by the following Sophia commands which set up the frames and carry out the appropriate manipulations:

```
> &kde 5:
> chainSimpRot([[N,H,3,q1],[H,B,1,Pi/2-q2],[B,C,3,q3]]):
> rQ := N &ev [q4,q5,0]:
> rC := rQ &++ (R &** (B&>2)):
> vNCQ := N &to ([N,C] &ffv [rC,rQ]):
```

where again the &kde command was used to declare the time dependence of the generalized coordinates. The vector rQ describes the contact point, while rC describes the origin of the disks rest frame. The last step combines finding the vector ${}^{N}\boldsymbol{v}^{C(Q)}$ and the transformation of the result to the frame N in which the base plane is taken to be fixed. The result is:

$$[[\cos(q1)Rq3t + q4t, q5t + \sin(q1)Rq3t, 0], N]$$

To check this result note that the $\boldsymbol{n}_3$ component vanishes. This is expected since we implicitly constrained the motion to one in which contact would be maintained with the base plane. While the result is also obtainable by intuitive methods the problem is difficult enough to make one pleased to have a computer algebra tool for its solution. Later in this text it will be seen how to determine and solve the dynamical equations of motion for this well known example. In the so called nonholonomic problem the constraint of no slip implies that the above vector vanishes, which provided two relations among the five generalized speeds. Thus the problem has *5 geometrical degrees of freedom* but is not a classic 5 degree of freedom system.

Now consider a contact problem in which the determination of the contact point requires more than simple intuition. A parabolic cylinder is allowed to roll on a flat surface. The surface of the cylinder has an invariant description in the frame B fixed to the body. The position of the cylinder is defined by the orientation of the frame and the location of the frame origin. The contact situation is defined by the position of the contact point and the constraint that the cylinder must tangent the fixed plane at the point of contact. With reference to the figure let q_1 denote the position of the contact point in the frame N and q_2 the orientation of the frame B in the frame N. The position of the origin of the frame B, taken as the vertex point of the parabola, is denoted as (x, y) in the frame N. The parameter xi is used to describe the parabola in the frame B, thus a point on the parabola has coordinates (xi, kxi^2). This defines the surface position vector, $\boldsymbol{s} = x_i\boldsymbol{b}_1 + kx_i^2\boldsymbol{b}_2$. The five quantities, q_1, q_2, x, y and xi are related by the contact condition that

$$\boldsymbol{r}_B + \boldsymbol{s} - \boldsymbol{r}_Q = 0$$

and the tangency condition, i.e. three equations. Therefore in the case where slip is allowed we have two independent parameters while in the no slip case there is only

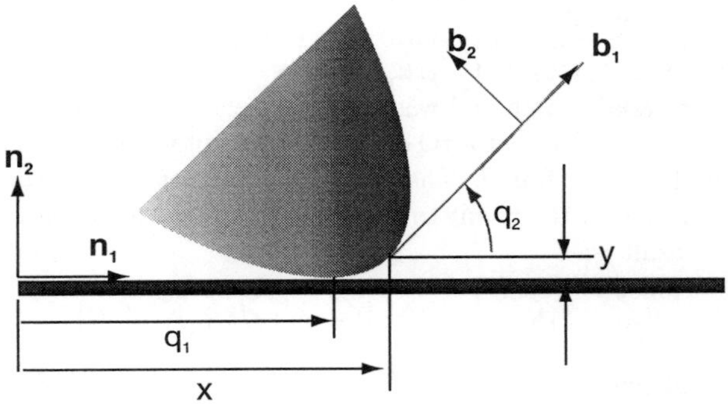

Figure 3.16: Rolling Parabolic Cylinder

one independent parameter. Using these conditions we can thus express x, y, and xi in terms of q_1 and q_2. This is done using the following Sophia-Maple statements:

```
> dependsTime(q1,q2,xi,x,y):
> chainSimpRot([[N,B,3,q2]]):
> s := B &ev [xi,k*xi^2,0]:
> ds := B &ev [1,2*k*xi,0]:
> rQ := N &ev [q1,0,0]:
> rB := N &ev [x,y,0]:
> vectEq := (rB &++ s &-- rQ):
> tangEq := ds &o (N&>2):
> ContactEqs := {vectEq &c 1,vectEq &c 2, tangEq};
> ContactSolution := solve(ContactEqs,{xi,x,y});
```

The declaration of time dependence will be used in deriving the non-slip condition. The second statement simply declares the relation of the cylinder frame to the reference frame. The vector $\boldsymbol{s}$ gives the surface coordinates in terms of the parameter xi. The vector $\boldsymbol{ds}$, in this simple case derived by inspection, is tangent to the parabola. The vectors $\boldsymbol{r}_Q$ and $\boldsymbol{r}_B$ are the position of the contact point and the vertex of the parabola, the latter also being the origin of the B frame. The next two steps derive the equations for contact and tangency, which are then gathered into the set 'ContactEqs' using the Evector component selection operators. Finally these are solved for the parameter value of the point of contact and the vertex position in terms of the position of the contact point and the orientation of the frame B. The results are:

$$y = \frac{\sin(q2)^2}{4\,k\cos(q2)}$$

$$x = \frac{2\sin(q2)\cos(q2)^2 + \sin(q2)^3 + 4\,q1\,k\cos(q2)^2}{4\,k\cos(q2)^2}$$

$$xi = -\frac{\sin(q2)}{2\,k\cos(q2)}$$

If a no slip condition is to be invoked we need the frame fixed velocity of the contact point in N at the contact point. This is easily found using our Sophia operator for frame fixed velocity and the standard Maple subs command, thus:

```
> vNBQ := [N,B] &ffv subs(ContactSolution,[rB,rQ])
```

which gives the result:

$$[[\frac{2\cos(q2)^3\,q1t\,k + q2t}{2\,k\cos(q2)^3}, 0, 0], N]$$

from which we see that the no-slip condition is given by:

$$\frac{2\cos(q2)^3\,q1t\,k + q2t}{2\,k\cos(q2)^3} = 0. \qquad (3.209)$$

Note that this is a condition imposed as a linear relation between the generalized velocities. In this planar problem the relation can be integrated to provide a standard holonomic constraint between the generalized coordinates.

The examples of this section provide a general approach toward the specification of contact problems, especially the means for dealing with conditions of non-slip rolling. More complex examples will be examined when we confront the general case of nonholonomic constraints as they arise in problems of dynamic rolling contact.

3.12 Problems

•**Problem 3.1** Figure 3.17 shows a robot arm consisting of three spherical and one cylindrical joint connect by means of three slender rods of length L. The end effector consists of a flat rectangular block of side L/2. Develop an expression for the position of the point Q on the corner of the end effector as a function of the joint variables. Use θ_j and ψ_j for each spherical joint. In setting up the joint coordinates take the indicated line as reference for the first angle θ_1 and use the resulting position of the line as reference for each of the following joints. The cylindrical joint only allows rotation about the axis by the angle ϕ.

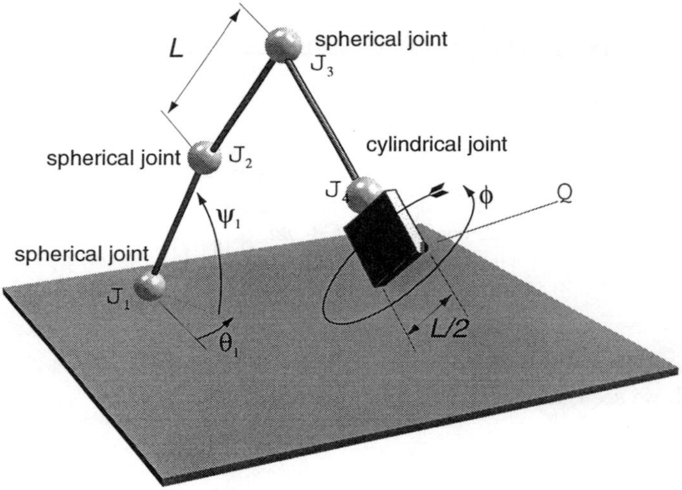

Figure 3.17: Jointed Robot Arm

•**Problem 3.2** Develop expressions for velocity and acceleration of the point Q in problem 3.17 given that $\psi_1 = \frac{kt}{1+t}$, $\psi_3 = \frac{kt}{2+t}$ and $\phi = t$ all other angles having the constant value 0.
•**Problem 3.3** Write the Kvector velocity for the point mechanism consisting of the two moving spherical joints J_2 and J_3 and the cylindrical joint J_4 in figure 3.17. Determine the tangent Kvectors to the configuration surface of the mechanism for the generalized coordinates ψ_j, θ_j, ϕ and using the time derivatives of these as generalized speeds.
•**Problem 3.4** The mechanism shown in figure 3.18 consists of two rotating disks and two rods. The larger disk of radius L is in forced rotation about an axis through its center at point O, the angle $\theta(t)$ being a given function of time. A rod of length L is attached to the edge of the disk at point P by a rotary joint which allows rotation

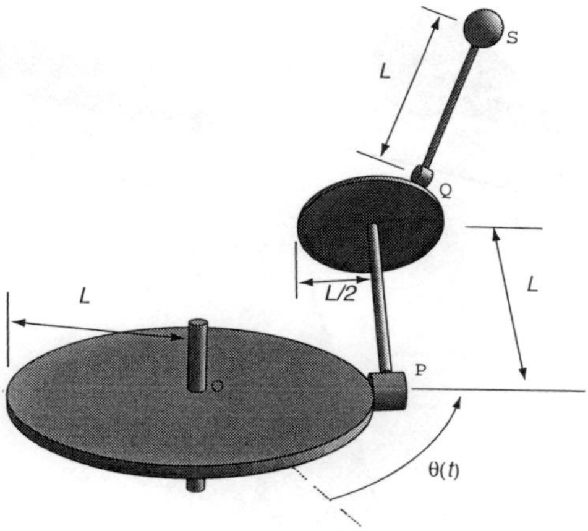

Figure 3.18: Jointed Disks

in a plane perpendicular to the radial line OP. The smaller disk of radius L/2 can rotate in a plane perpendicular to the rod. A second rod of length L is attached to the smaller disk at point Q by a rotary joint which allows rotation in a plane normal to the disk and passing through its center and the point Q. Choose suitable generalized coordinates and reference frames for describing the configuration of the mechanism. Use the velocity and acceleration decompositions of equations 3.40 through 3.43 to obtain the velocity and acceleration of point S.

•**Problem 3.5** Obtain the velocity Kvector for a point mechanism consisting of the points P, Q and S of figure 3.18 as described in problem 3.4. Choose at least two different forms of the kinematic differential equations, based on different choices of generalized speeds. Derive tangent Kvectors to the configuration surface for both choices. Verify the relations derived in the text between generalized speed transformations and tangent Kvectors basis to the configuration surface.

•**Problem 3.6** The orthonormal set, $\boldsymbol{a}_j$ provides a basis for Euclidian 4-space. The two vectors

$$\begin{aligned} \boldsymbol{s}_1 &= \boldsymbol{a}_1 + 3\boldsymbol{a}_2 + \boldsymbol{a}_3 \\ \boldsymbol{s}_2 &= \boldsymbol{a}_1 - 3\boldsymbol{a}_2 + \boldsymbol{a}_4 \end{aligned}$$

define a two dimensional hypersurface (plane) S. Find the operators $\boldsymbol{\Pi}_S$ and $\boldsymbol{\Pi}_{S*}$ which project vectors into the subspace S and its orthogonal complement S*. Use these to decompose the vector $\boldsymbol{v} = \boldsymbol{a}_1 + \boldsymbol{a}_2 + \boldsymbol{a}_3 + \boldsymbol{a}_4$ into parts $\boldsymbol{v}_1$ and $\boldsymbol{v}_2$ such that $\boldsymbol{v}_1$ is parallel to S and $\boldsymbol{v}_1 \cdot \boldsymbol{v}_2 = 0$. (Hint: develop a reciprocal basis expansion for $\boldsymbol{s}_j$.)

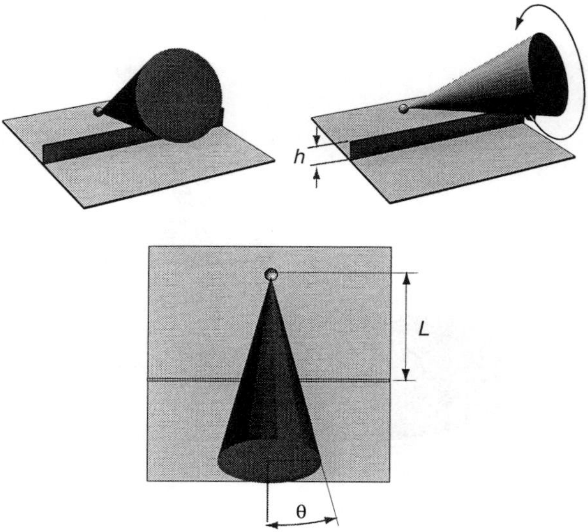

Figure 3.19: Rolling Cone

•**Problem 3.7** The vertex of a circular cone of half angle θ is attached to a surface by a universal joint as shown in figure 3.19. The surface of the cone is in contact with a rail of height h above the surface. The base of the rail is at a perpendicular distance L from the attachment point. It is assumed that the cone rolls with out slipping. Express the geometric configuration in terms of suitable generalized coordinates so as to express the position of an arbitrary point on the surface of the cone. In carrying out this task determine the position of the point of contact between the cone surface and the rail. Obtain the conditions for rolling with no slip in terms of your generalized coordinates and parameters. Check your solution by examining limiting cases, e.g. when the rail height tends to zero.

•**Problem 3.8** Figure 3.20 shows a two dimensional mechanism consisting of disk in forced rotation and two pinned links. Consider a mechanism consisting of the points P, Q and S. Attach the indicated frames to the disk A, and B and C the links of length L. Define the generalized coordinates q_1 and q_2 such that the relations $\cos q_1 = \boldsymbol{b}_1 \cdot \boldsymbol{n}_1$ and $\cos q_2 = \boldsymbol{c}_1 \cdot \boldsymbol{n}_1$, where N is the fixed reference frame. Also $\cos\theta(t) = \boldsymbol{a}_1 \cdot \boldsymbol{n}_1$, where the frame A is fixed in the disk and $\theta(t)$ is a given function of time. Define two sets of generalized speeds, one, labeled u and based on the generalized velocities for the given choice of generalized coordinates, the other labeled w and based on the relation: $\boldsymbol{v}^{<S} = w_1\boldsymbol{c}_1 + w_2\boldsymbol{c}_2$. Determine the transformation equations between the generalized speeds w and u, and the configuration surface tangent Kvectors. Provide a kinematic interpretation for each set of generalized speeds and tangent vectors in terms of the possible motions of the mechanism.

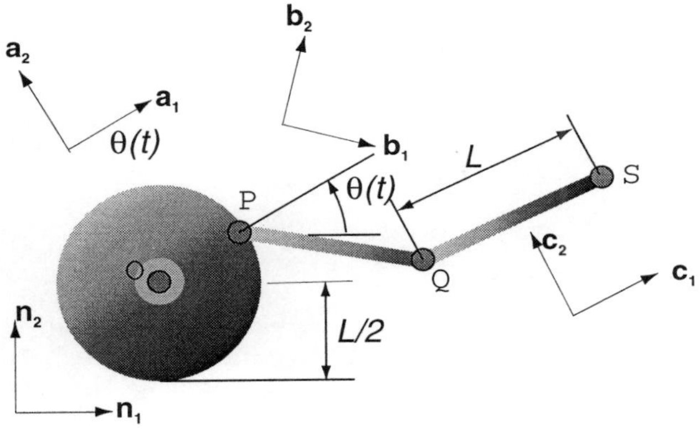

Figure 3.20: Simple Linkage

Chapter 4

The Dynamics of Point Mass Mechanisms

As we move about performing our everyday human activities we automatically carry out the solution of complex problems that are still beyond our best abilities at formal modeling. Grasping objects, the prediction and hopefully the avoidance of collisions with moving bodies, the movement of arms and legs are all examples of our 'genius' in the solution of problems of motion. The abstract task of modeling this activity provides us with tools for controlling and understanding our world. We certainly 'know' intuitively that it is not possible to produce all of the arbitrary family of kinematic motions described in the last chapter. To carry out a motion we need to take account of 'force', 'mass' and 'constraint'. It is the object of the present chapter to 'model' these everyday conceptions for a mechanism that consists of constrained point masses. After a brief discussion of some of the concepts involved we derive equations that govern the motions allowed by such a model. The major part of this derivation involves the elimination of the unknown constraint forces, which can then be determined after the motion has been calculated. To come to grips with the action of constraint forces we require a model for the realization of our common experience of how constraints function in typical mechanical devices. This is given by *D'Alembert's Principle of Constraint.* This principle, in combination with our previous development of the geometry of configuration and constraint and the Newtonian Principle, is used to obtain a form of the equations of motion. Under certain additional restrictions these lead in turn to Lagrange's Equations. The chapter concludes with a discussion of energy principles for point mass mechanisms.

4.1 Newtonian Dynamic Motions

Up to this point the main effort has been aimed at the general techniques of describing configuration. The concept of motion has involved the idea of how configuration may depend on an organized change of parameters controlling the generalized coordinates and the state of the overall constraints. We have used the variable 't' for the purpose of tracing a continuous change in a mechanism's possible configurations due to the alteration of the constraining geometry. Such a change in configuration describes the 'mapping' of a continuous interval of the one dimensional parameter 't' into a curve in the multidimensional configuration space of the mechanism. This type of change of configuration has been called a 'kinematic motion', mainly to distinguish it from the possible change in configuration that are restricted to a 'fixed' constraining geometry. This case can be considered as the mapping of a one dimensional parameter into a curve which is confined to a hypersurface that is in turn embedded in the full configuration space. *To distinguish these two types of change, the first has been called a kinematic motion and the second a test motion.* In the latter case the notation 'λ' has sometimes been used to label the curve points in the hypersurface.

A clear understanding of the above may be obscured by the intrusion of our everyday concept of time, which in a sense is equivalent to motion, i.e. we measure time by the observation of the motion of a 'clock'. At this point in the development of our mechanical theory the parameter 't', which we are calling the 'time', only plays a bookkeeping role. That is time has been a label of the possible geometric configurations of what has been called a mechanism. Two ways in which the mechanism's configuration can alter have been distinguished by the concept of time independent and time dependent constraint.

A simple example of this is the position of a bead on a wire. Assume the wire is fixed at one point and rotating in a plane. The bead is constrained to the wire, hence the bead has a position coordinate describing where it is on the wire. The wire is allowed to have a number of orientations described by an angle parameter. The full configuration space in this case is the three dimensional space of all positions. A convenient picture of a somewhat equivalent configuration space consists of all possible positions of the bead at all possible angular orientations of the wire. A test motion is a continuous change of configuration with the wire angle fixed. In a kinematic motion, both the position on the wire and the angle change, and both are continuous functions of the same single parameter.

The Newtonian principles, discussed briefly in the first chapter, allow us to distinguish a third type of 'motion', which in fact has the advantage of being very close to what we observe in our everyday reality. This type of motion is defined as a subclass of the set of all possible kinematic motions of a mechanism. It will be called a *Newtonian motion.* A Newtonian motion involves the concepts of mass, force, and an inertial observer. In this chapter we will consider *dynamic point mechanisms,* which are point mechanisms where each index point is assigned a scalar quantity and a vectorial quantity. The scalar quantity is called the *point inertial mass,* the vectorial quantity the *applied force.* The applied force may be a function of the configuration, the inertial mass and the velocity. We will also allow it to depend on the 'time', however implicit in this is the existence of another dynamical system which is in some sense separable from the one under study, and which is influencing it. A pendulum clock on a ship will be influenced by the motion of the vessel, but aside from its use by the crew in navigation, the clocks motion will have little influence on the vessel's path. Thus the specification of a *dynamic point mechanism* requires a set of *mass values* for the points and a set of given *applied force laws* that provide the *force vectors* at each point of the mechanism as a function of its configuration, velocity and time.

A *Newtonian motion* is then a kinematic motion, which in addition obeys the Newtonian law that the mass times the acceleration of each point is equal to the sum of the applied and constraint forces at that point. This law is taken to work for any *inertial observer.* An inertial observer sees the motion of a mechanism consisting of a single point under no forces as a straight line, where the point particle moves equal distances in equal times. It is easily verified that there is an infinite number of possible inertial observers related by the fact that they may all be moving with constant velocities in relation to one another.

Constraint forces are considered as a *result* of the presence of the constraints, that is the geometric restrictions that define possible kinematic motions. Thus the *solution of a dynamic problem* requires the determination of all possible Newtonian motions and constraint forces for a given mechanism. The unique specification of constraint forces requires more than the Newtonian principles. The additional principle, which may be called the *constraint principle of D'Alembert,* is in our terminology the statement that the projection of constraint forces onto the configuration hypersurface must vanish!

The viewpoint of classical mechanics is to accept the above as providing a model of a restricted part of observable reality. The fact that this part of reality coincides with the time and length scales of an overwhelming part of human experience and technology makes it well worth our effort to study the class of *Newtonian Dynamic Motions.* The full connection of this model with 'reality' is left to philosophers, theologians, politicians and psychologists.

4.2 The Dynamical Point Mechanism

A *dynamical point mechanism* is defined as a set of index points, each with position $\boldsymbol{r}^{<L}$, where a positive quantity, m_L, is associated with each point. The *momentum* associated with the point is defined as

$$\boldsymbol{P}^{<L} = m_L \boldsymbol{v}^{<L}. \tag{4.1}$$

During the course of time each index point changes position as measured by the *inertial observer* N, according to the Newtonian Law:

$$\frac{{}^N d}{dt}\boldsymbol{P}^{<L} - \boldsymbol{R}^{<L} - \boldsymbol{C}^{<L} = 0 \tag{4.2}$$

In the most common case the mass associated with each particle will remain constant during the course of a motion. The Newtonian Law then takes the form

$$m_L \frac{{}^N d}{dt}\boldsymbol{v}^{<L} - \boldsymbol{R}^{<L} - \boldsymbol{C}^{<L} = 0 \tag{4.3}$$

where $\boldsymbol{R}^{<L}$ for each point L is a *given* vectorial quantity called the *applied force*, and $\boldsymbol{C}^{<L}$ is a vectorial quantity called the *material constraint force.* The material constraint force is assumed to obey certain conditions, based on physical plausibility and experience, which will be set forth below. These conditions are needed in order to uniquely specify these quantities which must be found as part of the solution of the dynamical problem. The *dynamical problem* is to determine all possible kinematic motions of the mechanism that satisfy the Newtonian equations.

The notation developed for mechanisms in the last chapter will be used, hence the column array of position vectors of all points of the mechanism will be denoted by $\boldsymbol{r}^{<}$, and all velocities by $\boldsymbol{v}^{<}$, applied forces by $\boldsymbol{R}^{<}$ and material constraint forces by $\boldsymbol{C}^{<}$. The masses will sometimes be represented by a K diagonal matrix, $m^{<}$. The (L, L) entry denotes the mass of the Lth body, m_L. With this convention terms such as $m^{<}\boldsymbol{a}^{<}$ indicates a column with components of the form $m_L\boldsymbol{a}^{<L}$, so we can write the entire system of equations in the form:

$$m^{<}\frac{{}^{N}d\boldsymbol{v}^{<}}{dt} - \boldsymbol{R}^{<} - \boldsymbol{C}^{<} = 0. \tag{4.4}$$

With $\boldsymbol{P}^{<}$ taken as a column of momentum vectors the equations can also be stated in the more basic form:

$$\frac{{}^{N}d\boldsymbol{P}^{<}}{dt} - \boldsymbol{R}^{<} - \boldsymbol{C}^{<} = 0. \tag{4.5}$$

This notion of a point dynamical mechanism covers a large class of problems considered in classical mechanics, for example, the motions of celestial bodies considered as points with inverse square gravitational forces acting pair-wise between the bodies, or pendulums and other simple mechanical systems in which the masses are concentrated at certain locations. The importance of applied forces is that they are taken as being known quantities or functions of the mechanisms configuration. In some cases they may also depend on kinematic quantities such as the velocities of the mechanism's mass points. We will also follow the standard convention of calling the mechanism's mass points by the term *particles.*

•Illustration

Figure 4.1 depicts a mechanical device consisting of a 'collar' that can slide along a shaft which is forced to rotate about the point $\mathcal{O}$ in the plane defined by $\mathcal{O}$ and two of the prime observer's reference triad vectors, $\boldsymbol{a}_1$ and $\boldsymbol{a}_2$. The collar is also connected to the point $\mathcal{O}$ by a linear spring of *equilibrium* length l, and *stiffness* k. The spring develops a restoring force proportional to the stiffness times the difference between its stretched and equilibrium lengths. The rotation angle of the shaft is assumed as known as a function of the time, thus the quantity $\theta(t)$ is given as part of the specification of the mechanism. We can 'model' this device as a dynamical point mechanism by assuming that the mass of the spring and other components are small compared to the mass of the collar, and that the relative size of the collar is such as to allow us to represent its position by a single point, $\mathcal{P}$. We will assume that the point $\mathcal{O}$, is fixed in an inertial reference frame, i.e. that the prime observer is also an

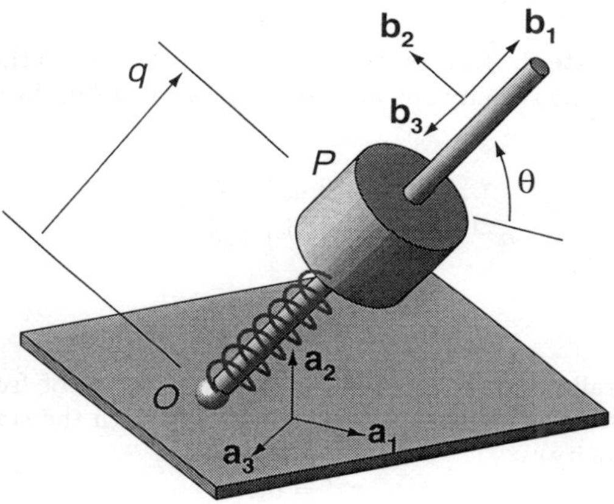

Figure 4.1: A simple example of a point dynamical mechanism

inertial observer. This is made into a dynamical mechanism by the assignment of a positive number, the mass, to the point $\mathcal{P}$. The Newtonian motions of this mechanism are kinematical motions which also satisfy the Newtonian equation of motion.

All kinematical motions are defined by noting that the mechanism has one degree of geometric freedom, which we will take as

$$q = |\boldsymbol{r}^{\mathcal{OP}}|. \tag{4.6}$$

The triad b is defined by the counter-clockwise rotation of $\boldsymbol{a}_1$ into the vector $\boldsymbol{b}_1$ by the angle $\theta(t)$. For this simple example, where only one point of the mechanism has a mass, we can identify $\boldsymbol{r}^<$ with $\boldsymbol{r}^{\mathcal{OP}}$. The configuration space is thus three dimensional position space and the configuration hypersurface is the straight line radiating out from $\mathcal{O}$. The spring exerts a force

$$\boldsymbol{R} = -k(q-l)\boldsymbol{b}_1 \tag{4.7}$$

on the particle representing the collar. In addition a constraint force $\boldsymbol{C}$ also acts on the collar as a result of the constraining shaft. This force is to be determined as part of the solution of the problem of determining all possible Newtonian motions. Thus we have to find solutions to the vectorial equation of Newtonian motion

$$-m\boldsymbol{a} + \boldsymbol{R} + \boldsymbol{C} = 0, \tag{4.8}$$

with the condition provided by D'Alembert's principle of material constraint, that

$$\boldsymbol{C} \cdot \boldsymbol{\beta} = 0, \tag{4.9}$$

where $\boldsymbol{\beta}$ is any vector that is tangent to a test motion. We have a case where the angular velocity relation between the triads a and b is 'simple', hence

$$\boldsymbol{\omega}^{(1)} = \dot{\theta}\boldsymbol{b}_3 \tag{4.10}$$

can be used to calculate the time derivatives of b with respect to the inertial observers frame. Thus we find the velocity and acceleration, expressed in the notation of chapter 2, to be

$$\boldsymbol{v}^{<1} = [\dot{q}, q\dot{\theta}, 0]b, \tag{4.11}$$

$$\boldsymbol{a}^{<1} = [\ddot{q} - q\dot{\theta}^2, q\ddot{\theta} + 2\dot{q}\dot{\theta}, 0]b. \tag{4.12}$$

We take

$$u = \dot{q} = {}^{\mathcal{B}}\boldsymbol{v}^{\mathcal{P}} \cdot \boldsymbol{b}_1 = \boldsymbol{v}^{1>} \cdot \boldsymbol{b}_1 \tag{4.13}$$

as the single generalized speed needed for this one degree of freedom mechanism. Inspection of the velocity expression then provides us with the single tangent vector (Kane's partial velocity)

$$\boldsymbol{\tau}^{<1} = \boldsymbol{b}_1, \tag{4.14}$$

where the symbol τ has been used to indicate the fact that in this case the tangent vector is derived from the derivative of the position with respect to the generalized coordinate. In this simple case the kinematic differential equation is just $\dot{q} = u$. We also have the part of the tangent vector that is tangent to a kinematic path, which has an out of instantaneous configuration surface component, given by

$$\boldsymbol{\tau}_t^{<1} = q\dot{\theta}\boldsymbol{b}_2, \tag{4.15}$$

which of course vanishes if θ is constant so that kinematic and test paths coincide.

The condition that the projection of the constraint force onto a tangent vector vanishes shows that the constraint force must be in the direction of the frame vector $\boldsymbol{b}_2$, hence taking the magnitude of the constraint force as λ, we have

$$\boldsymbol{C} = \lambda\boldsymbol{b}_2. \tag{4.16}$$

The following section of this chapter contains a detailed discussion of the direction of constraint forces.

D'Alembert's principle of constraint is then applied by taking the scalar product of the Newtonian equation with the tangent vector $\boldsymbol{\tau}^1$, which gives an equation for q that is free of the constraint force. Thus we obtain

$$-m(\ddot{q} - q\dot{\theta}^2) - k(q - l) = 0. \tag{4.17}$$

Given a particular function $\theta(t)$ and 'initial values' for q and $u = \dot{q}$ this equation can be integrated to produce a Newtonian motion. The study of all possible Newtonian motions involves the examination of how paths differ for different initial values, for a fixed set of defining parameters such as the stiffness and the rotation angle function. Knowing a Newtonian path we can now determine the constraint force by taking the dot product of the Newtonian equation with all possible vectors that are orthogonal to the single test vector. In this case this orthogonal vector is just $\boldsymbol{b}_2$, hence we find, with the assumed form of the constraint force, that:

$$\lambda = m(q\ddot{\theta} + 2\dot{q}\dot{\theta}). \tag{4.18}$$

In the next section we apply computer algebra to this example. •

4.2.1 Application of Sophia

The last illustration can easily be handled by the Sophia routines. First a listing of Sophia statements which resolve the problem are given with attached comments in italics. This is followed by further explanation emphasizing some important points needed to understand the statements and choice of variables. It should be realized that Sophia's Kvector mechanism is designed to handle multi-body mechanisms. The choice has been made to follow the kind of approach used in such cases even in this simple example which involves only one body and which could easily be treated in the fashion of chapter one. The statements are:

```
> &rot [A,B,3,theta]:
```

This sets up the direction cosine matrices between frames A and B.

```
> dependsTime(q1,u1,theta):
```

The fact that q1, u1 and theta are time dependent is recorded for use with the frame based differentiation commands.

```
> kde := {q1t = u1}:
```

The kinematic differential equations are recorded as a substitution set with the simple choice of generalized speed as generalized velocity.

```
> r1 := q1 &** (B&>1):
```

The Evector position of the mass particle is assigned the name r1.

```
> v1 := &simp subs(kde,A &fdt r1):
```

A combined step in which the frame based time derivative with respect to the inertial frame A is calculated, the q1t generalized coordinate derivative is replaced using the kinematic differential equations and an Evector simplification operation is carried out.

```
> p1 := m &** v1:
```

The Evector scalar multiplication operation is carried out to obtain the Evector for the particle momentum.

```
> vK := &KM [v1]:
```

The Kvector velocity is formed, which in this one particle case consists of only one Evector.

```
> tau := KMtangents(vK,u,1):
```

The Sophia command KMtangents is used to extract the list of tangent vectors. Note that this procedure expects subscripted quantities for the generalized speeds, this is the reason we use u1 rather than u. For the one particle system with one degree of freedom only one tangent vector is put on the tau list.

```
> taut := vKtime(vK,u,1):
```

The part of the velocity expansion due to the time dependent constraint imposed by the given rod motion

```
> pK := &KM [p1]:
```

Simply forms the Kvector for the particle momentum.

```
> ptK := &Ksimp subs(kde,A &Kfdt pK):
```

The inertial force Kvector is formed directly using the frame based Kvector differentiation operator with respect to the inertial frame A.

```
> R1 := (-k*(q1-l)) &** (B&>1):
```

The generalized active force Evector.

```
> RK := &KM [R1]:
```

The generalized active force Kvector.

```
> C1 := (lambda1) &** (B&>2):
> C2 := (lambda2) &** (B&>3):
> CK := &KM [C1 &++ C2]:
```

Evector and Kvector expressions are formed for the unknown constraint forces, the quantities lambda1 and lambda2 are to be determined.

```
> KaneEqs := tau &kane ( ((-1) &*** ptK) &+++ RK &+++ CK):
```

The &kane operator forms the projections of the sum of inertial active and constraint forces on the set of tangent vectors to give the list of expressions which when set to zero form the dynamical or Kane equations.

```
> gammaSK :=  [ &KM [B&>2], &KM [B&>3] ]:
```

This set of two vectors spans the orthogonal complement of the tangent space, hence the space of the constraint forces.

```
>  ConstraintEqs := gammaSK&kane ( ((-1) &*** ptK) &+++ RK &+++ CK):
```

Finally projections of the force summation onto these vectors provide equations for the constraint forces.

The variable theta must be declared as time dependent, also q1 and u1 are used for the generalized coordinate and velocity as the multi-body routines expect numerical attachments for these quantities. In this problem with only one body we could avoid the Kvectors, however it was desired to show a Sophia solution that would follow the pattern used for the more complex cases. Note the substitution operations which replace q1t by u1. This is absolutely necessary for the determination of the tangent vectors. While it is optional in the inertial force it is desirable to put the final dynamical equations in a form based on such a substitution. This neatly divides the dynamical equations as a first order set for the derivatives of the generalized speeds and the kinematical equations as a first order set for the derivatives of the generalized coordinates. The set of vectors, gammaSK define the orthogonal complementary space to the tangent space[1]. In this case this involves two vectors. The label gammaSK is used because Maple will give an error message when you try to assign something to the name gamma which is a system constant. It is a general problem with computer algebra systems that so many routines must be built in that it is almost inevitable that some such conflicts can occur and care should be taken in assigning names. The reader will for example note that D is not used as a name for a frame or anything else as this letter is used for a differential operator by Maple. The Sophia operator **&kane** takes a list of Kvectors, what we call an SKvector, on the left and a Kvector on the right. In the above case the right side is constructed as the negative of the momentum derivative plus the active and constraint forces, all as Kvectors. The result is a list that consists of the fat dot product of each of the vectors in the list with the Kvector. In the form above it gives a list terms. Kane calls each term the sum of the generalized inertia forces and the generalized active forces associated with a particular generalized speed. For this problem only one term results, which when set equal to zero gives the dynamical equation of motion. The constraint equations are found by projection onto the orthogonal complement of the tangent space. This is constructed by inspection in the present case and it is relatively obvious that it is spanned by the Kvectors contained in the set gammaSK. All this having been said let us now examine some of the results obtained when the above statements are used in an actual Maple-Sophia session. The explicit angular velocity expression was not used in the Maple calculations, however it is easily obtained, thus

```
> omegaAB := A &aV B;
```

$$[[0, 0, \mathit{thetat}], B]$$

which corresponds to the previously hand calculated result $\frac{d\theta}{dt}\boldsymbol{b}_3$. The list of expressions forming Kane's equations has just one member in this case as the dimension of the tangent space is one, i.e.

$$[\mathit{thetat}^2 m\, \mathit{q1} - m\, \mathit{u1t} - k\, q1 + k\, l]$$

which as the reader can verify is the same as the previously obtained result. In the Sophia calculation we have taken account of the two dimensional nature of the constraint, hence we have two equations, which output as two members of a list

$$[-2\, \mathit{thetat}\, m\, \mathit{u1} - m\, \mathit{thetatt}\, \mathit{q1} + \mathit{lambda1}, \mathit{lambda2}]$$

Each of these terms must be equated to zero to find the constraint forces. Doing this shows that the previous equation for the constraint force in the $\boldsymbol{b}_2$ direction is recovered while as expected from the nature of the posed problem the $\boldsymbol{b}_3$ component vanishes.

[1]The use of the geometric identification of tangent and cotangent vectors, as discussed in chapter 3, is used implicitly.

4.3 The Determination of Constraint Forces

Without constraints, a mechanism consisting of K particles would require $3K$ position coordinates to uniquely define its configuration. The imposition of constraints in effect restrict the motions of the systems by the specification of its geometric possibilities. The materials that impose the constraints could be looked upon as control devices which compute and apply the forces needed to maintain the particular set of allowed motions. In some sense this is what is done, as internal atomic structures of materials adjust to supply the electromagnetic forces that prevent the constraining devices from breaking apart. The Newtonian Motion that takes place under constraint is underdetermined unless we provide some statement of just how the constraint forces operate. This *extra* information is provided by D'Alembert's Principle, which states that the projection of constraint forces onto tangent Kvectors to test motions must vanish. The classical formulations of this goes under the name of the Principle of 'Virtual Work', our test motions having some connection with what are called virtual displacements. The problem with stating the way constraint forces operate as a principle of this form is that it moves our mental view away from the physical reasoning about why it provides a reasonable hypothesis for the actual mechanism behind such forces. What D'Alembert's reasoning was is an interesting study for the historian of mechanics, but it is a worthwhile exercise to produce ones own set of justifications. First of all the *principle* is not a law of nature in the same sense as Newton's principle of determinacy, but is a convenient hypothesis as to how materials exert constraining forces on bodies. It is quite possible to think of situations in which the principle would not hold and still not violate any fundamental law of physics. If the constraints are not themselves moving it is quite reasonable to assume that the forces operating on a particle will not set it in motion and will be normal to the object when it is in motion. If the constraints move it would be hard to expect them to use 'knowledge' of the past or future position of the system to exert constraint forces. Thus the D'Alembert Principle assumes that only the 'instantaneous' geometry can be responsible for the direction of the constraint force. A counterexample might be created by a constraining surface in which motions of the constrained particle set up 'waves' that led to constraint forces that depended on the history of the constraining objects motion. Such a situation might be dealt with by a more exacting theory that dealt with the constraint as a continuum or it might approximate its behavior by some other hypothesis than D'Alembert's for the action of constraints. While such speculation has its interest, there is no doubt that the D'Alembert Principle applies with great accuracy to a large number of constraint situations and has stood the test of several centuries of experience.

The statement of the principle, in terms of our 'big' dot product of Kvectors, is simple and elegant, that is if the constraint Kvector is given by $\boldsymbol{C}^{<}$ then

$$\boldsymbol{C}^{<} \bullet \boldsymbol{\beta}_j^{<} = 0, \tag{4.19}$$

where $\boldsymbol{\beta}_j^{<}$ is any of n independent Kvectors that are tangent to a test motion. Thus the

n generalized coordinates and speeds can be found as functions of time by taking the 'big' dot product of the Newtonian equation with the tangent vectors, the constraints being eliminated by the above equation.

There still remains the problem of finding the constraint forces. In practical problems of design this may be the main reason for carrying out computations. There also may be cases where we have another type of material constraint, than one that simply prevents motion in some direction. An example we will study is that of dry friction, where the force retards but does not restrict motion. In this case the retarding force's magnitude depends on the magnitude of constraint forces that do so restrict the motion. In other words we may have forces operating which are functions of constraint forces which in turn absolutely restrict motion. The full specification of such a problem requires the introduction of the unknown constraint forces explicitly into the equations of motion. The remainder of this section is devoted to the technical considerations needed to obtain explicit expressions for constraint forces that obey D'Alembert's Principle in a point mechanism.

We assume a K point mechanism, the configuration of which is determined by what we have called a Kvector, i.e., a column array with K ordinary position vectors as components. It should be evident that a Kvector is a $3K$ dimensional object, i.e., we can find $3K$ linearly independent Kvectors. If the mechanism is constrained to have n geometric degrees of freedom, its configuration is confined to an n dimensional surface in the $3K$ dimensional space of position Kvectors. When the configuration changes with time this surface may move and deform. Test motions are paths through a fixed surface; kinematic motions are paths through a family of configuration surfaces. D'Alembert's Principle states that the projection of the constraint forces on to the instantaneous surface, thus to test motions, must vanish. Thus at any point (q, t), with q representing the n generalized coordinates, we can examine a $3K$ dimensional vector space which encompasses all tangents vectors to all paths, constrained or not, that pass through the point.

This geometric picture can be used in a number of ways to develop methods for determination of the constraint forces. What constitutes the simplest approach will depend on the specific problem as well as the methods being used. Thus different approaches will be suitable for hand calculation, numerical methods and computer algebra. For this reason it is important to have a flexible store of methods available. As we proceed we will study several approaches. First we give one that depends on the concept of orthogonality.

4.3.1 Orthogonal Complements

We can now divide this vector space into two parts that are independent. One part consists of all vectors which are linear combinations of the n independent vectors tangent to the instantaneous hypersurface determined by the constraints. The other part, sometimes called the orthogonal complement, consists of linear combinations of all vectors that are orthogonal to the hypersurface. The dimension of this space will be given by $n' = 3K - n$. The orthogonality is of course with respect to our big dot product and convention for identifying tangent and cotangent vectors. D'Alembert's principle is then equivalent to the statement that the constraint force Kvector at the point (q, t) is a linear combination of basis vectors for the orthogonal complement of the tangent space to the instantaneous configuration hypersurface. This automatically implies that the dot product of the constraint force Kvector with any vector tangent to a test motion will vanish. In addition it tells us that in general we must find the n' expansion coefficients that represent the constraint force Kvector. This is easily done in principle. First the constraint forces are eliminated by taking the n scalar products of Newton's equations with the n independent vectors tangent to all possible test motions. This allows the complete determination of the n generalized coordinates and speeds, hence the full Newtonian path of the mechanism. Now the n' scalar products of the Newtonian equations with a basis for the space orthogonal to the hypertangent plane will give n' equations for the expansion coefficients. If some of part of the constraint force vector is needed to develop specific expressions for forces tangent to the configuration surface, such as the case with dry friction, then one must find a set of vectors in the orthogonal complement which will provide the needed components.

To give a formal statement of the above procedure, let β represent an arbitrary set of n tangent Kvectors to the tangent plane to the instantaneous configuration hypersurface, and let β' denote a basis set for the $n' = 3K - n$ dimensional orthogonal complement. It is understood that a particular point on the instantaneous configuration hypersurface is under consideration, thus all expansion coefficients will be functions of (q, t). Our condition implies that

$$\boldsymbol{\beta}_j^{<} \bullet \boldsymbol{\beta}_{j'}^{<} = 0, \tag{4.20}$$

where a prime on the subscript is used to indicate a basis vector in the orthogonal complement. The constraint force then has the representation:

$$\boldsymbol{C}^{<} = \sum_{j'=1}^{n'} \lambda_{j'} \boldsymbol{\beta}_{j'}^{<}. \tag{4.21}$$

The Newtonian path is obtained by solving the n equations:

$$(-m^{<}\boldsymbol{a}^{<} + \boldsymbol{R}^{<} + \boldsymbol{C}^{<}) \bullet \boldsymbol{\beta}_j^{<} = 0, \tag{4.22}$$

using $\boldsymbol{C}^{<} \bullet \boldsymbol{\beta}_j^{<} = 0$. The constraint expansion coefficient functions, $\lambda(q, t)$, are found by solving the n' equations:

$$(m^{<}\boldsymbol{a}^{<} - \boldsymbol{R}^{<}) \bullet \boldsymbol{\beta}_{j'}^{<} = \sum_{i'=1}^{n'} \lambda_{i'} \boldsymbol{\beta}_{i'}^{<} \bullet \boldsymbol{\beta}_{j'}^{<}. \tag{4.23}$$

When dry friction is present we may have to solve some or all of the second set of equations along with the first set. In practice there are a number of techniques for reducing the effort that a straight forward attack might require.

4.3.2 Cotangent Vector Summation

We now examine another approach, which does not use the concept of orthogonality, for determining the constraint forces. Thus the distinction between tangent and cotangent vectors is maintained. Because of this generality it is the natural starting point for the development of of 'optimal' algorithms for obtaining constraint forces. The 'big dot' is now interpreted as a linear functional between forces (cotangent vectors) and velocities or velocity time derivatives (tangent vectors). Velocity Kvectors that are tangent to the configuration hypersurface can be expanded as the linear sum of the n independent tangent Kvectors in the set τ. Let $\acute{\tau}$ denote a basis for the $3K$ dimensional space of tangent Kvectors for all unconstrained motions that includes the subset τ. The complement of the set τ will be denoted by $\hat{\tau}$. It will span a space of dimension $p = 3K - n$. Note that any set of linearly independent vectors that include τ will be sufficient, that is without further specification the choice $\hat{\tau}$ is *not unique.* It should also be clear that any set of n linear combinations of the Kvectors τ can be used to span the tangent hypersurface to the configuration surface. The transformation of basis kVectors was discussed in chapter three.

The constraint forces are defined as those forces which maintain the given configuration. If there were no motion whatever allowed, then each mass particle would be constrained by a force which could be expanded into three components in some reference frame. Therefore the space of all possible constraint forces acting on a K particle system will have dimension $3K$. If the forces acting on the multibody system are represented by Kvectors, the space of all constraint forces will require a $3K$ dimensional set of spanning Kvectors. Denote this set by $\hat{\gamma}$. Therefore the most general constraint force Kvector must have the form:

$$\boldsymbol{R}_c^< = \sum_{k=1}^{3K} \hat{c}_k \hat{\gamma}_k^<. \tag{4.24}$$

D'Alembert's principle has the general form that the constraint force's power functional on *any* tangent Kvector that is tangent to the configuration hypersurface must vanish. We continue to use the 'big dot' notation for this functional. Therefore

$$\boldsymbol{R}_c^< \bullet \boldsymbol{\tau}_j^< = \sum_{k=1}^{3K} \hat{c}_k \hat{\gamma}_k^< \bullet \boldsymbol{\tau}_j^< = 0, \tag{4.25}$$

where $\boldsymbol{\tau}_j^<$ is any of the n tangent Kvectors in the set τ. This provides n linear relations among the cotangent vectors γ. Therefore only $p = 3K - n$ of the set $\hat{\gamma}$ are linearly independent under the constraint condition or D'Alembert's principle. Choose p linear combinations of vectors in the set which are independent. Note that this is also a process which can be done in many ways, i.e. it is not unique. Any set that spans the p dimensional hyperplane defined by this set will suffice. Now complete this set with a further n linear combinations forming the set $\acute{\gamma}$. Together the sets of p Kvectors γ and n Kvectors $\acute{\gamma}$ span the space of all possible cotangent vectors. The p dimensional space or hyperplane spanned by γ is the *null* space of the linear functional '$\bullet$' with any of the n tangent Kvectors in the set τ. We may now express D'Alembert's principle by the somewhat picturesque statement: *The constraint forces live in the null space of the power functional with the tangent hyperplane to the configuration surface.* Or simply put that

$$\boldsymbol{R}_c{}^{<} = \sum_{k=1}^{p=3K-n} f_k \gamma_k^{<}, \tag{4.26}$$

where $\gamma_k^{<}$ is one of the p vectors in the set γ. Note that now there are only p quantities f_k to determine as D'Alembert's principle has been used to eliminate n terms. These are found by using Newton's equations, i.e.

$$\dot{\boldsymbol{P}}^{<} \bullet \acute{\boldsymbol{\tau}}_s^{<} = \sum_{k=1}^{p=3K-n} f_k \gamma_k^{<} \bullet \acute{\boldsymbol{\tau}}_s^{<} + \boldsymbol{R}_a{}^{<} \bullet \acute{\boldsymbol{\tau}}_s^{<}, \tag{4.27}$$

where $\acute{\boldsymbol{\tau}}_s^{<}$ is a member of the set of p tangent Kvectors $\acute{\tau}$. Therefore there are p equations to determine the p quantities f. The above process contains a number of steps which are not unique. Therefore it implies a number of alternative paths. For example instead of producing a basis set γ we could have kept the relations 4.25 as a set of equations for the c_k. The identification of a cotangent vector with a tangent Kvector and the consequent use of '$\bullet$' as a dot product leads to the developments of the previous section. It is emphasized once more that the 'best' method depends on both the problem and the means adopted for its solution. Several alternatives will be demonstrated in the remainder of this text, but over all emphasis will be given to the orthogonal complement method.

4.3.3 Internal Constraint Forces

It may be helpful to see how one frequently occurring situation, the constraint where particles are kept at fixed distances from one another, is dealt with using the formalism of Kvectors. To see this we examine the situation shown in the figure where two particles are separated by a rigid rod. Generalized coordinates q_1, q_2 and q_3 are chosen as the position of the first particle in the inertial frame N. The orientation of the rod is given by the angles q_4 and q_5 representing the rotation of the frame N into the frames A and B as shown. Thus the system has 5 degrees of freedom, while without the constraint is would have 6. Therefore there are 5 independent tangent vectors that span the tangent space. The orthogonal complement is spanned by only one vector and this vector must represent the constraint force in the rod. From elementary mechanics it should be clear that this has the form of an equal and opposite reaction between the two particles. Therefore the Kvector that spans the one dimensional space orthogonal to the configuration tangent space is an column of two ordinary vectors, with components $\boldsymbol{b}_1$ and $-\boldsymbol{b}_1$. If the formalism is working correctly this vector should have a vanishing fat dot product with all five of the

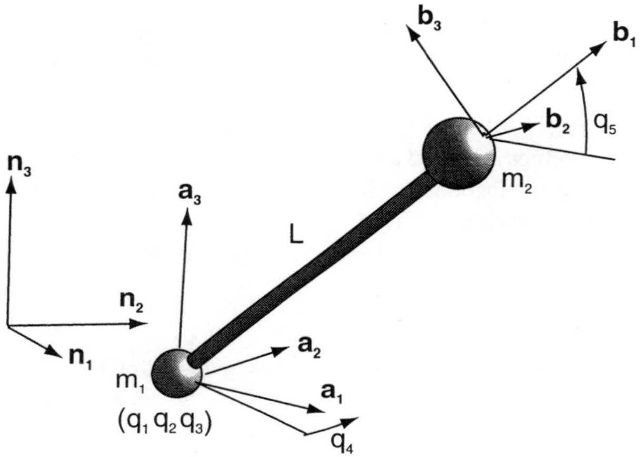

Figure 4.2: Distance Constrained Particles

tangent vectors! It is an easy exercise to check this with Sophia. The following Sophia statements determine the list or 'SkVector', tau, of the five tangent vectors that span the tangent space:

```
> chainSimpRot([[N,A,3,q4],[A,B,2,q5]]):
> &kde 5:
```

This sets up the reference frames, where B is fixed in the moving rod, and declares the time dependence and simple form of the kinematic differential equations. The next step is to define the vector $\boldsymbol{r}_A$ and $\boldsymbol{r}_2$, the displacement from the origin of the N system to the first particle and the displacement from the first to the second particle. Frame based differentiation is then used to find the velocities of these points which are gathered into the Kvector vK. Finally the KMtangents procedure is used to find the list of tangent vectors:

```
> rA := N &ev [q1,q2,q3]:
> r2 := rA &++ (B &ev [L,0,0]):
> v1 := N &fdt rA:
> v2 := N &fdt r2:
> vK := &Ksimp  subs(kde,&KM [v1,v2]):
> tau := KMtangents(vK,u,5):
```

The result is:

$$\begin{aligned}
\tau_1^< &= [[[1,0,0],N],[[\cos(q_4)\cos(q_5),-\sin(q_4),\cos(q_4)\sin(q_5)],B],2] \\
\tau_2^< &= [[[0,1,0],N],[[\sin(q_4)\cos(q_5),\cos(q_4),\sin(q_4)\sin(q_5)],B],2] \\
\tau_3^< &= [[[0,0,1],N],[[-\sin(q_5),0,\cos(q_5)],B],2]
\end{aligned}$$

$$\tau_4^< = [[[0,0,0],N],[[0,\cos(q_5)Ł,0],B],2]$$
$$\tau_5^< = [[[0,0,0],N],[[0,0,-Ł],B],2]$$

The space of constraints, i.e. the orthogonal complement to the tangent space, is spanned by a single Kvector. Our assumptions about action and reaction leads us to the assumed form:

```
> gamma1 := &KM [ (B&>1), ( (-1) &** (B&>1) )]:
```

If our 'fat' dot product is indeed a reflection of D'Alembert's principle of constraint this vector should be orthogonal to the tangent space. The verification is obtained by a simple Maple 'for' statement, thus

```
> for j from 1 to 5 do print( simplify(gamma1 &O tau[j]) ) od:
```

which prints out five zeros.

•Illustration

The main issue in this example is to see how constraint forces can be found in a case where several particles make up the mechanism. For this purpose the point dynamical mechanism shown in Figure 4.3 is analyzed. The mechanism consists of

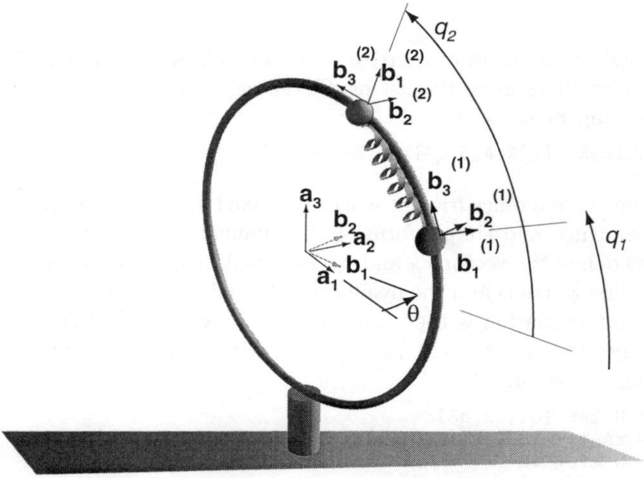

Figure 4.3: Two beads on a rotating hoop.

two beads allowed to slide on a rotating circular wire of radius l. The beads are connected by a spring which produces a restoring force proportional to the angle between them as subtended at the center of the hoop. The proportionality constant or 'stiffness' is given by k. The mass of each bead is m. The rotation angle, relative to the frame of the inertial observer, is given as a function of time, i.e., $\theta(t)$. Thus the constraint is time dependent. To carry out the kinematic analysis it is useful to use three sets of standard triads in addition to the 'a' triad of the inertial observer. The triad 'b', which rotates with the hoop, is defined by the condition that $\boldsymbol{a}_3 = \boldsymbol{b}_3$, and that $\boldsymbol{a}_1 \cdot \boldsymbol{b}_1 = \cos\theta(t)$. The two triads, b^L, with $L = 1, 2$, rotate in the plane of the hoop so as to maintain alignment with each of the beads. Thus they are chosen so that $\boldsymbol{b}_1^{(L)} \cdot \boldsymbol{b}_1 = \cos q_L$, where q_L is the angle between the $\boldsymbol{a}_1\boldsymbol{a}_2$ plane and the particle at point $\mathcal{P}_L$. The configuration of the system is determined by q_1 and q_2, hence they can be used as generalized coordinates for the mechanism. Thus the configuration Kvector is given by

$$\boldsymbol{r}^< = \begin{bmatrix} l\boldsymbol{b}_1^{(1)} \\ l\boldsymbol{b}_1^{(2)} \end{bmatrix} \tag{4.28}$$

The direction cosine matrices that relate the triads as obtained from the above description are:

$$R_{ba} = \begin{bmatrix} \cos\theta & \sin\theta & 0 \\ -\sin\theta & \cos\theta & 0 \\ 0 & 0 & 1 \end{bmatrix}, \tag{4.29}$$

$$R_{(L)b} = \begin{bmatrix} c_L & 0 & s_L \\ 0 & 1 & 0 \\ -s_L & 0 & cL \end{bmatrix}, \tag{4.30}$$

$$R_{(L)a} = \begin{bmatrix} c_\theta c_L & s_\theta c_L & s_L \\ -s_\theta & c_\theta & 0 \\ -c_\theta s_L & -s_\theta s_L & c_L \end{bmatrix} \tag{4.31}$$

Direct differentiation of the configuration expression, $\boldsymbol{r}^<$, or use of the expressions for relations between velocity for different observers leads to the result that for a kinematic motion:

$$\boldsymbol{v}^< = \dot{q}_1 \begin{bmatrix} l(-s_1\boldsymbol{b}_1 + c_1\boldsymbol{b}_3) \\ 0 \end{bmatrix} + \dot{q}_2 \begin{bmatrix} 0 \\ l(-s_2\boldsymbol{b}_1 + c_2\boldsymbol{b}_3) \end{bmatrix} + \begin{bmatrix} l\dot{\theta}c_1\boldsymbol{b}_2 \\ l\dot{\theta}c_2\boldsymbol{b}_2 \end{bmatrix}. \tag{4.32}$$

In this case the tangent vectors to the instantaneous hyperplane also are related to the generalized coordinates by differentiation of the configuration with respect to each coordinate. Therefore it is reasonable to use the notation $\boldsymbol{\tau}_j^<$ for these Kvectors, which by inspection of the above expression are immediately seen to be:

$$\boldsymbol{\tau}_1^< = l \begin{bmatrix} -s_1\boldsymbol{b}_1 + c_1\boldsymbol{b}_3 \\ 0 \end{bmatrix}, \tag{4.33}$$

$$\boldsymbol{\tau}_2^< = \begin{bmatrix} 0 \\ -s_2\boldsymbol{b}_1 + c_2\boldsymbol{b}_3 \end{bmatrix}. \tag{4.34}$$

In this case the the term $\boldsymbol{\tau}_t^<$ is nonvanishing and is obtained, also by inspection, as

$$\boldsymbol{\tau}_t^< = l\dot{\theta} \begin{bmatrix} c_1\boldsymbol{b}_2 \\ c_2\boldsymbol{b}_2 \end{bmatrix}. \tag{4.35}$$

The accelerations of the beads are also found by further differentiation of the coordinate expressions. Expressed as components, i.e. as the quantities $a_j^{<L}$ for the jth component of the acceleration of the Lth particle in the b system where

$$\boldsymbol{a}^{<L} = a_1^{<L}\boldsymbol{b}_1 + a_2^{<L}\boldsymbol{b}_2 + a_3^{<L}\boldsymbol{b}_3, \tag{4.36}$$

we find

$$a_1^{<L} = -\ddot{q}_L l s_L - \dot{q}_L\dot{\theta} l c_L - \dot{\theta}^2 l c_L, \tag{4.37}$$
$$a_2^{<L} = \ddot{\theta} l c_L - \dot{\theta}^2 l s_L - \dot{\theta} q_L l s_L \tag{4.38}$$
$$a_3^{<3} = \ddot{q}_L l c_L - \dot{q}_L\dot{\theta} l s_L. \tag{4.39}$$

The spring is arranged so that the force developed between the particles will vanish when the subtended angle has the value ϕ Therefore the Kvector representing the force associated with the spring is:

$$\boldsymbol{R}^{<} = \begin{bmatrix} k(q_2 - q_1 - \phi)\boldsymbol{b}_3^{(1)} \\ -k(q_2 - q_1 - \phi)\boldsymbol{b}_3^{(2)} \end{bmatrix}. \tag{4.40}$$

Examination of the above expressions for $\boldsymbol{\tau}_j^{<}$ shows that they can also be written in the form

$$\boldsymbol{\tau}_1^{<} = l\begin{bmatrix} \boldsymbol{b}_3^{(1)} \\ 0 \end{bmatrix}, \tag{4.41}$$

$$\boldsymbol{\tau}_2^{<} = l\begin{bmatrix} 0 \\ \boldsymbol{b}_3^{(2)} \end{bmatrix}. \tag{4.42}$$

These two Kvectors form a basis for the tangent hyperplane, therefore taking the big dot product with the equation of motion leads to two second order differential equations for the generalized coordinates q_j. The constraint forces are now discussed using the orthogonality method. The constraint force must be orthogonal (under the big dot product) to $\boldsymbol{\tau}_1^{<}$ and $\boldsymbol{\tau}_2^{<}$. The straightforward way of determining the constraints is to find 4 additional Kvectors which are orthogonal to the two tangent vectors and to expand the constraint Kvector in terms of them. In this case we can reduce the work, as it is possible to find such vectors by inspection and physical intuition. The particle at $\mathcal{P}_1$ can be expected to be subject to a constraint force which is in the plane that passing through the point and which is perpendicular to the constraining wire. Hence the most general constraint force acting on the particle would involve a linear combination of the ordinary vectors $\boldsymbol{b}_1^{(1)}$ and $\boldsymbol{b}_2$. These would account for the constraint force in and perpendicular to the plane of the wire hoop. Similar considerations apply to the second particle. Therefore we can postulate that, in terms of Kvectors associated with the full 6 dimensional configuration space for this two particle system, the following four vectors are a basis set for all vectors that are orthogonal to the two tangent vectors defining the instantaneous tangent hyperplane:

$$\boldsymbol{\tau}_{1'}^{<} = \begin{bmatrix} \boldsymbol{b}_1^{(1)} \\ 0 \end{bmatrix}, \tag{4.43}$$

$$\boldsymbol{\tau}_{2'}^{<} = \begin{bmatrix} 0 \\ \boldsymbol{b}_2^{(1)} \end{bmatrix}, \tag{4.44}$$

$$\boldsymbol{\tau}_{3'}^{<} = \begin{bmatrix} \boldsymbol{b}_1^{(2)} \\ 0 \end{bmatrix}, \tag{4.45}$$

$$\boldsymbol{\tau}_{4'}^{<} = \begin{bmatrix} 0 \\ \boldsymbol{b}_2^{(2)} \end{bmatrix}. \tag{4.46}$$

This set of four vectors not only provide the needed basis, but they have the fortunate property of being orthonormal under the big dot product, which leads to particularly simple uncoupled equations for the components of the constraint force Kvector. This latter vector must exist in the space spanned by the set $\boldsymbol{\tau}_{j'}^{<}$, hence has the form:

$$\boldsymbol{C}^{<} = \sum_{j'=1}^{j'=4} \lambda_{j'} \boldsymbol{\tau}_{j'}^{<}. \tag{4.47}$$

If the differential equations for q_1 and q_2 are solved the Kvector for $m^{<}\boldsymbol{a}^{<}$ can be computed and the four equations obtained by taking their big dot product with the constraint Kvector. The details are left as an exercise.

4.4 Kane's Equations

The use of tangent vectors that are derived from configuration representations in terms of generalized coordinates is at the heart of the subject of analytical mechanics. D'Alembert deserves the credit for recognizing that the constraint forces for most mechanical systems will be orthogonal to the instantaneous configuration; though in his time it would have been most unusual to express this, as we have done, in terms of the geometry of surfaces in many dimensional spaces. Therefore it was natural to arrive at the idea calculating something that has become known as virtual work to express the expected behavior of idealized constraint forces. The only problem with the idea of virtual work is that its main utility and indeed the reason for considering it at all is due to the orthogonality condition. The name work and the connection of work with the concept of energy can be confusing to the newcomer to the subject and, in the author's opinion, can even lead the professional into error. Many writers even appear to take the vanishing of virtual work itself as a basic principle of nature rather than the expression of a constitutive equation that describes the material behavior of the constraining parts of a mechanism. By distancing ones thoughts from the geometry and its insights the concept of virtual work can limit the use of the problem solver's mechanical and geometric intuition. The development of Lagrange's equations, discussed in the following section, can be carried out with increased insight using the geometric tool of tangent vectors, and better yet forms can be found which apply to basis of tangent vectors that are not derivable by differentiation of configuration expressions with respect to generalized coordinates. This insight, in my opinion, has been made needlessly complex by the use of the theory of Pfaffian forms and quasi-coordinates.

The rest of this section is devoted to connecting the formulation presented in this text with the terminology of Kane. It was seen in the last section, in the discussion of calculation of constraints, that the big dot product of Newton's equations of motion with n independent tangent vectors provided equations for the generalized coordinates that were free of constraint forces. Once the generalized coordinates are determined as functions of time by the integration of these equations, it was seen that the constraint forces could be found by the use of a complementary basis of vectors which are orthogonal to the tangent vectors. Kane has put the equations so obtained into a form which is independent of the terminology of the $3K$ dimensional configuration spaces. Each component of a generalized tangent basis Kvector is called a partial velocity. Thus in terms of our particle mechanism, each point in the mechanism has its own three dimensional set of partial velocity vectors, each member of the set being associated with one of the generalized speeds, u_j. The term generalized speed is also due to Kane. What we have termed the big dot product of a tangent vector associated with a generalized speed and Newton's equations of motion is the sum over all bodies in the mechanism of the ordinary dot product of Kane's partial velocity with the equation of motion of each body. Kane decomposes this into two parts, one due to the applied forces, the other due to the inertia or acceleration terms. The first is called the generalized active force, the second the generalized inertial force. To see this in detail start with the equation of motion in the Kvector representation and take the big dot product of this with any one of the n independent generalized tangent vectors associated with one of the n generalized speeds. This gives

$$(-m^{<}\tfrac{^A d}{dt}\boldsymbol{v}^{<} + \boldsymbol{R}^{<} + \boldsymbol{C}^{<}) \bullet \boldsymbol{\beta}_j^{<} = 0, \tag{4.48}$$

which if we define the *generalized active force* as

$$F_j = \boldsymbol{\beta}_j^{<} \bullet \boldsymbol{R}^{<}, \tag{4.49}$$

and the *generalized inertia force* as

$$F_j^* = -m^{<}\boldsymbol{a}^{<} \bullet \boldsymbol{\beta}_j^{<} = -\dot{\boldsymbol{P}}^{<} \bullet \boldsymbol{\beta}_j^{<}, \tag{4.50}$$

can be written as

$$F_j^* + F_j = 0. \tag{4.51}$$

Equations 4.51, which depend on the satisfaction of D'Alembert's principle so that $\boldsymbol{C}^{<} \bullet \boldsymbol{\beta}_j^{<} = 0$, will be referred to as *Kane's Equations.* If there are n independent tangent vectors this provides a set of n first order differential equations for the generalized speeds, the kinematic differential equations, providing another n equations linking the generalized speeds to the generalized coordinates. The complications that arise when there are less than n independent basis tangent vectors is treated in detail in the discussion of nonholonomic systems later in this text. It is a worthwhile exercise for the reader, who does not see why n independent tangent vectors are needed, to prove that less than n independent first order equations for the derivatives of the generalized speeds would result in such a case. In other words, to examine the case where linear relations might exist among the n kinematic differential equations.

To obtain agreement with Kane's terminology, the Lth component of the Kvector $\beta_j^{<}$ will be called the *partial velocity* of the Lth particle or point of the mechanism associated with the jth generalized speed of the mechanism. Sometimes the partial velocity will be indicated by the notation $\boldsymbol{v}^{<L}j$, or even ${}^{B}\boldsymbol{v}^{\mathcal{P}}{}_{j}$ to emphasize interest in a partial velocity connected with the relative motion of a point $\mathcal{P}$ with respect to a body B. This establishes the connection between the Kvector notation and the notation used by Kane and his colleagues. In particular the texts of Kane and Levinson and of Kane, Litkins and Levinson, with their many interesting problems and insights, should be readily accessible to readers of this work.

The utility of Kane's equations is increased by careful kinematic analysis. Insights into the geometry can aid the analyst in the choice of useful definitions for the generalized speeds. The generalized speeds can be defined in advance of a commitment to particular sets of generalized coordinates. The process of choosing generalized speeds is independent of the process of choosing generalized coordinates. In effect the analyst has increased freedom in formulating a problem. The generalized speeds and coordinates are connected by the appropriate kinematic differential equations. The increased flexibility can help in obtaining 'simple' forms of the equations of motion. Equally important, it can help focus interest on the most meaningful variables for a particular problem. Naturally the ability to perform the 'best' possible analysis requires practice and experience. In addition Kane's formalism, hence the formalism of Kvectors used in this text, is very general. It applies to a wide spectrum of particle and rigid body problems that are often treated by diverse methods. In the remainder of this chapter we will also see how Kane's equations reduce in certain special situations to the formalism of classical analytical mechanics, i.e., to the equations of Lagrange and the mechanics of conservative systems. Following chapters extend the work of this chapter from particle mechanisms to mechanisms that also contain rigid bodies and to problems involving constraints between the velocities of different parts of a system. The latter involve what are called nonholonomic constraints.

It is emphasized that the equations of motion can also be expressed in terms of the momentum Kvector, $\boldsymbol{P}^{<}$. The components of this vector are simply the momenta of the particles, usually given by the product of the particle mass and velocity. The generalized inertia force is then given by:

$$F_j^* = -\frac{{}^{A}d}{dt}\boldsymbol{P}^{<} \bullet \beta_j^{<}. \tag{4.52}$$

When it is clear which frame represents an inertial frame, this will also be written as:

$$F_j^* = -\dot{\boldsymbol{P}}^{<} \bullet \beta_j^{<}. \tag{4.53}$$

•Illustration

To appreciate how Kane's formulation of the equations governing a Newtonian motion can be expressed in the notation of Kvectors, we will examine a relatively complete description of a point mechanism. As shown in Figure 4.4, a rod is attached to a pin joint at a point $\mathcal{O}$, this point being fixed in the inertial frame A. The collar at point $\mathcal{P}_1$ is free to slide along the rod. A pendulum, of length l, with a point mass m_2 attached to one end, is fixed to the collar by another pin joint. The collar can be considered a point with mass m_1. The entire mechanism is subjected to a uniform gravitational field, with gravitational acceleration g. To obtain the equations

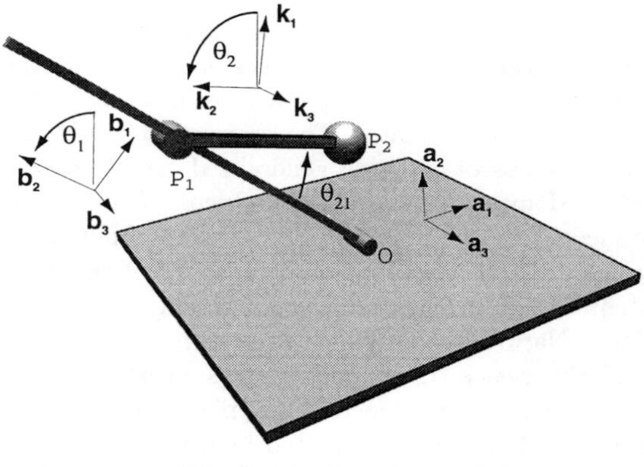

Figure 4.4: Sliding double pendulum.

of motion of this constrained mechanism we take $\mathcal{O}$ as the reference point of the inertial observer with standard triad a. The vector $\boldsymbol{a}_2$ is aligned in the direction counter to the gravitational acceleration. We choose a second standard triad, b, with $\boldsymbol{b}_2$ aligned along the rod and such that $\boldsymbol{b}_2 \cdot \boldsymbol{a}_2 = \cos\theta_1$, i.e., the last expression defines θ_1 as the angle made by the rod with the vertical. We will also make use of a third standard triad, k, where $\boldsymbol{k}_2$ is aligned along the pendulum in the direction $\mathcal{P}_2\mathcal{P}_1$. The angle θ_2 describes the movement of this pendulum from the vertical direction, hence we have the relation $\boldsymbol{k}_2 \cdot \boldsymbol{a}_2 = \cos\theta_2$. The symbol q will be used to designate the distance from $\mathcal{O}$ to $\mathcal{P}_1$. All quantities needed to formulate the equations have now been defined. The alert reader may realize that the current mechanism is in effect a disguised form of a mechanism used in an earlier example. Given the above definitions of the triads a, b and k it is a straightforward manner to find the direction cosine matrices that relate them. The problem at hand is essentially planar in its geometry, though it is convenient to occasionally use the vectors $\boldsymbol{a}_3$, $\boldsymbol{b}_3$ or $\boldsymbol{k}_3$, to represent angular velocity vectors, hence we need only consider 2×2 direction cosine matrices.These are given by:

$$R_{ba} = \begin{bmatrix} c_1 & s_1 \\ -s_1 & c_1 \end{bmatrix}, \tag{4.54}$$

$$R_{ka} = \begin{bmatrix} s_2 & c_2 \\ -c_2 & s_2 \end{bmatrix}, \tag{4.55}$$

$$R_{kb} = \begin{bmatrix} s_{21} & c_{21} \\ -c_{21} & s_{21} \end{bmatrix}. \tag{4.56}$$

In these expressions terms such as s_2 stand for $\sin\theta_2$, while s_{21} indicates $\sin(\theta_2 - \theta_1)$. As the reader may expect, θ_1, θ_2 and q provide generalized coordinates, i.e., they totally define the geometry of the constrained system. It is tempting to base the analysis on these quantities, however this would suppress the freedom given us by the notion of generalized speeds. Therefore even if these quantities seem like the best choice for generalized coordinates it is useful to delay making a final decision and instead to proceed with a kinematical analysis based on an expedient choice of generalized speeds. In choosing generalized speeds it is helpful to keep in mind that one faces the usually formidable task of calculating acceleration components, as well as the problem of determining a suitable set of basis vectors for the instantaneous configuration hypersurface. In the present case we have a two particle mechanism. The mechanism is essentialy planar. The geometry is determined by three parameters and the configuration space is 4 dimensional. The instantaneous configuration hypersurface is three dimensional. It is left as an exercise for the reader to specify and use position vectors to derive a set of coordinate based tangent vectors. Taking the route of kinematical analysis based on picking three independent generalized speeds note that the velocity of the point $\mathcal{P}_1$ in the frame of the inertial observer can be expressed in terms of projections onto the triad b. Therefore use two generalized speeds to specify the velocity ${}^{A}\boldsymbol{v}^{\mathcal{P}_1} = \boldsymbol{v}^{<1}$ as:

$$\boldsymbol{v}^{<1} = u_1\boldsymbol{b}_1 + u_2\boldsymbol{b}_2. \tag{4.57}$$

It is then possible to use the principles of chapter 2 to calculate the velocity ${}^{A}\boldsymbol{v}^{\mathcal{P}_2}$. This calculation would be easier if we had an expression for the angular velocity of the pendulum, i.e. if we knew the value of ${}^{A}\boldsymbol{\omega}^{K}$. As we have the freedom to choose an additional generalized speed, it appears reasonable to use this to specify this angular velocity, i.e. we take

$${}^{A}\boldsymbol{\omega}^{K} = u_3\boldsymbol{a}_3. \tag{4.58}$$

This gives the result

$$ {}^{A}\boldsymbol{v}^{\mathcal{P}_2} = {}^{A}\boldsymbol{v}^{\mathcal{P}_1} - l u_3 \boldsymbol{b}_3 \times \boldsymbol{k}_2. \tag{4.59}$$

Knowing the velocity of each point we are in a position to write the velocity Kvector, which is:

$$\boldsymbol{v}^{<} = \begin{bmatrix} u_1\boldsymbol{b}_1 + u_2\boldsymbol{b}_2 \\ u_1\boldsymbol{b}_1 + u_2\boldsymbol{b}_2 + l u_3\boldsymbol{k}_1 \end{bmatrix}. \tag{4.60}$$

A set of basis vectors for the instantaneous configuration hypersurface is found by inspecting this expression for coefficients of the generalized speeds. Thus

$$\beta_1^{<} = \begin{bmatrix} \boldsymbol{b}_1 \\ \boldsymbol{b}_1 \end{bmatrix}, \tag{4.61}$$

$$\beta_2^{<} = \begin{bmatrix} \boldsymbol{b}_2 \\ \boldsymbol{b}_2 \end{bmatrix}, \tag{4.62}$$

$$\beta_3^{<} = \begin{bmatrix} 0 \\ l\boldsymbol{k}_1 \end{bmatrix}. \tag{4.63}$$

To determine the generalized active force, as defined by Kane, we need the Kvector for the applied forces. In the problem as set forth, the only applied force is due to the gravitational field. Therefore the applied force Kvector is:

$$\boldsymbol{R}^{<} = \begin{bmatrix} -m_1 g\boldsymbol{a}_2 \\ -m_2 g\boldsymbol{a}_2 \end{bmatrix}. \tag{4.64}$$

With the definitions:

$$m_2 = m, \tag{4.65}$$

$$m_1 + m_2 = M, \tag{4.66}$$

the three generalized active forces are obtained by taking the big dot product of the applied force Kvector with each of the basis tangent vectors, therefore,

$$F_1 = \boldsymbol{R}^{<} \bullet \beta_1^{<} = -Mg\sin(\theta_1), \tag{4.67}$$

$$F_2 = \boldsymbol{R}^{<} \bullet \beta_2^{<} = -Mg\cos(\theta_1), \tag{4.68}$$

$$F_3 = \boldsymbol{R}^{<} \bullet \beta_3^{<} = -mLg\sin(\theta_2). \tag{4.69}$$

To calculate what Kane has called the generalized inertia force requires an expression for the acceleration Kvector. From the calculational point of view this is the most involved step, and the place where one might make interesting comparisons regarding various methods of approaching the problem of equation formulation in mechanics. A direct determination of the acceleration can be obtained by differentiation, with respect to the inertial frame A, of the representation of the velocity Kvector as a sum over basis tangent vectors. That is we differentiate:

$$\boldsymbol{v}^{<} = u_1\beta_1^{<} + u_2\beta_2^{<} + u_3\beta_3^{<}, \tag{4.70}$$

with the operator

$$\frac{^A d}{dt}() = (\dot{\ }). \tag{4.71}$$

Therefore the acceleration Kvector has the form:

$$\boldsymbol{a}^< = \sum_{j=1}^{3} (\dot{u}_j \boldsymbol{\beta}_j^< + u_j \dot{\boldsymbol{\beta}}_j^<). \tag{4.72}$$

The generalized inertia forces are obtained by taking the big dot product of the Kvector, $m^< \boldsymbol{a}^<$ with each of the basis tangent vectors, so in this case the 3 generalized inertia terms are given by

$$F_j^* = -(m^< \boldsymbol{a}^<) \bullet \boldsymbol{\beta}_j^<, \tag{4.73}$$

where

$$m^< \boldsymbol{a}^< = \begin{bmatrix} m_1 \boldsymbol{a}^{<1} \\ m_2 \boldsymbol{a}^{<2} \end{bmatrix}. \tag{4.74}$$

In carrying out this calculation scalar products of the form $\boldsymbol{k}_j \cdot \boldsymbol{b}_i$ involving the three triads a, b, and k occur. These can easily be read off from the direction cosine matrices, which have already been set down. In addition scalar products of the form $\boldsymbol{k}_j \cdot \dot{\boldsymbol{b}}_i$ occur. With the notation that $\boldsymbol{\omega}^{(1)}$ refers to the angular velocity of the triad b in A and $\boldsymbol{\omega}^{(2)}$ to the angular velocity of the triad k in A one obtains the relations:

$$\dot{\boldsymbol{b}}_1 = \boldsymbol{\omega}^{(1)} \times \boldsymbol{b}_1 = \dot{\theta}_1 \boldsymbol{b}_2, \tag{4.75}$$
$$\dot{\boldsymbol{b}}_2 = \boldsymbol{\omega}^{(1)} \times \boldsymbol{b}_2 = -\dot{\theta}_1 \boldsymbol{b}_1, \tag{4.76}$$
$$\dot{\boldsymbol{k}}_1 = \boldsymbol{\omega}^{(2)} \times \boldsymbol{k}_1 = \dot{\theta}_2 \boldsymbol{k}_2, \tag{4.77}$$
$$\dot{\boldsymbol{k}}_2 = \boldsymbol{\omega}^{(2)} \times \boldsymbol{k}_2 = -\dot{\theta}_2 \boldsymbol{k}_1. \tag{4.78}$$

Thus, after some calculation, one obtains the generalized inertia forces as:

$$\begin{aligned} -F_1^* &= M\dot{u}_1 - lmc_{21}\dot{u}_3 - u_2 M\dot{\theta}_1 \\ &\quad -mls_{21}\dot{\theta}_2 u_3, \end{aligned} \tag{4.79}$$
$$\begin{aligned} -F_2^* &= M\dot{u}_2 + lms_{21}\dot{u}_3 + u_1 M\dot{\theta}_1 + \\ &\quad -mlc_{21}\dot{\theta}_2 u_3 \end{aligned} \tag{4.80}$$
$$\begin{aligned} -F_3^* &= -mlc_{21}\dot{u}_1 + mls_{21}\dot{u}_2 + ml^2\dot{u}_3 \\ &\quad +ml\dot{\theta}_1 s_{21} u_1 + ml\dot{\theta}_1 c_{21} u_2, \end{aligned} \tag{4.81}$$

where terms such as $s_{21} = \sin(\theta_2 - \theta_1)$, and $c_1 = \cos\theta_1$. The generalized active forces can then be expressed as

$$F_1 = -Mgs_1, \tag{4.82}$$
$$F_2 = -Mgc_1, \tag{4.83}$$
$$F_3 = -mgls_2. \tag{4.84}$$

Using these results we obtain three first order differential equations for the quantities $\dot{u}_i$. To complete the formulation we must still set down the three kinematic differential equations which related the rates of change of a suitable set of generalized coordinates to the generalized speeds. It is at this stage that we must make some commitment to a set such coordinates. We have already used θ_1 and θ_2 as convenient labels for the angles made by the rod and the pendulum with the vertical, and if we add the distance of the collar, q, from the pin at $\mathcal{O}$ we have three perfectly acceptable candidates for the role of generalized coordinates. The important point to note is that the way we set up the problem, using generalized speeds, did not depend on this choice. We have the freedom to choose other possible sets of generalized coordinates if it is convenient to do so! To emphasize our choice it is natural to use the notation:

$$q_1 = \theta_1, \tag{4.85}$$

$$q_2 = \theta_2, \tag{4.86}$$

$$q_3 = q. \tag{4.87}$$

A generally useful procedure for deriving the kinematic equations is to use the chosen coordinates to arrive at the velocity expressions which defined the generalized speeds. This could be done by setting down vectorial expressions for the positions of the points of the mechanism and then differentiating them, or by use of the facts about the relations of velocities with respect to different reference frames derived in chapter 2. The first approach also forces us to check on the validity of the choice, i.e., we can check to see if we can uniquely determine the geometric configuration from the particular choice of generalized coordinates. On the other hand it is usually easier to follow the second route. Doing so we use the relation between the velocities of a point as seen by an observer in one frame when it is moving in another. To help the reader we use the explicit point referral notation, thus we have

$$^{A}\boldsymbol{v}^{\mathcal{P}_1} = {}^{A}\boldsymbol{v}^{B(\mathcal{P}_1)} + {}^{B}\boldsymbol{v}^{\mathcal{P}_1}. \tag{4.88}$$

Recall that the first term on the left represents the velocity the point would have at its instantaneous position if it were at rest in the frame B, while the second term is the velocity of the point to an observer at rest in frame B. Using this we have the result that

$$^{A}\boldsymbol{v}^{\mathcal{P}_1} = \boldsymbol{v}^{<1} = \boldsymbol{\omega}^{(1)} \times q_3\boldsymbol{b}_2 + \dot{q}_3\boldsymbol{b}_2 = -q_3\dot{q}_1\boldsymbol{b}_1 + \dot{q}_3\boldsymbol{b}_2. \tag{4.89}$$

Also we can use the concept of simple angular velocity discussed in chapter 2 to see that

$$\boldsymbol{\omega}^{(2)} = \dot{q}_2\boldsymbol{a}_3. \tag{4.90}$$

Referring to our choice of generalized speeds and comparing these results with the equations used to define these quantities we can directly arrive at the *kinematic differential equations.* Thus

$$\dot{q}_1 = -\frac{u_1}{q_3}, \tag{4.91}$$

$$\dot{q}_2 = u_3, \tag{4.92}$$

$$\dot{q}_3 = u_2. \tag{4.93}$$

•

4.4.1 Inertial Frames of Reference

The Newtonian idea of an inertial frame of reference is that in such a frame a particle experiencing no applied forces will move with uniform velocity in a straight line. A basic postulate of Newtonian mechanics is that any frame moving with a constant rectilinear motion with respect to an inertial frame is also an inertial frame. Therefore there are an infinite number of inertial frames. This does not mean that it is easy to find a true inertial frame! The surface of the earth is certainly not a suitable candidate, as the earth circulates about the sun and rotates about its own axis, not to mention the motion of the entire solar system through the moving rotating galaxy. From another point of view *any sufficiently small region of space can be taken as an inertial frame for a sufficiently small period of time.* The bullet shot from a rifle follows a parabolic path in the approximately constant gravity of the earth's surface, however for an observation on the scale of a few micro-seconds it can be taken as an inertial frame. In practice this is the case for all our candidates for an inertial frame, the real question being one of the involved time and space scales. It is a matter of some considerable interest to quantify these ideas. The classic example is the so called Foucault's pendulum. In summary the earth's surface is certainly not an exact inertial frame of reference. Just what is an inertial frame depends on the time and accuracy of the observations. For most engineering calculations a frame fixed with respect to our measuring instruments or laboratory is sufficient. A frame with origin at the earth's center and rotating about the sun, but not spinning with the earth, is the ultimate approximation to an inertial frame for atmospheric navigation. Astronomers might have to use a frame that is fixed with respect to the rotation of the solar system or the milky way galaxy or even the local group of galaxies. To examine the equations of motion under the assumption of a fixed frame at the earth's center we use five frames. The first, N, is the approximate inertial frame with origin fixed at the earth's center. The frame N1 is fixed in the rotating earth, hence the angular velocity ${}^{N}\boldsymbol{\omega}^{N1} = \omega_e \boldsymbol{n}_3$. The frame S is fixed at the location of the pendulum support on the surface of the earth located at ϕ degrees south of the north pole, with $\boldsymbol{s}_1$ pointing south and $\boldsymbol{s}_2$ pointing east. Its origin is at a distance R from the origin of N at the earth's center, i.e. it is assumed that the earth is a perfect sphere. The pendulum of mass m is supported at a point a distance h above the surface and is suspended by a string of length L. Its configuration is determined by two angular variables, q_1 and q_2. The support string is at rest in a frame B. This frame is related to the earth surface fixed frame S by an auxiliary frame A. We proceed from S to A by a positive rotation q_1 about the $\boldsymbol{s}_1$ axis, while to go from A to B we make a positive rotation q_2 about the $\boldsymbol{a}_2$ axis. This is all expressed in the Sophia statement:

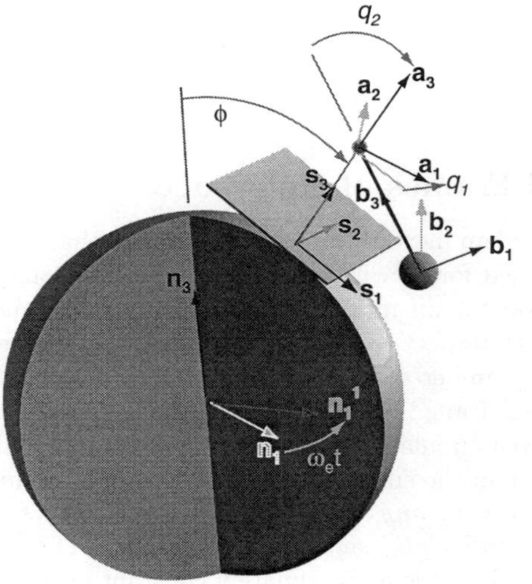

Figure 4.5: Pendulum on Earth's Surface

```
> chainSimpRot([[N,N1,3,we*t],[N1,S,2,phi],[S,A,1,q1],[A,B,2,q2]]):
```

Note that in figure 4.5 the angle $q_2 < 0$. This two degree of freedom problem requires two generalized speeds, which in this case will not be taken as simply being the generalized velocities or derivatives of the generalized coordinates. We will see that the velocity of the pendulum relative to the surface frame S contains only two components when expressed in the frame B fixed in the support string. This should be evident as we are taking the support string as having a fixed length. To proceed we first declare the time dependence of the generalized velocities and speeds. We then express the vectorial distance $\boldsymbol{r}$ of the pendulum mass from the earth's center. The vector $\boldsymbol{s}$ is taken as the distance of the mass from surface origin of frame S. We also calculate the velocity $\boldsymbol{v}$ of the mass point in the inertial frame and $\boldsymbol{v}_s$ the velocity relative to the surface origin point. With some steps to simplify expressions this is implemented by the following Sophia statements:

```
> dependsTime(q1,q2,u1,u2):
> r := ( (R+h) &** (S&>3) ) &-- ( L &** (B&>3) ) :
> v := &simp ( N &fdt r):
> s := ( (h) &** (S&>3) ) &-- ( L &** (B&>3) ):
> vs := B &to (S &fdt s):
```

The last result shows that $\boldsymbol{v}_s$ is given by the expression:

$$[[-q2t\, L, q1t\, \cos(q2)L, 0], B]$$

which as expected has a vanishing $\boldsymbol{b}_3$ component. The generalized speeds are defined by the expression:

$$\boldsymbol{v}_s = u_1\boldsymbol{b}_1 + u_2\boldsymbol{b}_2.$$

To obtain the kinematic differential equations we use Sophia-Maple to first form inverse kinematic differential equations for u_j by comparing the calculated form of $\boldsymbol{v}_s$ with the above form. Solving these for the derivatives of the generalized coordinates give the kinematic differential equations:

```
> ikde := {u1 =( vs &c 1),u2 = (vs &c 2)}:
> kde := solve(ikde,{q1t,q2t}):
```

The next step is to use the subs command to express the velocity $\boldsymbol{v}$ in terms of the generalized speeds and to find the set of two independent tangent vectors. Despite the fact that we are dealing with only one mass point the use of the Sophia routines still requires the use of the Kvector formalism. The steps needed to calculate the tangent vectors then takes the form:

```
> v := subs(kde,v):
> vK := &KM [v]:
> tau := KMtangents(vK,u,2):
```

The next step is to form the momentum Kvector and the applied force Kvector. The latter consists of the gravitational force. This points in the negative $\boldsymbol{s}_3$ direction. Because the length of the pendulum is a very small fraction of the radius of the earth it is appropriate to use the local gravitational acceleration approximation for the magnitude of this force. It is left as an exercise for the interested reader to replace this by a more exact expression for the Newtonian gravitational law. The generalized inertial force is obtained from the derivative of the momentum vector with respect to the assumed inertial frame, thus

```
> pK := m &*** vK:
> Rg := S &ev [0,0,-m*g]:
> RgK := &KM [Rg]:
> pKt := subs(kde, N &Kfdt pK):
```

Note that in the last step the kinematic differential equations are used to replace the time derivatives of generalized coordinates resulting from the frame based derivative process. This is not strictly necessary but it is required if the equations of motion are to take the standard form of a set of 2n first order equations in the derivatives of the generalized speeds and velocities for an n degree of freedom system. To proceed we use the &kane operator which takes as input the set tau of tangent vectors and the applied force and momentum time derivative Kvectors. We use the symbol GAF for the set of generalized active forces and MGIF for minus the generalized inertia forces. Thus:

```
> GAF := tau &kane RgK:
> MGIF := tau &kane pKt:
```

The equations of motion are then formed. The additional step is then taken of putting them in standard state space form by solving for the derivatives of the generalized speeds,

```
> eqs := {GAF[1]=MGIF[1],GAF[2]=MGIF[2]}:
> eqs := solve(eqs,{u1t,u2t}):
```

We don't show an explicit expression for this complicated expression. In fact most of the complexity is contained in terms multiplied by the square of the earths angular velocity which is the order of 10^{-10}. If these terms are neglected and the notation s_j and c_j are used to represent $\sin(q_j)$ and $\cos(q_j)$ the dynamical equations take the approximate form:

$$u_{1t} = c_1\, s_2\, g + \frac{s_2\, {u_2}^2}{c_2\, L} - 2\, s_2\, \omega_e\, \sin(\phi) u_2 + 2\, c_2\, \cos(\phi) \omega_e\, c_1\, u_2$$

and

$$u_{2t} = -s_1\, g - \frac{u_2\, s_2\, u_1}{c_2\, L} + 2\, \omega_e\, s_2\, u_1\, \sin(\phi) - 2\, c_2\, c_1\, u_1\, \omega_e\, \cos(\phi)$$

Together with the kinematic differential equations this completes the needed relations for the problem. As a further step we obtain expressions for the projection of the motion into the plane tangential to the earth, i.e. the plane of $\boldsymbol{s}_1$-$\boldsymbol{s}_2$. This is easily done by the following steps:

```
> rsS := S &to s:
> xs := rsS &c 1:
> ys := rsS &c 2:
```

or

$$x_s = -\sin(q_2) L,$$

and

$$y_s = \sin(q_1)\cos(q_2) L.$$

The derivation of these results has followed the formulation given in the text of Kane and Levinson. These authors examine numerical solutions of the equations assuming that either the earth center or the surface base point is fixed in an inertial frame. Comparison of the results with experimental data shows that the earth center frame is a superior approximation to an inertial frame. Even so the surface frame is adequate for most calculations involving the behavior of technical devices. In the final analysis just what constitutes an inertial frame is a question of the length and time scales that describe the mechanism under study.

4.5 Acceleration Components

A major part of the algebraic labor in the derivation of equations governing Newtonian motions consists of the development of expressions for the acceleration components with respect to the reference frame of the inertial observer. In carrying out this task, using the apparatus of generalized speeds, one soon learns that searching for 'optimal choices' is well worth the expansion of some effort. In the special case where the tangent vectors are the derivatives of the position vectors with respect to some set of generalized coordinates is possible to obtain formal expressions for the projection of the accelerations onto this set of *coordinate tangent vectors.* These expressions use the *Euler-Lagrange* operator, involving derivatives of a scalar quantity (essentially the kinetic energy) with respect to the generalized coordinates, velocities and finally the time. The complexity of the operations make it difficult to decide, in any objective manner, the question of comparative difficulty between using the Euler-Lagrange operator and the direct method of tangent vector projection that is basic to Kane's formulation. The approach to formulating equations of motion by the use of the Euler-Lagrange operator does have considerable historical and theoretical interest, and has been the object of most of the mathematically oriented studies of mechanics under the names of Lagrangian and Hamiltonian mechanics. Understanding modern mechanics requires a clear appreciation of the relationships between these various points of view. From the viewpoint of generality, Newton's formulation covers the widest range of situations. Once D'Alembert's principle of constraint is introduced the idea of projection onto the instantaneous constraint surface becomes a useful technique. If one restricts consideration to tangent vectors which can be derived from differentiation of position expressions, and if the n degree of freedom system possesses n independent vectors, then it is possible to use the *Lagrangian* formulation. Certain modifications are possible which still allow the use of a Lagrangian formulation when these latter conditions don't apply. However for the most part the Lagrangian formulation must be considered to be more restrictive. The full Lagrangian formulation is most interesting for the study of systems for which it is possible to develop 'force' expressions by the differentiation of scalar functions. These *potential energy* functions will be discussed later in this chapter.

In the calculation that follows, we first obtain a direct expression for the projection of the part of the Kvector representing the acceleration of a single point of the mechanism, on to the part of an arbitrary member of the tangent Kvector that represents the point in question. In our notation this is the quantity $\boldsymbol{a}^{<L} \cdot \boldsymbol{\tau}_j^{<L}$. Note that this is the ordinary dot product between vectors in Euclidian three dimensional space. This will then be manipulated into the Euler-Lagrange form. In carrying out this manipulation, use is made of our 'knowledge' of the fact that the square of the velocity is the essential scalar function that we seek. Why this is so can be motivated by examination of a simple, single particle system without constraints. In the initial part of our calculation we will use coordinate based tangent vectors of the form: $\boldsymbol{\tau}_j^{<L} = \frac{{}^{A}\partial}{\partial q_j}\boldsymbol{r}^{<L}$. The key step in this calculation is to use the expansion for the velocity in terms of the basis tangent vectors, i.e. to use the expansion:

$$\boldsymbol{v}^{<L} = \sum_{j=1}^{j=n} \boldsymbol{\tau}_j^{<L}\dot{q}_j + \frac{{}^{A}\partial}{\partial t}\boldsymbol{r}^{<L} \tag{4.94}$$

which immediately gives the important relationship that:

$$\frac{{}^{A}\partial}{\partial \dot{q}_j}\boldsymbol{v}^{<L} = \boldsymbol{\tau}_j^{<L}. \tag{4.95}$$

Therefore using the definition of acceleration in terms of a time derivative of the velocity we can write:

$$\boldsymbol{a}^{<L} \cdot \boldsymbol{\tau}_j^{<L} = \frac{{}^{A}d}{dt}\boldsymbol{v}^{<L} \cdot \frac{{}^{A}\partial}{\partial \dot{q}_j}\boldsymbol{v}^{<L}. \tag{4.96}$$

We now want to express the right side of the above equation as differentiation operations on a scalar function. The form of the expression suggests that we try some quadratic function of the velocities for this role. It turns out that the function to consider is $\boldsymbol{v}^{<L} \cdot \boldsymbol{v}^{<L}$, which the knowledgeable reader will see is related to the yet to be defined kinetic energy of the mechanism. Thus we start with the relation that:

$$\frac{{}^{A}\partial}{\partial \dot{q}_j}(\boldsymbol{v}^{<L} \cdot \boldsymbol{v}^{<L}) = 2\boldsymbol{v}^{<L} \cdot \frac{{}^{A}\partial}{\partial \dot{q}_j}\boldsymbol{v}^{<L} \tag{4.97}$$

If we take the total time derivative of the last expression we obtain the result that:

$$\frac{{}^{A}d}{dt}(\boldsymbol{v}^{<L} \cdot \frac{{}^{A}\partial}{\partial \dot{q}_j}\boldsymbol{v}^{<L}) = \frac{{}^{A}d}{dt}\boldsymbol{v}^{<L} \cdot \frac{{}^{A}\partial}{\partial \dot{q}_j}\boldsymbol{v}^{<L} + \boldsymbol{v}^{<L} \cdot \frac{{}^{A}d}{dt}\frac{{}^{A}\partial}{\partial \dot{q}_j}\boldsymbol{v}^{<L} \tag{4.98}$$

Using equation 4.96, the first term on the right hand side of this is just the projection of the acceleration on the part of the tangent plan $T_n(q,t)$ that is involved with the Lth particle. Therefore we have the result that:

$$\frac{{}^{A}d}{dt}(\frac{{}^{A}\partial}{\partial \dot{q}_j}(\frac{1}{2}\boldsymbol{v}^{<L} \cdot \boldsymbol{v}^{<L})) = \boldsymbol{a}^{<L} \cdot \boldsymbol{\tau}_j^{<L} + \boldsymbol{v}^{<L} \cdot \frac{{}^{A}d}{dt}\frac{{}^{A}\partial}{\partial \dot{q}_j}\boldsymbol{v}^{<L}. \tag{4.99}$$

The last term in the above can be further evaluated using the expression for the expansion of $\boldsymbol{v}^{<L}$. Thus

$$\frac{{}^{A}\partial}{\partial \dot{q}_j}\boldsymbol{v}^{<L} = \frac{{}^{A}\partial}{\partial \dot{q}_j}(\sum_{k=1}^{k=n} \frac{{}^{A}\partial}{\partial q_k}\boldsymbol{r}^{<L}\dot{q}_k + \frac{{}^{A}\partial}{\partial t}\boldsymbol{r}^{<L}) = \frac{{}^{A}\partial}{\partial q_j}\boldsymbol{r}^{<L} \tag{4.100}$$

showing that:

$$\boldsymbol{v}^{<L} \cdot \frac{{}^{A}d}{dt}\frac{{}^{A}\partial}{\partial \dot{q}_j}\boldsymbol{v}^{<L} = \boldsymbol{v}^{<L} \cdot \frac{{}^{A}\partial}{\partial q_j}\boldsymbol{v}^{<L} = \frac{{}^{A}\partial}{\partial q_j}(\frac{1}{2}\boldsymbol{v}^{<L} \cdot \boldsymbol{v}^{<L}). \tag{4.101}$$

Now a slight rearrangement of terms provides the main result, i.e. that:

$$\boldsymbol{a}^{<L} \cdot \boldsymbol{\tau}_j^{<L} = (\frac{{}^{A}d}{dt}\frac{{}^{A}\partial}{\partial \dot{q}_j} - \frac{{}^{A}\partial}{\partial q_j})(\frac{1}{2}\boldsymbol{v}^{<L} \cdot \boldsymbol{v}^{<L}). \tag{4.102}$$

This result shows that for the Lth particle the projection of the acceleration on the Lth component of the coordinate basis tangent Kvector can be expressed in terms of differentiation operations on a pure scalar function. The differentiation operations are carried out using the Euler–Lagrange operator. This result is the major stepping stone to the derivation of Lagrange's equations.

The above applies only to tangent vectors that form a coordinate basis. It is useful to also have a result for an arbitrary basis, e.g. for a basis made up of what Kane has called partial velocities, $\boldsymbol{\beta}_j^{<L}$. This is an easy step if one recalls that the partial velocities are linear combinations of the coordinate basis vectors. Thus we have the relation:

$$\boldsymbol{\beta}_k^{<L} = \sum_{j=1}^{j=n} \boldsymbol{\tau}_j^{<L} W_{jk}^{\beta\tau} \tag{4.103}$$

so that all we need do is multiply 4.102 by $W_{jk}^{\beta\tau}$ and sum over j. Therefore

$$\sum_{j=1}^{j=n} W_{jk}^{\beta\tau} \boldsymbol{a}^{<L} \cdot \boldsymbol{\tau}_j^{<L} = \sum_{j=1}^{j=n} (\tfrac{^{A}d}{dt}\tfrac{^{A}\partial}{\partial \dot{q}_j} - \tfrac{^{A}\partial}{\partial q_j})(\frac{1}{2}\boldsymbol{v}^{<L} \cdot \boldsymbol{v}^{<L}) W_{jk}^{\beta\tau}, \quad (4.104)$$

so that simple rearrangement of terms show that

$$\boldsymbol{a}^{<L} \cdot \boldsymbol{\beta}_k^{<L} = \sum_{j=1}^{j=n} (\tfrac{^{A}d}{dt}\tfrac{^{A}\partial}{\partial \dot{q}_j} - \tfrac{^{A}\partial}{\partial q_j})(\frac{1}{2}\boldsymbol{v}^{<L} \cdot \boldsymbol{v}^{<L}) W_{jk}^{\beta\tau} \quad (4.105)$$

which is the desired result.

•Illustration

The mechanism shown in Figure 4.6 will be used to show how the Euler-Lagrange operator can be used to compute acceleration components. The mechanism consists

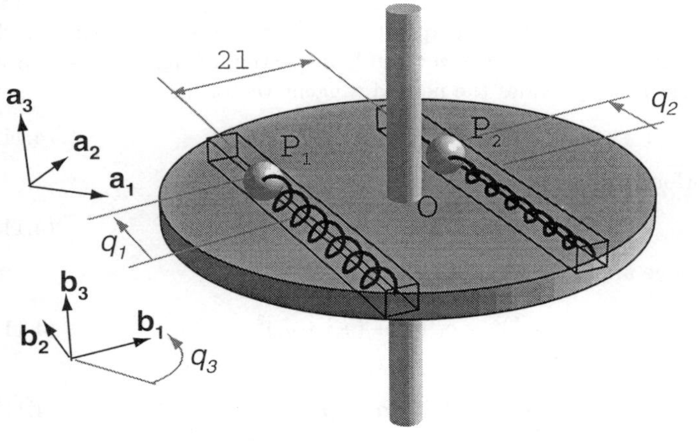

Figure 4.6: Acceleration components of particles.

of a wheel with two slots which are parallel to a diameter and a distance l from the center. Masses are connected to springs in each slot. In the equilibrium position the masses are centered in the respective slots. We will continue working with this problem in the next section, where we will be concerned with the forces generated by motion against the springs. For the moment we concentrate on the kinematic features. A standard triad b rotates with the wheel. For generalized velocities we take the velocities of each mass relative to a triad rotating with the wheel and the angular velocity of the wheel, i.e. we choose:

$$u_1 = {}^{B}\boldsymbol{v}^{\mathcal{P}_1} \cdot \boldsymbol{b}_2, \quad (4.106)$$

$$u_2 = {}^{B}\boldsymbol{v}^{\mathcal{P}_2} \cdot \boldsymbol{b}_2, \tag{4.107}$$
$$u_3 = {}^{A}\boldsymbol{\omega}^{B} \cdot \boldsymbol{b}_3. \tag{4.108}$$

The simplest derivation of acceleration components using the Euler-Lagrange operator requires a coordinate basis. Therefore we take q_1 and q_2 as the displacements of the masses from their center positions, and q_3 as the angular position of the wheel, so that the kinematic equations are: $u_j = \dot{q}_j$. With these choices we have:

$$ {}^{A}\boldsymbol{\omega}^{B} = u_3 \boldsymbol{b}_3. \tag{4.109}$$

To obtain the velocity of the masses with respect to the inertial observer we could write down expressions for the vector distances and differentiate. Instead we again use the ideas of chapter two, so that:

$$\boldsymbol{v}^{<L} = {}^{A}\boldsymbol{v}^{B(\mathcal{P}_L)} + {}^{B}\boldsymbol{v}^{\mathcal{P}_L}. \tag{4.110}$$

The result is that

$$\boldsymbol{v}^{<L} = -q_L u_3 \boldsymbol{b}_1 + (l u_3 + \dot{q}_L)\boldsymbol{b}_2. \tag{4.111}$$

For our demonstration first calculate the quantity $\boldsymbol{a}^{<L} \cdot \boldsymbol{\tau}_3^{<1}$ directly. Then use the Euler-Lagrange operator applied to $(1/2)\boldsymbol{v}^{<1} \cdot \boldsymbol{v}^{<1}$. Inspection of the velocity expression for the coefficient of u_3 provides the needed tangent vector, i.e.,

$$\boldsymbol{\tau}_3^{<1} = -q_1 \boldsymbol{b}_1 + l \boldsymbol{b}_2. \tag{4.112}$$

It is then a straightforward matter to calculate the result that:

$$\boldsymbol{a}^{<1} \cdot \boldsymbol{\tau}_3^{<1} = 2 q_1 u_1 u_3 + {q_1}^2 \dot{u}_3 + l \dot{u}_1 + l^2 \dot{u}_3. \tag{4.113}$$

For the Euler-Lagrange approach we need:

$$\Psi = \frac{1}{2}(\boldsymbol{v}^{<1})^2 = {q_1}^2 {\dot{q}_3}^2 + (\dot{q}_3 l + \dot{q}_1)^2. \tag{4.114}$$

To verify that:

$$\left(\frac{{}^{A}d}{dt}\frac{{}^{A}\partial}{\partial \dot{q}_3} - \frac{{}^{A}\partial}{\partial q_3}\right)\Psi = \boldsymbol{a}^{<1} \cdot \boldsymbol{\tau}_3^{<1}, \tag{4.115}$$

calculate that,

$$\frac{\partial}{\partial \dot{q}_3}\Psi = {q_1}^2 \dot{q}_3 + \dot{q}_3 l^2 + \dot{q}_1 l, \tag{4.116}$$
$$\frac{\partial}{\partial q_3}\Psi = 0, \tag{4.117}$$
$$\frac{{}^{A}d}{dt}\frac{\partial}{\partial \dot{q}_3}\Psi = \boldsymbol{a}^{<1} \cdot \boldsymbol{\tau}_3^{<1} = 2 q_1 u_1 u_3 + {q_1}^2 \dot{u}_3 + l \dot{u}_1 + l^2 \dot{u}_3, \tag{4.118}$$

where we have used the kinematic differential equations. Thus both the direct route and the Euler-Lagrange route lead to the same final result. In this simple example the labor is about equal for both approaches. It is another matter if we use a non-coordinate basis, in which case the direct route appears to have the advantage. The direct route is at the heart of Kane's approach to mechanics, and is the central theme of this book.

4.6 The First Form of Lagrange's Equations

The subject of 'analytical' mechanics seeks to reduce, mechanics to the study of certain scalar functions. In the eighteenth century Lagrange discovered a means to express D'Alembert's principle of constraint in terms of such a scalar function, which has become known as the Lagrangian. In his own book on the subject Lagrange emphasized the analytic, as opposed to geometric, character of his contribution by avoiding the use of *any diagrams or figures.* The trend in 20th century mechanics is to integrate geometric and analytical thinking. Thus the explicit use of higher dimensional geometry to interpret both the physical geometric configuration and the analytical theory. Kane's recognition and explicit use of what he calls 'partial velocities', the components of our tangent Kvectors, provides a revealing way to look at the connections between the analytical and the vectorial theory. The link comes from the last sections demonstration. The components of each tangent Kvector are vectors tangent to the motion of each of the systems particles for a motion associated with the generalized velocity related to the tangent Kvector. The Euler-Lagrange operator applied to the square of the particles velocity provides the dot product of the particle acceleration with this partial velocity vector. The Newtonian motion of a particle mechanism is governed by Kane's equations. The generalized inertia force term that appears in these equations consists of a sum over the mechanism's particles of the scalar products of the acceleration with the tangent Kvector components, each multiplied by the particle mass. In other words the Euler-Lagrange operator can be used to obtain the generalized inertia force in Kane's equations. In the case where the tangent vectors are coordinate tangent vectors we obtain the classical so called first form[2] of Lagrange's equations. It is easy to extend this to the general or noncoordinate based tangent Kvector case by use of the theory of transformations of tangent Kvectors developed in chapter 3.

Many readers will recognize the that a function which involves the square of the particle velocity is related to the so called *kinetic energy.* The significance of this quantity can be appreciated by examining the equation of motion of a single particle,

$$m\tfrac{d\boldsymbol{v}}{dt} = \boldsymbol{R}, \tag{4.119}$$

The ordinary dot product of $\boldsymbol{v}$ with both sides of this equation, assuming fixed mass, gives

$$m\tfrac{d\boldsymbol{v}}{dt}\cdot\boldsymbol{v} = \tfrac{d}{dt}(\frac{1}{2}m\boldsymbol{v}^2) = \boldsymbol{R}\cdot\boldsymbol{v}. \tag{4.120}$$

The term on the right hand side is defined as the *power* due to the applied force, i.e. we can write

$$\mathcal{P} = \boldsymbol{R}\cdot\boldsymbol{v}. \tag{4.121}$$

[2]There is no universal agreement about the terminology

The *kinetic energy* is defined as the bracketed term on the left,

$$\mathcal{K} = \frac{1}{2} m \boldsymbol{v}^2. \tag{4.122}$$

Thus the rate of change of kinetic energy equals the applied power. If no force, $\boldsymbol{R}$, is applied, or if the force is orthogonal to the Newtonian path we can make the statement that the kinetic energy is *conserved,* or constant for the motion. Such would be the case for a free particle moving in a straight line at constant velocity. If the power could also be represented by the total time derivative of a scalar function of the position, say $-\mathcal{V}$, we would have the interesting relationship that

$$\tfrac{d}{dt}(\mathcal{K} + \mathcal{V}) = 0, \tag{4.123}$$

which implies the existence of a constant $\mathcal{E}$ such that

$$\mathcal{E} = \mathcal{K} + \mathcal{V}. \tag{4.124}$$

Thus, if the power is indeed representable as the total time derivative of $\mathcal{V}$, then the quantity $\mathcal{E}$ would be conserved or constant throughout the motion. The reader may recognize $\mathcal{E}$ as the energy of the mechanical system (particle plus force) and the above result as the 'law of conservation of mechanical energy'. If the function $\mathcal{V}$ can be found it implies the result that

$$\mathcal{P} = \boldsymbol{R} \cdot \boldsymbol{V} = -\tfrac{d\mathcal{V}}{dt} = -(\tfrac{\partial}{\partial x_1}\mathcal{V}\tfrac{dx_1}{dt} + \tfrac{\partial}{\partial x_2}\mathcal{V}\tfrac{dx_2}{dt} + \tfrac{\partial}{\partial x_3}\mathcal{V}\tfrac{dx_3}{dt}), \tag{4.125}$$

where it is assumed that $\mathcal{V}$ is not an explicit function of time. As

$$\boldsymbol{r} = \mathbf{x} = x_1\mathbf{a}_1 + x_2\mathbf{a}_2 + x_3\mathbf{a}_3$$

and

$$\boldsymbol{v} = \tfrac{d\boldsymbol{r}}{dt},$$

this can be put into the form:

$$\boldsymbol{R} \cdot \boldsymbol{v} = -\nabla\mathcal{V} \cdot \boldsymbol{v}. \tag{4.126}$$

This shows that if the applied force is equal to the gradient of a scalar function we have the conservation law that the energy is constant for the motion of a particle. Observe that as the constraint forces are normal to the velocity they make no contribution to the power. This provides some intuitive idea of the relation of a particles kinetic energy to its motion. Now we continue with the development of the relationship between the generalized inertial force and the Euler-Lagrange operator. The generalized inertia force associated with the jth generalized speed, was defined as

$$F_j^* = -m^{<}\boldsymbol{a}^{<} \bullet \boldsymbol{\tau}_j^{<} = -\sum_{L=1}^{K} m^L \boldsymbol{a}^{<L} \cdot \boldsymbol{\tau}_j^{<L} = -\dot{\boldsymbol{P}}^{<} \bullet \boldsymbol{\tau}_j^{<}, \tag{4.127}$$

where $m^{<}$ is a diagonal matrix with the L, L entry being the mass of the Lth particle, i.e., $m^{<}_{IL} = m_L \delta_{IL}$. Each term in the sum is a scalar product of the type studied in the last section, hence it can be represented by the action of the Euler-Lagrange operator on the kinetic energy of the particle, i.e., as

$$F_j^* = -\left(\frac{{}^A d}{dt}\frac{{}^A\partial}{\partial \dot{q}_j} - \frac{{}^A\partial}{\partial q_j}\right)\sum_{L=1}^{K}(\mathcal{K}^{(L)}), \tag{4.128}$$

where $\mathcal{K}^{(L}$ denotes the kinetic energy of the Lth particle. If

$$\mathcal{K}^{(} = \sum_{L=1}^{K} \mathcal{K}^{(L)} = \frac{1}{2}(m^{<}\boldsymbol{v}^{<}) \bullet \boldsymbol{v}^{<}, \tag{4.129}$$

we have the important result that

$$F_j^* = -\left(\frac{{}^A d}{dt}\frac{{}^A\partial}{\partial \dot{q}_j} - \frac{{}^A\partial}{\partial q_j}\right)\mathcal{K}^{(}. \tag{4.130}$$

The *first form* of Lagrange's equations can now be written down by replacing the generalized inertia term in Kane's equations with the above and noting that the generalized active force is the big dot product with the coordinate tangent Kvector, i.e.

$$\left(\frac{{}^A d}{dt}\frac{{}^A\partial}{\partial \dot{q}_j} - \frac{{}^A\partial}{\partial q_j}\right)\mathcal{K}^{(} = \sum_{L=1}^{K} \boldsymbol{R}^{<L} \cdot \frac{{}^A\partial \boldsymbol{r}^{<L}}{\partial q_j}. \tag{4.131}$$

The derivative of the position vector is of course the same as $\boldsymbol{\tau}^{<L}_{\cdot}$ The conventional terminology for this generalized active force term, when the tangent vectors are also coordinate basis vectors, is *generalized force.*

To complete the discussion of the first form of Lagrange's equations we apply the theory of transformations of basis tangent vectors to obtain expressions for the generalized inertia force in terms of the Euler-Lagrange operator. The notation for generalized inertia and active forces must now reflect the fact that more than one tangent basis to the constraint hypersurface is being considered. Therefore we write

$$F_j^{*\tau} = -(m^{<}\boldsymbol{a}^{<}) \bullet \boldsymbol{\tau}_j^{<}, \tag{4.132}$$

$$F_j^{*\beta} = -(m^{<}\boldsymbol{a}^{<}) \bullet \boldsymbol{\beta}_j^{<}, \tag{4.133}$$

where the convention is followed that τ refers to the coordinate basis associated with q. To obtain the desired relation we sum over the product of both sides of the first of the above relations with the transformation matrix $W^{\tau\beta}ji$ so that

$$\sum_{j=1}^{n} F_j^{*\tau} W_{ji}^{\tau\beta} = -\sum_{j=1}^{n}(m^{<}\boldsymbol{a}^{<}) \bullet \boldsymbol{\tau}_j^{<} W_{ji}^{\tau\beta} = -(m^{<}\boldsymbol{a}^{<}) \bullet \boldsymbol{\beta}_i^{<} = F_i^{*\beta}. \tag{4.134}$$

A similar result holds for the generalized active force, hence we have the generalized first form of Lagrange's equations

$$\sum_{j=1}^{n}\left[\left(\frac{{}^A d}{dt}\frac{{}^A\partial}{\partial \dot{q}_j} - \frac{{}^A\partial}{\partial q_j}\right)\mathcal{K}^{(}\right] W_{ji}^{\tau\beta} = \boldsymbol{R}^{<} \bullet \boldsymbol{\beta}_i^{<}. \tag{4.135}$$

The representation of a mechanical system in terms of a single scalar function of the n geometric degrees of freedom has considerable attraction, and has dominated the theoretical development of the subject for nearly two hundred years. It has also had a major influence on the development of theoretical structures in modern physics and even subjects such as control theory. Therefore it must be understood and mastered by any serious student of the subject. It is also clear that the viewpoint of generalized speeds, active and inertia forces coupled with generalized tangent vector basis, provide the problem solver with considerable flexibility. Suitably used, they can increase the organization and decrease the labor involved in solving complex mechanical systems. In the section that follows we consider an extensive illustration of the use of the various approaches discussed.

4.7 An Example Mechanism The Model Cable Car

The purpose of this section is to demonstrate the theory by examining a specific problem in some detail. Despite the simplicity of the model the full equations of motion are still relatively complicated, making it meaningful to examine different approaches to arriving at equations of motion. In addition the opportunity will be taken to discuss a number of related matters such as the 'art' of modeling and the use of dimensional reasoning for eliminating unnecessary parameters from the governing equations of a mechanism. The problem will also be carried through to a numerical solution. This section is also somewhat of a digression from the main theory. It is meant to give the reader some hints as to how to apply the theory and to carry through an analysis of the resulting equations. Those interested mainly in the theoretical development of the subject can either skip or skim this section for what is of interest to them.

The goal is to discuss the mechanics of a car, suspended from a cable. It is assumed that the car can swing and that the cable can deform. We will also assume that the cars suspension point is at rest for our first try at the problem. If the cars suspension point is moving along the cable at constant velocity we can choose an inertial observer with the same velocity. Thus the assumption of the suspension point being at rest for an inertial observer does include the case of uniform motion. To reduce the system to a point mass mechanism we will assume the deflection of the cable can be replaced by a linear spring with equilibrium length l. Energy lost by wave motion in the cable could be simulated by a viscous damper. The suspension mechanism is assumed to have mass m_1, which includes the mass of the cable. The car is represented by a mass m_2 located at the position of its center of mass (center of mass will be discussed in chapter 5), a distance l from the suspension mechanism. The same length l is used as for the equilibrium spring length for simplicity. Thus we replace the complex cable car system by a mass suspended by a linear spring and constrained to move in the vertical direction, and a second mass attached to a massless rigid rod of length l. While this is far from the original system it does contain some of its essential features.

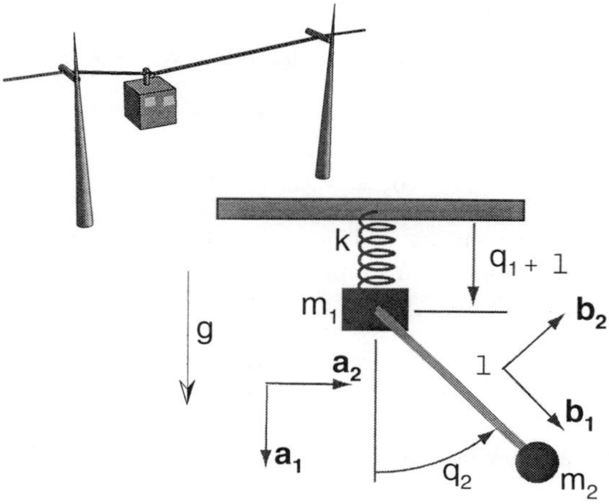

Figure 4.7: Cable car model

Another point of view is that one can not expect to understand the full complex system without an understanding of this simple model. An effective discussion of the problem starts with enumerating some essentials. Thus it might be expected that important frequencies or time scales can be related to the subsystems formed by the spring and masses, or the pendulum and mass. Based on these assessments it is possible to 'scale' the variables and seek various non-dimensional parameters that describe the qualitative behavior of the model. One such parameter might be the ratio of the natural periods of motion of the spring-mass to the pendulum. An experienced analyst will have some ideas about such matters from simply examining the physical situation. Until such experience is gained it is useful to have concrete equations to deal with.

As a first approach to the problem, start by taking as the generalized coordinates q_1, the displacement of the suspension mass from its equilibrium length, and q_2, the angle of the suspension rod with the vertical direction. The inertial triad a is chosen with $\boldsymbol{a}_1$ in the direction of the gravitational acceleration. A second triad b is oriented so that $\boldsymbol{b}_1$ points from the suspension mass to the mass representing the car. With these orientations the in plane part of the direction cosine matrix is

$$\begin{bmatrix} c_2 & s_2 \\ -s_2 & c_2 \end{bmatrix}. \tag{4.136}$$

The position vectors of the two points of the mechanism can be found by inspection of figure 4.7 to be

$$\boldsymbol{r}^{<1} = (q_1 + l)\boldsymbol{a}_1, \tag{4.137}$$
$$\boldsymbol{r}^{<2} = (q_1 + l)\boldsymbol{a}_1 + l\boldsymbol{b}_1. \tag{4.138}$$

From the use of the concept of 'simple' angular velocity we see that ${}^A\boldsymbol{\omega}^B = \dot{q}_2\boldsymbol{b}_3$, hence that

$$\dot{\boldsymbol{b}}_1 = \dot{q}_2\boldsymbol{b}_2, \tag{4.139}$$
$$\dot{\boldsymbol{b}}_2 = -\dot{q}_2\boldsymbol{b}_1. \tag{4.140}$$

In this simple case these relations can be easily obtained from the direction cosine matrix. Using these relations it is a simple exercise to find the velocities with respect to the inertial frame, which are:

$$\boldsymbol{v}^{<1} = \dot{q}_1\boldsymbol{a}_1, \tag{4.141}$$
$$\boldsymbol{v}^{<2} = \dot{q}_1\boldsymbol{a}_1 + \dot{q}_2 l\boldsymbol{b}_2. \tag{4.142}$$

The kinetic energy can now be calculated, thus

$$\mathcal{K}^{(} = \frac{1}{2}(m^{<}\boldsymbol{v}^{<}) \bullet \boldsymbol{v}^{<} \tag{4.143}$$
$$= \frac{1}{2}m_1\dot{q}_1{}^2 + \frac{1}{2}m_2(\dot{q}_1{}^2 + l^2\dot{q}_2{}^2 - 2s_2 l\dot{q}_1\dot{q}_2)$$

The applied force and the coordinate tangent Kvectors are also needed to apply the first form of Lagrange's equations. These quantities provide the information needed to calculate the generalized applied forces. In standard texts on Lagrangian mechanics these are called generalized forces. The tangent Kvectors, or partial velocities can be calculated directly from the position vectors, or they can be obtained by inspection of the velocity expressions. The result is that

$$\boldsymbol{\tau}_1^{<1} = \frac{{}^A\partial}{\partial q_1}\boldsymbol{r}^{<1} = \boldsymbol{a}_1, \tag{4.144}$$
$$\boldsymbol{\tau}_2^{<1} = \frac{{}^A\partial}{\partial q_2}\boldsymbol{r}^{<1} = 0, \tag{4.145}$$
$$\boldsymbol{\tau}_1^{<2} = \frac{{}^A\partial}{\partial q_1}\boldsymbol{r}^{<2} = \boldsymbol{a}_1, \tag{4.146}$$
$$\boldsymbol{\tau}_2^{<2} = \frac{{}^A\partial}{\partial q_2}\boldsymbol{r}^{<2} = l\boldsymbol{b}_2. \tag{4.147}$$

The active forces on the particles arise from the gravitational field and the spring. Internal constraints will be eliminated automatically, hence they need not be considered. The resulting applied force vectors are

$$\boldsymbol{R}^{<1} = -kq_1\boldsymbol{a}_1 + m_1 g\boldsymbol{a}_1, \tag{4.148}$$
$$\boldsymbol{R}^{<2} = m_2 g\boldsymbol{a}_1. \tag{4.149}$$

It is now possible to calculate the generalized active force terms as:

$$F_1^{\tau} = \boldsymbol{R}^{<} \bullet \boldsymbol{\tau}_1^{<} = (m_1 + m_2)g - kq_1, \tag{4.150}$$

$$F_2^{\tau} = \boldsymbol{R}^{<} \bullet \boldsymbol{\tau}_2^{<} = -m_2 g l s_2. \tag{4.151}$$

Since the coordinate basis of Kvectors is associated with the chosen generalized coordinates these terms may also be called the generalized forces. The Euler-Lagrange operator is now used to calculate the generalized inertia terms, which are:

$$F_1^{*\tau} = \left(\frac{{}^A d}{dt}\frac{{}^A\partial}{\partial \dot{q}_1} - \frac{{}^A\partial}{\partial q_1}\right)\mathcal{K}^{()} \tag{4.152}$$

$$= (m_1 + m_2)\ddot{q}_1 - m_2 l s_2 \ddot{q}_2 - m_2 l c_2 \dot{q}_2{}^2,$$

$$F_2^{*\tau} = \left(\frac{{}^A d}{dt}\frac{{}^A\partial}{\partial \dot{q}_2} - \frac{{}^A\partial}{\partial q_2}\right)\mathcal{K}^{()} \tag{4.153}$$

$$= m_2 l (l\ddot{q}_2 - s_2 \ddot{q}_1).$$

The Euler-Lagrange approach provides two second order differential equations for the generalized coordinates:

$$(m_1 + m_2)\ddot{q}_1 - m_2 l s_2 \ddot{q}_2 - m_2 l c_2 \dot{q}_2{}^2 = \tag{4.154}$$

$$-kq_1 + (m_1 + m_2)g,$$

$$l\ddot{q}_2 - s_2\ddot{q}_1 = -g s_2. \tag{4.155}$$

For comparison the equations are now derived by the direct method of calculating the generalized inertial terms from the accelerations. Straightforward differentiation of the velocity expressions using the relations for the derivatives of the frame vectors, $\boldsymbol{b}_j$, gives the result:

$$m^{<}\boldsymbol{a}^{<} = \begin{bmatrix} m_1 \dot{u}_1 \boldsymbol{a}_1 \\ m_2 \dot{u}_1 \boldsymbol{a}_1 + m_2 l \dot{u}_2 \boldsymbol{b}_2 - m_2 l u_2^2 \boldsymbol{b}_1 \end{bmatrix}, \tag{4.156}$$

where u_j has been used to represent generalized speeds in keeping with the spirit of this approach where one seeks a set of first order equations in the generalized speeds and coordinates for the mechanism. The kinematic equations are the relations $\dot{q}_j = u_j$. It is easily verified that the resultant equations are the same as was found by the Euler-Lagrange operator, i.e. by the use of the first form of Lagrange's equations.

The derived equations are nonlinear and difficult to deal with by pure analytic means. The next step in a methodical approach to this problem is to gain some insight by simplifying the the equations. One approach is to assume small displacements from an equilibrium

solution in which the mechanism is at rest. This may provide information about basic oscillation frequencies and some idea as to the couplings between basic motions. This can then be followed by numerical treatment, and or nonlinear analysis around various possible conditions. A practical problem that may have to be dealt with is that the form of the equations is as a set of second order equations. Techniques for nonlinear analysis and many numerical integration methods are based on having first order equations for the derivatives in terms of 'state' variables. While it is not hard to put the Lagrange equations into such a form the direct tangent Kvector method based on generalized speeds and coordinates leads naturally to a state space form. Therefore the analysis is postponed until we develop Kane's form of the equations of motion for the mechanism.

Also instead of developing expressions for the velocity in terms of the generalized coordinates we start with a choice of generalized speeds which are then connected to the generalized coordinates by a set of kinematic equations. This is to emphasize the freedom we have to choose generalized speeds that naturally fit the problem. So as to keep the notation from becoming overly complex we will use the same symbols, u_j for the new generalized speeds. As the mechanism has two geometric degrees of freedom, we may choose two independent generalized speeds. It is frequently useful to consider the representation of the velocity relative to the inertial observer in a reference frame that is fixed in one of the moving bodies. To see how this may work out choose generalized speeds such that

$$\boldsymbol{v}^{<2} = u_1\boldsymbol{b}_1 + u_2\boldsymbol{b}_2. \tag{4.157}$$

An expression for the velocity $\boldsymbol{v}^{<1}$ in terms of these generalized speeds is now required. To obtain this use the result from chapter two for the difference of two velocities for points fixed in a moving reference frame. This gives

$$\boldsymbol{v}^{<2} = \boldsymbol{v}^{<1} + l\,{}^A\boldsymbol{\omega}^B \times \boldsymbol{b}_1, \tag{4.158}$$

where ${}^A\boldsymbol{\omega}^B = \dot{q}_2\boldsymbol{b}_3$. Therefore it is seen that:

$$\boldsymbol{v}^{<1} = u_1\boldsymbol{b}_1 + u_2\boldsymbol{b}_2 - \dot{q}_2 l\boldsymbol{b}_2. \tag{4.159}$$

To obtain tangent vectors by inspection terms such as $\dot{q}_2$ must be eliminated from this expression. This elimination uses the kinematic differential equations for the particular choice of generalized speeds and coordinates. To obtain these note the result that

$$\boldsymbol{v}^{<1} = \dot{q}_1\boldsymbol{a}_1, \tag{4.160}$$

hence the vectorial relationship

$$\dot{q}_1\boldsymbol{a}_1 = u_1\boldsymbol{b}_1 + (u_2 - \dot{q}_2 l)\boldsymbol{b}_2. \tag{4.161}$$

Taking the dot product of this equation with $\boldsymbol{b}_j$ for $j = 1, 2$, gives the two equations

$$\dot{q}_1 = \frac{u_1}{c_1}, \tag{4.162}$$

$$\dot{q}_2 = \frac{s_2}{lc_2}u_1 + \frac{u_2}{l}, \tag{4.163}$$

which are the desired kinematic differential equations. Substitution for $\dot{q}_2$ in the expression for $\boldsymbol{v}^{<1}$ then gives an expression for the velocity Kvector which is linear in the generalized speeds. Therefore tangent basis Kvectors may be obtained by examination of the coefficients of the generalized speeds in this expression. The velocity Kvector is

$$\boldsymbol{v}^{<} = \begin{bmatrix} u_1(\boldsymbol{b}_1 - \tan q_2 \boldsymbol{b}_2) \\ u_1\boldsymbol{b}_1 + u_2\boldsymbol{b}_2 \end{bmatrix}, \tag{4.164}$$

and by picking out the coefficients of the generalized speeds the basis Kvectors are seen to be

$$\beta_1^{<} = \begin{bmatrix} \boldsymbol{b}_1 - \tan q_2 \boldsymbol{b}_2 \\ \boldsymbol{b}_1 \end{bmatrix}, \tag{4.165}$$

$$\beta_2^{<} = \begin{bmatrix} 0 \\ \boldsymbol{b}_2 \end{bmatrix}. \tag{4.166}$$

This analysis can also be carried out more directly by starting with the particle displacement vectors and differentiating them relative to the inertial observers frame, as was done in the first approach to the problem. Comparison with the forms containing the new generalized speeds then give the kinematic differential equations. In any case the most convenient form for $\boldsymbol{v}^{<}$ is:

$$\boldsymbol{v}^{<} = \begin{bmatrix} \dot{q}_1\boldsymbol{a}_1 \\ \dot{q}_1\boldsymbol{a}_1 + l\dot{q}_2\boldsymbol{b}_2 \end{bmatrix}. \tag{4.167}$$

Which, using the kinematic differential equations, can be written as:

$$\boldsymbol{v}^{<} = \begin{bmatrix} u_1 \sec q_2 \boldsymbol{a}_1 \\ u_1\boldsymbol{b}_1 + u_2\boldsymbol{b}_2 \end{bmatrix}, \tag{4.168}$$

the tangent vectors taking the form

$$\beta_1^{<} = \begin{bmatrix} \sec q_2 \boldsymbol{a}_1 \\ \boldsymbol{b}_1 \end{bmatrix}, \tag{4.169}$$

$$\beta_2^{<} = \begin{bmatrix} 0 \\ \boldsymbol{b}_2 \end{bmatrix}. \tag{4.170}$$

Direct differentiation, using the relations for $\dot{\boldsymbol{b}}_i$, gives the acceleration Kvector as:

$$\boldsymbol{a}^{<} = \begin{bmatrix} (\dot{u}_1 \sec q_2 + u_1\dot{q}_2 \tan q_2 \sec q_2)\boldsymbol{a}_1 \\ (\dot{u}_1 - \dot{q}_2 u_2)\boldsymbol{b}_1 + (\dot{u}_2 + \dot{q}_2 u_1)\boldsymbol{b}_2 \end{bmatrix}, \tag{4.171}$$

hence the result that:

$$m^{<}\boldsymbol{a}^{<} = \begin{bmatrix} m_1(\dot{u}_1 \sec q_2 + u_1\dot{q}_2 \tan q_2 \sec q_2)\boldsymbol{a}_1 \\ m_2(\dot{u}_1 - \dot{q}_2 u_2)\boldsymbol{b}_1 + m_2(\dot{u}_2 + \dot{q}_2 u_1)\boldsymbol{b}_2 \end{bmatrix}. \tag{4.172}$$

The quantity $\dot{q}_2$ in these expressions is to be replaced in the final result with its value in terms of the generalized speeds from the kinematic differential equations. Evaluation of

$$m^< \boldsymbol{a}^< \bullet \boldsymbol{\beta}_j^< = -F_j^{*\beta} = F_j^\beta = \boldsymbol{R}^< \bullet \boldsymbol{\beta}_j^<, \tag{4.173}$$

leads to the two equations (Kane's Equations)

$$(m_1 \sec^2 q_2 + m_2)\dot{u}_1 + m_1 \sec^2 q_2 \tan q_2 \dot{q}_2 u_1 \tag{4.174}$$
$$-m_2 \dot{q}_2 u_2 = (m_1 g - k q_1)\sec q_2 + m_2 g \cos q_2,$$
$$m_2 \dot{u}_2 = -m_2 g \sin q_2 - m_2 \dot{q}_2 u_1. \tag{4.175}$$

As $\dot{q}_2$ can be replaced by nondifferentiated terms from the kinematic differential equations, these are two first order equations for the derivatives of u_j. The fortunate circumstance that the equations are almost in the form $\dot{u}_j = \ldots$, is not automatic, and in the present case is connected with the fact that our generalized speeds were chosen to represent the velocity relative to the inertial observer in a single 'body fixed' frame. In the general case one can not expect this happy circumstance for 'free'.

In summary the full set of equations for the problem consist of the dynamic and kinematic differential equations in the variables q_j and u_j. These are now given in explicit form for the readers convenience:

$$\dot{u}_1 = \frac{f_1}{f_2} + g \cos q_2, \tag{4.176}$$
$$\dot{u}_2 = -g \sin q_2 - \frac{1}{l}(u_2 + u_1 \tan q_2)u_1, \tag{4.177}$$
$$\dot{q}_1 = \sec q_2 u_1, \tag{4.178}$$
$$\dot{q}_2 = \frac{1}{l}(u_1 \tan q_2 + u_2), \tag{4.179}$$

where

$$f_1 = (\tan q_2 u_1 + u_2)(m_2 u_2 - m_1 \sec^2 q_2 \tan q_2 u_1) \tag{4.180}$$
$$-klq_1 \sec q_2,$$
$$f_2 = l(m_1 \sec^2 q_2 + m_2). \tag{4.181}$$

Though the model was based on a considerable simplification , the governing equations are certainly not simple! In the next section we will see how one can obtain an 'energy integral' which can help in analyzing the mechanism. Such integrals, while sometimes helpful, are not generally available for mechanisms with dry friction and other nonconservative forces. In fact as noted in the formulation of this problem it is probably required to represent the loss of energy due to wave motion on the cable by a nonconservative term. It is therefore important to have tools which can directly help us with equations of the above form. To at least 'hint' at some of the steps one can take in this task, we continue with the analysis of the mechanism.

The above system of equations can be considered as the description of all possible trajectories of a point in a four dimensional *state space,* with a point given by the values of u_j and q_j. If such a point is represented by a column vector of the form $(x_1, x_2, x_3, x_4)^T$, with $x_1 = q_1, x_2 = q_2, x_3 = u_1$ and $x_4 = u_2$, the equations may be cast in the form:

$$\tfrac{dx}{dt} = F(x). \tag{4.182}$$

Every possible state of the system is represented by some point in this four dimensional space. Given a particular state, or equivalently some point in the space of states, the above equation provides all the information needed to determine the evolution of the system in time. Thus a point in the state space determines a trajectory or curve which passes through the point. All possible states that can be obtained from the given state lie on this curve. Special points are *equilibrium points,* those for which $F(x) = 0$. It is of interest to study *linearized* systems in the vicinity of such equilibrium points. In such studies interest is in questions of *stability* in the sense of asking if small disturbances will take the system trajectory far from equilibrium or not. The important stable equilibrium point for the present system is the case where the car lies at rest below the suspension point, stretching the cable as a result of the systems total weight. In such a case the x of the above form must vanish, i.e. the generalized velocities, $\dot{q}_i$ and the rates of change of the generalized speeds $\dot{u}_i$, go to zero. This condition easily leads to the static equilibrium result that

$$(m_1 + m_2)g = kq_1. \tag{4.183}$$

Linearized equations of motion about this equilibrium point are developed by assuming that:

$$q_1 = \frac{(m_1 + m_2)g}{k} + \xi_1, \tag{4.184}$$

$$q_2 = \xi_2, \tag{4.185}$$

where ξ_j are to be considered small. These expressions are then substituted into the equations of motion, and terms quadratic and smaller in ξ are dropped. This gives the result that:

$$\ddot{\xi}_1 + \frac{k}{m_1 + m_2}\xi_1 = 0, \tag{4.186}$$

$$\ddot{\xi}_2 + \frac{g}{l}\xi_2 = 0. \tag{4.187}$$

This corresponds with our original *intuition* about the problem, i.e. that the basic motion will be built around the two important frequencies of the oscillation due to the mass of the car and cable interacting with the effective system stiffness and the pendulum motion of the car. It is somewhat surprising that to there is no interactive effect for the disturbance modes at the linear level. The validity of this for an actual system, as opposed to our model, must be checked by experiment or against a more adequate model.

The next natural step in the analytical examination is to deal with the question of 'scaling' and parameter reduction. This has the two fold purpose of reducing the labor of integrating the system and of determining what parameters should be used to examine different ranges of qualitative behavior. The subject is a complicated one that makes considerable use of experience and hence does not lend itself easily to an objective exposition. It relates to aspects of dimensional analysis, perturbation theory and similarity solutions of differential equations using group theoretical concepts. References are given to pertinent literature in the appendix B of this text.

For the problem as posed, the solution for any set of initial conditions would depend on 9 parameters and the time. These are the two masses m_1 and m_2, the stiffness k, the pendulum support length, which is also taken as the spring equilibrium length, l, the gravitational acceleration g, the initial displacements of the pendulum support point $q_1(0)$, the initial angle of the pendulum $q_2(0)$, and the two initial generalized speeds. Clearly it is desirable to reduce the number of parameters to a minimum, for even if we could obtain a full solution of the problem in analytic form the task of interpretation would indeed be formidable. In effect the existence of numerical computational techniques and fast computers provide us with the equivalent of a full solution for any particular set of parameters. Therefore reducing the dimension of the parameter space for the mechanism is an important practical part of the task of 'understanding' the system.

There are a number of ways of going about this task. A popular method is to identify critical scales for the variables. This can be done on physical and intuitive grounds, or by substitution of arbitrary scales into the equations of motion. In the latter approach the scales are picked so as to simplify the equations while preserving the generality of the problem. This approach is popular in the study of fluids and has led to the use of certain 'dimensionless' numbers to describe fluid flow regimes. Examples which may be familiar to the reader are the Mach number which identifies compressibility effects and the Reynolds number for viscous friction effects. Another approach, popular in the theoretical physics literature, is to use our freedom to set a certain number of independent units. Careful study shows equivalence between these methods, i.e. setting units to desirable values in effect is equivalent to the identification of dimensionless parameters through normalization. Since it appears less often in the mechanics literature and is very useful when dealing with large awkward systems of equations we will examine our cable car problem using the unit setting or as it is sometimes called the natural units method.

The rational behind this approach is that the choice of mass, length and time as fundamental units is entirely arbitrary. For a mechanics problem we are free to choose both the type of unit and the units particular realization. That is even if we choose 'length' as a fundamental unit we still have the choice of 'scale'. From a purely abstract view centimeters are as 'good' as feet! Therefore it is natural to choose units for a particular problem that conform to the problem's physical magnitudes and geometry. The limitation is that only three arbitrary units may be chosen. In addition the units must be *independent* in the sense that all needed physical magnitudes that are present in the problem can be represented in terms of the chosen set. Once such a set is chosen it is relatively simple to convert between the suitable problem units and the normal measurement systems used in engineering and science.

This idea is now applied to the cable car problem with its nine parameters. The theory says the problem can be reduced to six parameters. Take as a unit of length, the length of the pendulum l. This means that all lengths will be measured in units of the particular pendulum length. In particular it means that in the new system of units the length of the pendulum is 1. Therefore one of the nine parameters is now eliminated. To convert back to normal scientific units take any quantity in the problem solution that would have dimensions of length and multiply it by the value of l in the particular problem. Thus the normal dimensional value of the displacement q_1 is obtained by multiplying it by l, in the desired unit system, e.g. centimeters. Another parameter in the problem is the mass of the cable car, m_2. By measuring all masses in units of m_2 we have the desirable result that for our problem we can take $m_2 = 1$, so we are down to seven parameters. There still remains an independent unit at our disposal. So far we have produced a mass and a length scale, but it appears that we are staying with a *mass, length, time* system of units. We depart from this in our third choice by deciding to measure all accelerations in terms of the gravitational constant, g. Thus in our problem we can take $g = 1$, reducing the number of parameters to six. To convert back to standard units it would be useful to have the unit of time in our new natural system. This is easy to accomplish by the use of *dimensional equations.* Suppose we pick *mass, length and time* as our base physical quantities. Then the dimensional equation for the force unit would be written as:

$$[F] = MLT^{-2}. \tag{4.188}$$

The square brackets indicate that this is a dimensional equation, and it is standard to use M, L, and T, to indicate the units of mass, length and time. To obtain the conversion factor from a mass, length, acceleration system to a mass, length time system use the dimensional equation

$$[A] = LT^{-2}, \tag{4.189}$$

substitute, $A = g$, and $L = l$. and solve for T, which gives:

$$T = \sqrt{\frac{l}{g}}. \tag{4.190}$$

Not surprisingly our natural choice of problem units gave us a time scale which is near the period of a linear pendulum of length l in a gravitational acceleration field of magnitude g. This provides the conversion between time in standard units and in the natural units. If in the solution of our problem we wish to find out what standard time applied to a time t_* in the natural units we simply multiply by $\sqrt{l/g}$. Thus the parameters m_2, l and g have been eliminated from the equations without loss of generality. But how do we choose a parameter such as the stiffness, k, given a value in standard units? Suppose that we have measured k in say kilogram, meter, seconds as base units. Call this value k_*. The linear spring force law shows that the dimensional equation for k is:

$$[k] = \frac{[F]}{[L]} \rightarrow \frac{m_2 g}{l}, \tag{4.191}$$

hence we have the relation that the k in the equation expressed in the new natural units is given by:

$$\bar{k} = \frac{k_* l}{m_2 g}, \tag{4.192}$$

and the notation $\bar{k}$ has been used to indicate that this can be interpreted as a dimensionless number. It has thus been demonstrated that the important parameter is $\bar{k}$ which can be looked upon as the ratio of the linear spring force to the weight. Similar reasoning gives $\bar{m} = m_1/m_2$ as another dimensionless parameter. The four remaining dimesionless parameters are derived from the initial conditions. The angle $q_2(0)$ is already in dimensionless form, and the initial displacement q_1 is made into a dimensionless quantity by dividing with l. The generalized speeds can be scaled with $\sqrt{gl}$.

From this point assume that the equations have been interpreted in the fashion that has been outlined, hence m will refer to the mass ratio and k to what was called $\bar{k}$ above. It should be noted that with this interpretation a single numerical solution covers *all* cases for which these new parameters have the same value, i.e. an infinite number of possible value of the full set of 9 parameters. In terms of the new scaled variables the equilibrium solution is given by

$$x_1 = \frac{1+m}{k}, \tag{4.193}$$

the state variable notation being used for the set of generalized coordinates and speeds.

To conclude this discussion we examine two numerical integrations of the equations. This was done by a variable step Runge-Kutta method using the off the shelf software package, Matlab. Four sets of curves are presented for each case. The integration was carried out for 100 time units, which in light of the above discussion corresponds to about 15 periods of the linear pendulum (this is because of the factor of 2π in the full linear period formula). The results are shown in Figures 4.8 and 4.9. The upper left set shows the generalized coordinates as functions of time, and the upper right set shows the generalized speeds, also as a function of time. From our linear analysis we expect at least a rough distinction to show up between what we might call the spring and pendulum modes. The lower left and right curve plot the generalized speeds against the generalized velocities as indicated in the figures. In

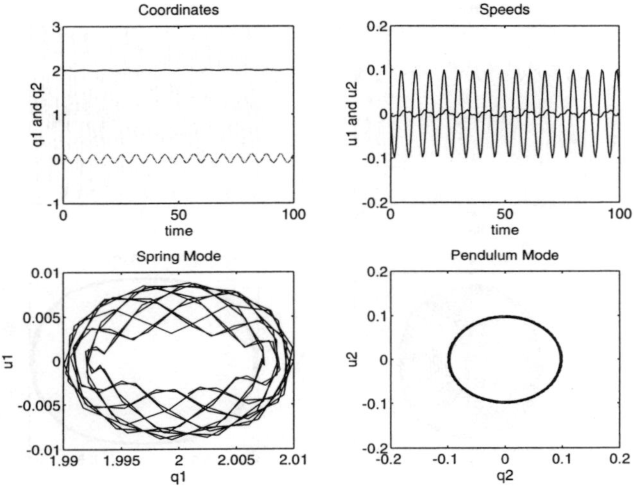

Figure 4.8: Cable car model near equilibrium.

both cases $m = k = 1$, hence the equilibrium solution is given by $q_2 = u_1 = u_2 = 0$, and $q_1 = 2$. Also in both cases initial conditions are chosen so that $u_1(0) = u_2(0) = 0$. The initial values of the generalized coordinates are chosen to examine what happens when the displacement is near its equilibrium value. Thus we take $q_1(0) = 2.01$ in both cases. In the first case the pendulum angle is started somewhat near the equilibrium value, so that $q_2(0) = 0.1$, while in the second case it is started with a relatively large value of 1.0. The first set of curves show that the periodic pendulum motion proceeds independent of the spring motion, The spring displacement remains small and almost follows the simple harmonic motion expected for a linear spring, except for a modulation due to the pendulums motion. In the second set we see a more dramatic effect on the spring mode, which is driven to quite large displacement values. We also see that the periodic pendulum motion is destroyed by the interaction. To pursue the full discussion of this problem would take us into the field of dynamic systems and chaos, hence stop at this point. The main purpose of this section has been to give the reader an idea how the somewhat abstract theory that has been set forth can be applied to actual problems. This includes false starts and searches for useful quantities to work with. The freedom of choosing generalized speeds as well as generalized coordinates increases ones flexibility in this task, but requires some practice and experience for full usefulness. Many commercial software packages now exist for setting up and integrating mechanical models of engineering systems. Their

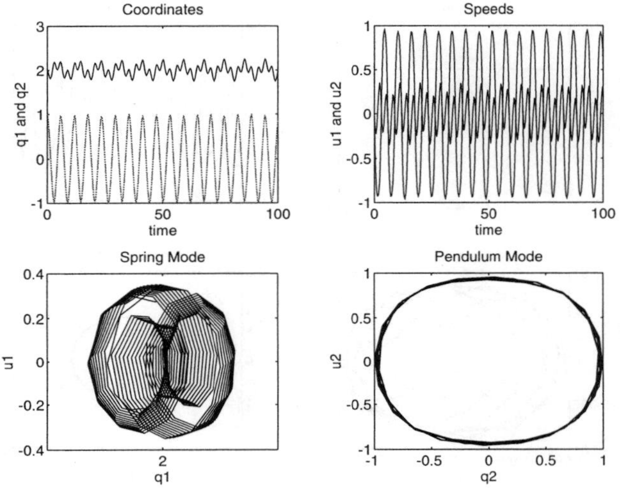

Figure 4.9: Cable car model with large pendulum angle.

workings are generally invisible to the user. In many cases they provide animations and involved graphical presentations of the results. They usually carry out the manipulations for elimination of constraints as part of a pointwise numerical process of solution. Thus they provide little deep theoretical insight and should be looked upon and used as easy 'experimental' simulations and tools for obtaining numbers in the process of design. The analytical methods set forth in this book can best be used to obtain 'understanding' of a mechanism. This includes understanding the simulations which are best looked upon as a type of experiment. Understanding means the abstraction of the essential features and working with a model which is simpler than the full system but exhibits those features.

4.8 Potential Energy and The Second Form of Lagrange's Equations

It was shown in the previous section that if the power associated with a mechanism consisting of a single particle is defined as the ordinary scalar product of the forces acting on the particle with the particles velocity, then in a Newtonian motion

$$\mathcal{P} = \tfrac{d}{dt}\mathcal{K}(\boldsymbol{v}). \tag{4.194}$$

If the force acting on the particle were derivable from a *potential*, i.e., if $\boldsymbol{R} = -\nabla V(\boldsymbol{r})$, in certain special cases it is also possible to write the power of a mechanism undergoing a Newtonian motion in the form

$$\mathcal{P} = -\tfrac{d}{dt}\mathcal{V}(\boldsymbol{r}), \tag{4.195}$$

in which case there is an immediate *first integral* of the Newtonian motion

$$\mathcal{K} + \mathcal{V} = \mathcal{E}_0. \tag{4.196}$$

The *energy* $\mathcal{E}_0$ is a constant, hence the above *conservation law* gives a relation between the velocity and the position, reducing the order of the differential equation system for the Newtonian motion from 6 to 5. The question of deciding on the existence of the function $\mathcal{V}$ is clearly of considerable interest. Elementary vector analysis gives an immediate *necessary* condition, as the curl of any gradient must vanish the applied force must satisfy the relation:

$$\nabla \times \boldsymbol{R}(\boldsymbol{r}, t) = 0. \tag{4.197}$$

In addition for the energy conservation integral to exist, the potential must not be an *explicit* function of t, or

$$\tfrac{\partial}{\partial t}\mathcal{V} = 0. \tag{4.198}$$

For a single particle, the first form of Lagranges equations can be constructed using the cartesian position expressions as generalized coordinates, i.e. $x_i = q_i = \boldsymbol{r} \cdot \boldsymbol{a}_i$. In this case, if a potential $\mathcal{V}(\boldsymbol{r}, t)$ exists, so that

$$F_i = \tfrac{\partial}{\partial x_i}\mathcal{V}, \tag{4.199}$$

it is possible to define the *Lagrangian*,

$$\mathcal{L}(\boldsymbol{r}, \boldsymbol{v}, t) = \mathcal{K} - \mathcal{V}, \tag{4.200}$$

in terms of which the equations for a Newtonian motion are:

$$F_i^* + F_i = (\tfrac{d}{dt}\tfrac{\partial}{\partial \dot{x}_i} - \tfrac{\partial}{\partial x_i})\mathcal{L} = 0. \tag{4.201}$$

Thus the Newtonian motions of a single particle mechanism subject to applied forces derivable from a potential is completely described by the scalar function $\mathcal{L}$. The next step is to extend this results, as far as possible, to the K body point mechanism.

Following the pattern used in the calculation for the single body, the big dot product of the Newtonian equation for the K particle system is:

$$\boldsymbol{v}^< \bullet \tfrac{d}{dt}(m^< \boldsymbol{v}^<) = \boldsymbol{v}^< \bullet \boldsymbol{R}_a{}^< + \boldsymbol{v}^< \bullet \boldsymbol{R}_c{}^<, \tag{4.202}$$

where we have indicated the applied and constraint forces by the subscripts a and c. As before we define the *power* associated with the mechanism by

$$\mathcal{P}^< = \tfrac{d}{dt}\mathcal{K}^<, \tag{4.203}$$

which from the above big dot product we can write as

$$\mathcal{P}^< = \boldsymbol{v}^< \bullet \boldsymbol{R}_a{}^< + \boldsymbol{v}^< \bullet \boldsymbol{R}_c{}^<. \tag{4.204}$$

Using the generalized speed expansion for the velocity

$$\boldsymbol{v}^< = \sum_{j=1}^{n} \dot{q}_j \boldsymbol{\tau}_j^< + \boldsymbol{\tau}_t^<, \tag{4.205}$$

it is seen that

$$\mathcal{P}^< = \sum_{j=1}^{n} \dot{q}_j F_j + \boldsymbol{\tau}_j^< \bullet \boldsymbol{R}_a{}^< + (\boldsymbol{\Pi}_* \boldsymbol{\tau}_t^< \bullet \boldsymbol{R}_c{}^<), \tag{4.206}$$

where $\boldsymbol{\Pi}_*$ is the projection operator for the subspace orthogonal to the configuration hypersurface. Now assume that we can find a function $\mathcal{V}^<(q,t)$, such that

$$\mathcal{P}^< = \tfrac{d}{dt}\mathcal{V}^<(q(t),t) = \sum_{j=1}^{n} \dot{q}_j \tfrac{\partial}{\partial q_j}\mathcal{V}^< + \tfrac{\partial}{\partial t}\mathcal{V}^<, \tag{4.207}$$

where the generalized coordinate set $q(t)$ describes a Newtonian motion. In the *special case* where

$$\boldsymbol{\tau}_j^< \bullet \boldsymbol{R}_a{}^< + (\boldsymbol{\Pi}_* \boldsymbol{\tau}_t^<) \bullet \boldsymbol{R}_c{}^< = 0, \tag{4.208}$$

so that

$$\tfrac{\partial}{\partial t}\mathcal{V}^< = 0 \tag{4.209}$$

we again have an *energy conservation law,* i.e.,

$$\tfrac{d}{dt}(\mathcal{K}^< + \mathcal{V}^<) = 0, \tag{4.210}$$

implying that

$$\mathcal{K}^< + \mathcal{V}^< = \mathcal{E}_0^<, \tag{4.211}$$

where $\mathcal{E}_0^<$ is the constant energy of the particular Newtonian motion. A necessary condition for the existence of $\mathcal{V}^<$ is easily derived from the requirement that

$$F_j = -\tfrac{\partial}{\partial q_j}\mathcal{V}^<. \tag{4.212}$$

For $\mathcal{V}^<$ to be a continuous function its mixed partial derivatives must satisfy the condition that

$$\frac{\partial^2}{\partial q_j \partial q_i}\mathcal{V}^< = \frac{\partial^2}{\partial q_i \partial q_j}\mathcal{V}^<. \tag{4.213}$$

This implies that the F_j must satisfy the $n-$ dimensional equivalent of the three dimensional vanishing curl of the applied force, i.e. that

$$\tfrac{\partial}{\partial q_i}F_j - \tfrac{\partial}{\partial q_j}F_i = 0, \tag{4.214}$$

which gives $n(n-1)/2$ necessary conditions on the generalized active forces. If any of these conditions fail to be satisfied, no potential exists. If they are all satisfied then one can attempt the construction of the potential function. As we shall see, the above condition plays an important part in the possible success of the construction.

To carry out the construction, first consider the case where $\mathcal{V}^<$ is not an explicit function of the time and fix attention on the instantaneous configuration hypersurface. Let the parameter s designate any 'test' motion, so that the set of generalized coordinates $q_j(s)$ describe a curve in the configuration manifold. As the path is in the configuration surface, only applied forces need be considered. Now

$$\boldsymbol{R}^< \bullet \tfrac{d}{ds}\boldsymbol{r}^< = \sum_{j=1}^{n} \boldsymbol{R}^< \bullet \boldsymbol{\tau}_j^< \tfrac{d}{ds}q_j = \sum_{j=1}^{n} F_j \tfrac{d}{ds}q_j, \tag{4.215}$$

A particular 'test path', $q(s)$, is designated by $\gamma(s)$. The *test work* associated with the test path is defined by

$$\mathcal{W}(s_1, s_2, \gamma) = -\int_{s_1;\gamma}^{s_2} \sum_{j=1}^{n} F_j \tfrac{d}{ds}q_j ds. \tag{4.216}$$

An infinitesimaly short test path (whatever that may be?) is frequently called a virtual displacement, and the work associated with such a path, virtual work. If $F_j = -\partial\mathcal{V}^</\partial q_j$, as t is held constant for a test path we have the result that

$$\mathcal{W}^< = \int_{s_1,\gamma}^{s_2} \tfrac{d}{ds}\mathcal{V}^< ds = \mathcal{V}^<, \tag{4.217}$$

hence that the test work $\mathcal{W}^<$ is independent of the path γ and depends only on the end points $q(s)$. This gives us a recipe for calculating $\mathcal{V}^<$. We simply calculate the test work between a *reference point,* $q(s_1)$ and a field point $q(s_2)$. Let the reference point be designated by the set q_{j0}, the field or general point by q_j and the integration variable by ξ, and choose a path where first we integrate from the point $q_{10}, q_{20}, \ldots, q_{n0}$, to the point $q_1, q_{20}, \ldots, q_{n0}$, with $q_2, \ldots, q_n$ held constant, then continue with q_2 as the variable, q_3 as the variable, until we reach the point $q_1, \ldots, q_n$. This gives the formula for the potential as

$$\begin{aligned} \mathcal{V}^< = &-\int_{q_{10}}^{q_1} F_1(\xi, q_{20}, \ldots, q_{n0})d\xi \\ &-\int_{q_{20}}^{q_2} F_2(q_1, \xi, \ldots, q_{n0})d\xi - \ldots \\ &-\int_{q_{n0}}^{q_n} F_n(q_1, q_2, \ldots, \xi)d\xi. \end{aligned} \tag{4.218}$$

It is left as an exercise for the reader to show that taking the partial derivatives with respect to the q_j of this result yields the generalized active forces as long as the necessary condition on the partial derivatives, discussed above, is satisfied. The above result also gives the potential when $\partial\mathcal{V}^{<}/\partial t$ does not vanish. In this case the q_{j0} can be functions of time and an arbitrary function of time can be added to the above result. In such a case we do not in general have the result that the total derivative of the potential is equal to the power. Note also that the existence of a potential requires the independence of path of the test work integral. The condition on mixed partial derivatives is necessary for this, but not sufficient. Only the successful demonstration of path independence can fully establish the result, though in most practical applications of the above result the potential does exist if the mixed partial derivative criteria is filled.

The special case in which an energy conservation law holds for the particle mechanism has considerable theoretical interest, however its significance for the practical integration of many particle mechanisms is relatively limited.

The existence of the potential allows us to write down the second form of Lagrange's equations for many particle systems. The Lagrange function is now defined as

$$\mathcal{L}^{<} = \mathcal{K}^{<} - \mathcal{V}^{<}, \tag{4.219}$$

and the Lagrange equations are

$$(\frac{d}{dt}\frac{\partial}{\partial \dot{q}_j} - \frac{\partial}{\partial q_j})\mathcal{L}^{<} = 0. \tag{4.220}$$

The elegance of this form is very appealing, all of the properties of the constrained mechanism being rolled up into one scalar function of the generalized coordinates, their time derivatives the generalized velocities and the time. In practice it is usually easier to stay with Kane's equations and the flexibility of defining useful generalized speeds and basis vectors for the configuration hypersurface which may not be coordinate vectors. There is some controversy about this in the literature and in the end the student of the subject must decide for himself what is the most suitable approach for a particular problem. In any case mastery of both approaches and their connections is needed by any serious user of mechanical theory.

A simple example concludes this chapter on many particle mechanisms. The next two chapters extend this theory to the case of mechanisms which contain rigid body components.

•Illustration

Two collars are connected by a linear spring and are constrained to slide along two parallel rods, a distance l apart. The equilibrium (unstretched) spring has length l. The configuration for this problem is shown in Figure 4.9. The vectors $\boldsymbol{b}_{1,2}$ are taken to be aligned and orthogonal to the spring. Take as generalized coordinates the distance q_1 of the first collar from a reference point on the wire, and $q2$ as the angle between the vectors $\boldsymbol{b}_1$ and $\boldsymbol{a}_1$. If h is the stretched length of the spring, the relation $h \sin q_2 = l$ must hold. Therefore the spring is stretched a distance

$$h - l = l(\csc q_2 - 1). \tag{4.221}$$

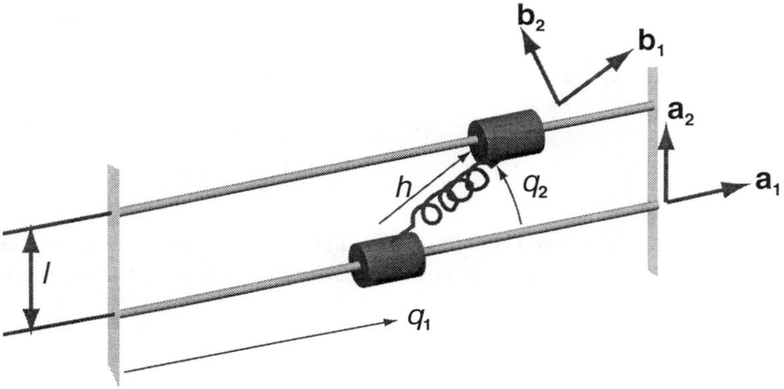

Figure 4.10: Spring connected constrained masses.

If the stiffness of the spring is k, the applied force Kvector is thus given by

$$\boldsymbol{R}^{<} = \begin{bmatrix} kl(\csc q_2 - 1)\boldsymbol{b}_1 \\ -kl(\csc q_2 - 1)\boldsymbol{b}_1 \end{bmatrix}. \tag{4.222}$$

The configuration of the system, i.e. the K displacement vector, is given by

$$\begin{bmatrix} q_1\boldsymbol{a}_1 \\ q_1\boldsymbol{a}_1 + l\cot q_2\boldsymbol{a}_1 + l\boldsymbol{a}_2 \end{bmatrix}. \tag{4.223}$$

Taking the derivative of this with respect to the inertial observer's frame, the velocity Kvector is seen to be

$$\begin{bmatrix} u_1\boldsymbol{a}_1 \\ u_1\boldsymbol{a}_1 - u_2 l(\cot^2 q_2 + 1)\boldsymbol{a}_1 \end{bmatrix}. \tag{4.224}$$

Where $u_i = \dot{q}_i$, and the coordinate basis tangent Kvectors are seen by inspection to be

$$\boldsymbol{\tau}_1^{<} = \begin{bmatrix} \boldsymbol{a}_1 \\ \boldsymbol{a}_1 \end{bmatrix}, \tag{4.225}$$

$$\boldsymbol{\tau}_2^{<} = \begin{bmatrix} 0 \\ -l(\cot^2 q_2 + 1)\boldsymbol{a}_1 \end{bmatrix}. \tag{4.226}$$

Thus the generalized active forces are:

$$F_1 = \boldsymbol{R}^{<} \bullet \boldsymbol{\tau}_1^{<} = 0, \tag{4.227}$$

$$F_2 = \boldsymbol{R}^{<} \bullet \boldsymbol{\tau}_2^{<} = kl^2(\csc q_2 - 1)(\cot^2 q_2 + 1)\cos q_2. \tag{4.228}$$

It is easily verified that

$$\tfrac{\partial}{\partial q_2}F_1 = \tfrac{\partial}{\partial q_1}F_2, \tag{4.229}$$

so that a potential may exist. Using the integral formula derived above, with $q_{10} = 0$, and $q_{20} = \pi/2$,

$$\begin{aligned} \mathcal{V}^{<} &= -\int_{\pi/2}^{q_2} kl^2(\csc\xi - 1)(\cot^2\xi + 1)\cos\xi d\xi \\ &= -kl^2(\csc^2 q_2 - 2\csc q_2 + 1) \\ &= -\frac{1}{2}k(h-l)^2. \end{aligned} \tag{4.230}$$

The last result could have been written down directly from the elementary theory of the linear spring. The Lagrangian is now easily constructed from the expressions for $\boldsymbol{v}^{<}$ and $\mathcal{V}^{<}$.

4.9 Problems

•Problem 4.1

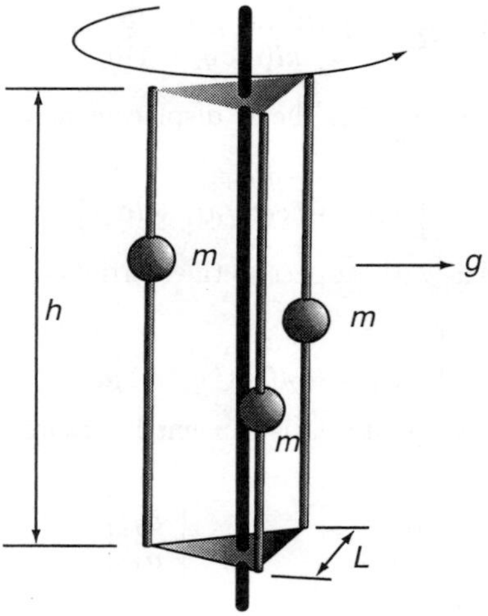

Figure 4.11: Problem 4.1

Three particles, each of mass 'm' are free to slide along the edges of a prismatic frame of height 'h' . The ends of the frame form two equilateral triangles of side 'L'. The frame is free to rotate about an axis which passes through the centroids of the triangles. This mechanism is placed in a uniform gravitational field with gravitational acceleration 'g'. Using appropriate generalized coordinates and speeds obtain both the kinematic and dynamic equations for friction free motion. Find a Kvector basis for the tangent space and for its orthogonal complement. Obtain the constraint forces that act on the frame at the positions of the particles. Assuming a coefficient of kinetic friction μ determine the modified equations of motion. The problem should be solved both by hand calculation and by use of Sophia-Maple.

•Problem 4.2

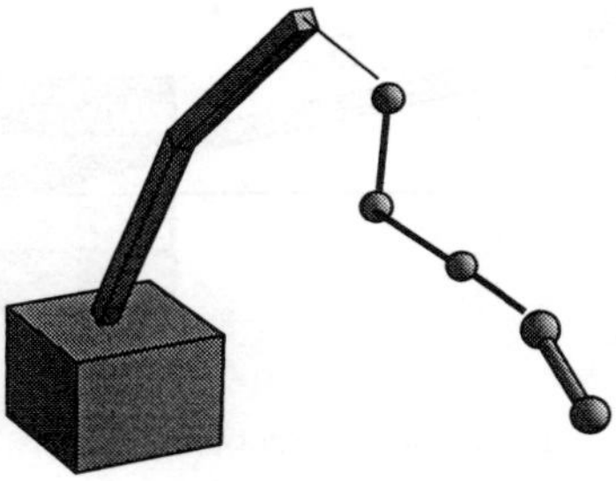

Figure 4.12: Problem 4.2

The diagram provides a simplified model for the study of the sport of 'bungy cord' jumping. The rope is modeled by three equal masses connected by springs. The jumper is modeled by a rigid rod with two attached mass points. Use this diagram to derive appropriate equations of motion to study this problem. It is not necessary to restrict your considerations to two dimensions! It might also be advisable to add dampers between the mass points.

•Problem 4.3

The shaft A rotates with angular velocity f(t). A mass m_1 is attached to the pin joint connecting shaft A to B. The mass m_2 attached to shaft B is guided to move in a channel aligned with the pin joint fixing the end of shaft A. Assume that shafts A and B have lengths L_A and L_B respectively and that the local uniform gravitational field is at an arbitrary angle with the channel. Determine the equation of motion and the constraint force on the channel wall assuming frictionless motion.

•Problem 4.4

The two rods are free to rotate at right angles to the connecting shaft. The center of the shaft moves along the path given by the vector $\boldsymbol{R}(t)$. Also the connecting shaft is constrained to be in a plane that is orthogonal to the vector $\boldsymbol{R}(t)$. The rod and shafts have negligible mass. The connecting rods and shaft all have length 2L. Four equal masses are attached to the ends of the rotating rods. Derive equations of for the motion of this mechanism using the projection method. Also derive equations for motion in a uniform gravitational field using a Lagrangian.

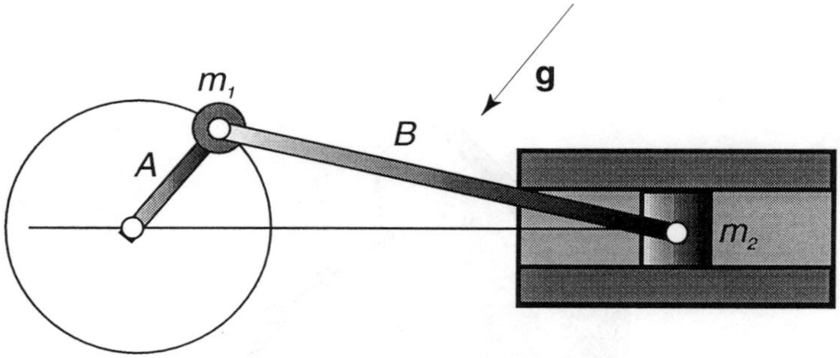

Figure 4.13: Problem 4.3

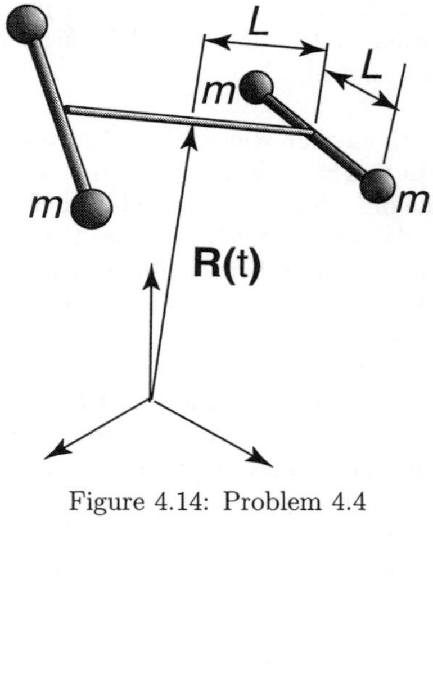

Figure 4.14: Problem 4.4

Chapter 5

Dynamics of a Rigid Body

Any competent physicist will tell you there are no such things as rigid bodies. Push one end of a long rigid rod and the other end moves. There does not appear to be any perceptible time delay. If the rod were made of steel, the motion of one end would propagate a wave, moving at about 6000 m/s, down the rod. The message is that, for bodies of reasonable scale (to humans) made of the relatively stiff materials used in constructing machines, the approximation of rigidity is reasonable and useful. This chapter examines the implications for Newtonian motion of the assumption that a body is rigid. To do this the dynamical properties that are intrinsic to a body and the way forces may act on it must be characterized. Previous work has shown that the configuration of a rigid body is entirely determined by the specification of the location of a point fixed in the body and the attitude of the body. The latter being set by a knowledge of the direction cosine matrix of a standard triad fixed in the body. Thus a total of six parameters entirely specify the configuration. This has deep consequences in regard to the nature of the force system and the specification of the dynamical and geometric properties. Classes of equivalent bodies and force systems exist which have the same dynamic characterization. Thus for a body having a general distribution of mass and acted on by a general distribution of force only certain functions of these distributions are important. Some examples of these functions are the total mass and the total force acting on the body.

5.1 Characterization of a Rigid Body

In the discussion of a reference frame in Chapter two the term body was used to denote a set of points to which was attached an observer equipped with a standard triad of vectors and some distinguished point from which to measure displacements. The body becomes *physical* in the Newtonian point of view when it has mass. Now characterize a *Newtonian rigid body* as a body in the sense of chapter two and a mass distribution over the points of the body. Attention is concentrated on one body in a mechanism, hence for convenience specialize the notation of chapter three to such a case. Thus simply refer to the body triad as b, rather then $b^{(L)}$, and let

$$\boldsymbol{r}^{<L} = \boldsymbol{\eta}, \quad (5.1)$$

$$\boldsymbol{r}^{L>} = \boldsymbol{\xi} \quad (5.2)$$

and

$$\boldsymbol{r}^{<L>} = \boldsymbol{r}. \quad (5.3)$$

The physical body is specified by giving its *mass distribution* $\rho(\boldsymbol{\xi})$. The total mass of a body is always positive, however there are situations in which it is useful to allow a local negative mass density. This arises in computations involving compound bodies. An example will be given in the discussion of the 'center of mass' for bodies. The

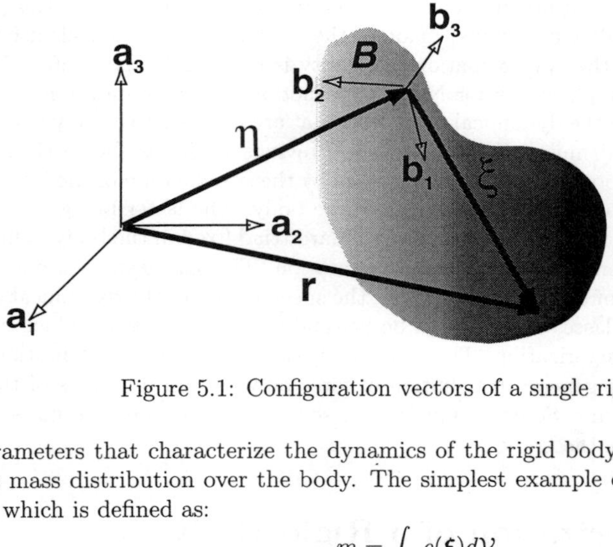

Figure 5.1: Configuration vectors of a single rigid body.

parameters that characterize the dynamics of the rigid body involve integrations of the mass distribution over the body. The simplest example of this is the *total mass* m, which is defined as:

$$m = \int_B \rho(\boldsymbol{\xi}) d\mathcal{V} \quad (5.4)$$

The integral is formally considered to be carried out over the entire range of $\boldsymbol{\xi}$, and to include the special cases where the mass in the body may be distributed over points, lines and surfaces as well as volumes. Thus point masses are included by allowing ρ to have delta functions such as $m_j\delta(\boldsymbol{\xi} - \boldsymbol{\xi}_j)$, as components. By use of these singular functions or distributions, the theory may include cases such as a massless rigid rod with point masses attached at each end or mass distributed over a two dimensional surface or a one dimensional filament.

It will soon be evident that what is required for the calculation of Newtonian motions is not the full density function, $\rho(\boldsymbol{\xi})$, but only its so called zeroth, first and second moments. The zeroth moment is given by the total mass, the other moments are defined by the integrals:

$$m_j = \int_B \rho \xi_j d\mathcal{V}, \tag{5.5}$$

$$m_{jk} = m_{kj} = \int_B \rho \xi_j \xi_k d\mathcal{V}, \tag{5.6}$$

where the subscripts on the components of $\boldsymbol{\xi}$ run from 1 to 3.

These integrals will be seen to relate to the position of the *mass center* and the *moment of inertia tensor* of the body. During a motion any position in the body is specified by its own position vector $\boldsymbol{\xi}$, which is fixed during the motion. Thus the reader should be careful to note that $\boldsymbol{\xi}$ ranges over the body in the above integrations. However any particular point in the body is described by one and the same $\boldsymbol{\xi}$ as the body moves. That is $\boldsymbol{\xi}$ *labels* the body points. Thus the above integrals will not be functions of time. They provide an invariant characterization of the bodies dynamical properties.

Forces act on the body, changing its state of motion. In the Newtonian viewpoint these force may be functions of position, velocity and time. In what follows the velocity dependence is suppressed from the notation. Again the idea of a distribution function will be invoked to describe the forces. Thus assume a *force distribution* function $\boldsymbol{f}(\boldsymbol{\xi}, t)$ which gives the force per unit volume at each point of the body. Point, line and surface distributions are included by the use of delta functions, in the same manner as was done with the mass. A consequence of the rigidity assumption will be that only certain integrals of the force distribution will be needed to determine the motion. These will be seen to be the *total force* and the *total torque about the bodies index point:*

$$\boldsymbol{R} = \int_B \boldsymbol{f} d\mathcal{V}, \tag{5.7}$$

$$\boldsymbol{T} = \int_B \boldsymbol{\xi} \times \boldsymbol{f} d\mathcal{V}. \tag{5.8}$$

It is now possible to derive suitable equations for the Newtonian motion in terms of the six quantities that determine the configuration of the rigid body. In some sense what follows can be looked upon as a demonstration that the dynamical behavior of any rigid bodies that have the same total mass, center of mass, moments of inertia, total force and total torque will be the same. Also note that nothing has been said about internal forces. In fact they are included in the above. Their elimination from consideration will be discussed after deriving the equations governing Newtonian motion.

•Illustration

It is instructive to apply the results of rigid body theory to the singular case of point mass distributions. In many cases one models actual bodies by composites made up of continuous and smooth mass distributions. Understanding how the discrete case fits into the continuous one can be helpful in the formulation of the equations for a particular model. For now, consider the purely discrete or singular case of three equal point masses, which are attached by three 'weightless' rods. The rods are attached to

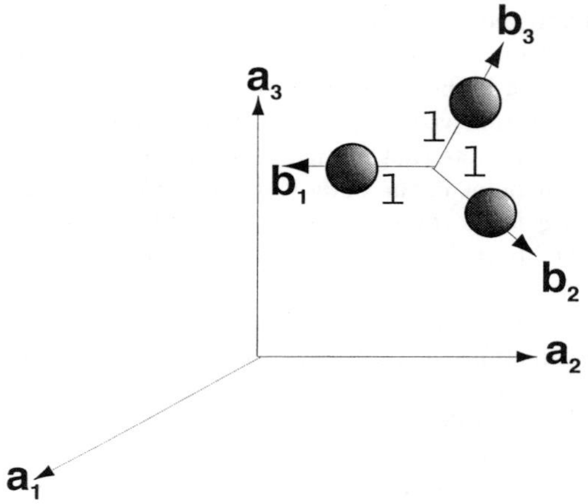

Figure 5.2: Three mass points with rigid connections.

one another at right angles from a common point, with the mass $m_0/3$ on each rod at a distance l from the attachment point. Therefore a standard triad b can be aligned with the rods. Taking the common point as origin, this forms a reference frame fixed in the body. If a position vector fixed in this frame is denoted by $\boldsymbol{\xi}$, the position of the mass points are given by $\boldsymbol{\xi} \cdot \boldsymbol{b}_i$, or with the 'coordinates', $\xi_i = \boldsymbol{\xi} \cdot \boldsymbol{b}_i$, by the three equations, $\xi_i = l$. Therefore the density distribution, defining the Newtonian dynamics, can be written as:

$$\rho = \frac{1}{3} m_0 [\delta(\xi_1 - l)\delta(\xi_2)\delta(\xi_3) + \delta(\xi_1)\delta(\xi_2 - l)\delta(\xi_3) + \delta(\xi_1)\delta(\xi_2)\delta(\xi_3 - l)], \tag{5.9}$$

where $\delta(x)$ is the delta distribution, sometimes called the delta function. For our purposes the reader only needs to know that in carrying out an integration the delta function has the 'sifting' property.

$$\int_{-\infty}^{+\infty} f(x)\delta(x - x_0)dx = f(x_0). \tag{5.10}$$

More detailed information can be found in any modern text on applied mathematics which renders this operation legitimate using the modern theory of distributions. From the definition of total mass,

$$m = \int_B \rho(\boldsymbol{\xi}) d\mathcal{V} = \int_{-\infty}^{+\infty} \int_{-\infty}^{+\infty} \int_{-\infty}^{+\infty} \rho(\boldsymbol{\xi}) d\xi_1 d\xi_2 d\xi_3 = m_0. \tag{5.11}$$

The three first order moments of this mass distribution are:

$$m_j = \int_B \xi_j \rho(\boldsymbol{\xi}) d\mathcal{V} = \frac{1}{3} m_0 l, \tag{5.12}$$

i.e. the three moments are equal to one another. The second order moments are:

$$m_{jk} = \int_B \xi_j \xi_k \rho(\boldsymbol{\xi}) d\mathcal{V} = \frac{1}{3} m_0 l^2 \delta_{jk}. \tag{5.13}$$

A force density of strength R_0/l is applied to the body along the line from the origin of the body coordinates to the point at $\xi_1 = l, \xi_2 = \xi_3 = 0$. The direction of the force is the same as that of an inertial observers $\boldsymbol{a}_3$ triad vector. To represent this force distribution use the 'box car' function:

$$B(x, x_1, x_2) = \begin{cases} 1 & \text{if } x_1 \leq x \leq x_2 \\ 0 & \text{if } x < x_1 \text{ or } x_2 < x. \end{cases} \tag{5.14}$$

The force distribution is:

$$\boldsymbol{f}(\boldsymbol{\xi}, t) = \frac{R_0}{l} \boldsymbol{a}_3 B(\xi_1, 0, l) \delta(\xi_2) \delta(\xi_3). \tag{5.15}$$

The total force is then

$$\boldsymbol{R} = R_0 \boldsymbol{a}_3, \tag{5.16}$$

and the total torque is

$$\boldsymbol{T} = \int_B \boldsymbol{\xi} \times \boldsymbol{f} d\mathcal{V} = \frac{R_0}{l} \int_0^l \xi_1 \boldsymbol{b}_1 \times \boldsymbol{a}_3 d\xi_1 = \frac{1}{2} R_0 l \boldsymbol{b}_1 \times \boldsymbol{a}_3. \tag{5.17}$$

As will soon be evident these parameters completely characterize the Newtonian behavior of the body.

5.2 Total Force and the Center of Mass

As indicated above, the vector $\boldsymbol{\eta}$ is the position of some *index point* in the rigid body relative to an inertial observer fixed at some origin. The vector $\boldsymbol{r}$ is the position of some *arbitrary point* which is *fixed* in the body. Again $\boldsymbol{r}$ is relative to the inertial observer's origin. The vector $\boldsymbol{\xi}$ is the displacement of the arbitrary point relative to the index point. It is *fixed* in the body B. The velocity of the index point will be denoted by $\boldsymbol{u}$ and the velocity of the arbitrary point by $\boldsymbol{v}$. Both these velocities are relative to the inertial observer A, i.e.

$$\frac{{}^{A}d\boldsymbol{r}}{dt} = \boldsymbol{v}, \tag{5.18}$$

$$\frac{{}^{A}d\boldsymbol{\eta}}{dt} = \boldsymbol{u}. \tag{5.19}$$

As $\boldsymbol{\xi}$ is fixed in the body,

$$\frac{{}^{B}d\boldsymbol{\xi}}{dt} = \mathbf{o}. \tag{5.20}$$

A more general form of Newton's law of motion that applies to a body with distributed mass, was introduced by Euler. Thus for a body with distributed mass:

$$\frac{{}^{A}d}{dt}\int_{B}\rho(\boldsymbol{\xi})\boldsymbol{v}d\mathcal{V} = \int_{B}\boldsymbol{f}d\mathcal{V} \tag{5.21}$$

The velocity $\boldsymbol{v}$ of any arbitrary point in the body can be expressed in terms of the velocity of the index point, $\boldsymbol{u}$, and the angular velocity of the body relative to the inertial observer, ${}^{A}\boldsymbol{\omega}^{B}$. Thus, dropping the superscripts from the latter, replace $\boldsymbol{v}$ in the equation of motion by:

$$\boldsymbol{v} = \boldsymbol{u} + \boldsymbol{\omega} \times \boldsymbol{\xi}. \tag{5.22}$$

With this replacement

$$\frac{{}^{A}d}{dt}\int_{B}\rho(\boldsymbol{\xi})\boldsymbol{u}d\mathcal{V} + \frac{{}^{A}d}{dt}\int_{B}\rho(\boldsymbol{\xi})\boldsymbol{\omega} \times \boldsymbol{\xi}d\mathcal{V} = \int_{B}\boldsymbol{f}d\mathcal{V}. \tag{5.23}$$

Now observe that $\boldsymbol{u}$, the velocity of the index point, is *not a function of* $\boldsymbol{\xi}$. The angular velocity, $\boldsymbol{\omega}$, is also independent of $\boldsymbol{\xi}$. The *total force* is defined as the integral of the force distribution over the body, i.e. as

$$\boldsymbol{R} = \int_{B}\boldsymbol{f}d\mathcal{V}. \tag{5.24}$$

Therefore using the definition of the total mass of the body as the integral of the mass distribution over the body

$$\frac{{}^{A}d}{dt}m\boldsymbol{u} + \frac{{}^{A}d}{dt}(\boldsymbol{\omega} \times \int_{B}\rho(\boldsymbol{\xi})\boldsymbol{\xi}d\mathcal{V}) = \boldsymbol{R}. \tag{5.25}$$

The second term on the left can be eliminated by proper choice of the body reference system, i.e. of the index point. The idea is to choose the index point so that the integral $\int_{B}\rho(\boldsymbol{\xi})\boldsymbol{\xi}d\mathcal{V}$, i.e., m_j vanishes. This leads to the concept of center of mass for the body. Thus given any arbitrary choice of index point, define the *center of mass* $\boldsymbol{\xi}^*$ as

$$\boldsymbol{\xi}^* = \frac{1}{m}\int_{B}\rho(\boldsymbol{\xi})\boldsymbol{\xi}d\mathcal{V}. \tag{5.26}$$

In terms of this definition the equation of motion is

$$\frac{{}^{A}d}{dt}m\boldsymbol{u} + \frac{{}^{A}d}{dt}\boldsymbol{\omega} \times m\boldsymbol{\xi}^* = \boldsymbol{R} \tag{5.27}$$

Now assume that the index point $\boldsymbol{\eta}$ is redefined so that it is at the center of mass of the body, in which case $\boldsymbol{\xi}^* = \mathbf{o}$, and the equation of motion reduces to the simpler form:

$$m\frac{^A d}{dt}\boldsymbol{u} = \boldsymbol{R}. \tag{5.28}$$

This result implies that the motion of the bodies center of mass depends only on the total force. It is interesting to note that any force distribution having a vanishing integral over the body will have *no effect on the motion of the center of mass* of the body. This must be considered to be a *result of the constraint of rigid motion.* Now make the fundamental assumption that the internal forces that hold the rigid body together must be such that their total integral vanishes, i.e. let $\boldsymbol{f}'(\boldsymbol{\xi})$ represent the internal forces that the particles of the body exert on one another. It is reasonable to require that the internal constraint forces do not cause any motion of the mass center. This implies that

$$\int_B \boldsymbol{f}'(\boldsymbol{\xi})d\mathcal{V} = \mathbf{o}. \tag{5.29}$$

In basic treatments of mechanics it has become popular to derive the equations of motion for a rigid body by summing over a large (infinite?) system of particles. In doing this it is usually assumed that the pair-wise forces between particles obey the so called third law of Newton, i.e. they are equal and opposite. It is also usually required that the force is alligned with the interparticle displacement vector. While this does lead to the correct equations of motion it fails to emphasize that the nature of the internal forces must be a result of the constraint of assuming a rigid body! As far as the motion of the center of mass is concerned, the important requirement for the internal forces is that they sum to zero. For the center of mass motion, any two *applied* force distributions are *equivalent* if they have the same integral over the body.

•Illustration

The center of mass of the three point mass rigid body, discussed at the end of the last section, is located at the point with coordinates:

$$\xi_j = \frac{1}{m}m_j = \frac{1}{3}l, \tag{5.30}$$

so that the displacement vector from the given origin to the center of mass point is:

$$\boldsymbol{\mu} = \frac{1}{3}l(\boldsymbol{b}_1 + \boldsymbol{b}_2 + \boldsymbol{b}_3). \tag{5.31}$$

The displacement of an arbitrary body point in the center of mass frame, denoted for the moment by $\boldsymbol{\xi}'$ is

$$\boldsymbol{\xi}' = \boldsymbol{\xi} - \frac{1}{3}l(\boldsymbol{b}_1 + \boldsymbol{b}_2 + \boldsymbol{b}_3), \tag{5.32}$$

so that the mass point at $\xi_1 = l$ in the new frame will be located at

$$\boldsymbol{\xi}' = \frac{2}{3}l\boldsymbol{b}_1 - \frac{1}{3}l(\boldsymbol{b}_2 + \boldsymbol{b}_3). \tag{5.33}$$

There may be some advantage in rotating to a new triad in which two of the triad vectors are parallel to the plane defined by the three mass points. The total applied force calculation will not be effected by the change in frame origin, however the total applied torque depends on the chosen origin, thus if $\boldsymbol{T}'$ denotes the torque calculated about the center of mass origin,

$$\begin{aligned} \boldsymbol{T}' &= \int_B \boldsymbol{\xi}' \times \boldsymbol{f} d\mathcal{V} \\ &= \int_B (\boldsymbol{\xi} \times \boldsymbol{f} - \boldsymbol{\mu} \times \boldsymbol{f} d\mathcal{V} \\ &= \boldsymbol{T} - \boldsymbol{\xi}^* \times \boldsymbol{R}. \end{aligned} \tag{5.34}$$

For the three particle configuration this gives

$$\boldsymbol{T}' = \frac{1}{6} R_0 l \boldsymbol{b}_1 \times \boldsymbol{a}_3 - \frac{1}{3} R_0 l (\boldsymbol{b}_2 + \boldsymbol{b}_3) \times \boldsymbol{a}_3. \tag{5.35}$$

The new coordinate system, or one obtained by rotation to take advantage of some symmetry of the configuration, is also fixed in the body, hence body positions referred to it will not change if the body position and orientation vary. The first moments of the mass distribution in the new system will vanish, indeed this was the motivation for the change. The second moments, m_{ij}, will also change but will not in general vanish. The way the second moments behave under displacement and rotation of the frame is discussed at some length in the remainder of this chapter.

•Illustration

Problems in technical mechanics frequently involve complex objects which can be build up from relatively simple bodies. The compound body object shown in figure 5.3 provides an example of this type of situation. The center of mass is most easily located by a two step process. First the center of mass of each component body is located, along with its mass. These can then be considered as a collection of point mass objects. The center of mass of this collection is then located. Using this approach it is also possible to use component elements to represent holes or voids. Thus to find the center of mass of a plate with a hole of a given radius, simply treat a compound body made up of a plate and a disk. The mass of the disk would be taken as negative with a value calculated according to its area relative to the plate. In the example of the figure it is assumed the object consists of 4 square plates of side L and a solid rod of length L and that each of these bodies has the mass m. It is also assumed that there are no holes in the bodies and that each object has a uniform mass distribution.

There is also a Sophia operator for calculating the position of the mass center of a set of mass points relative to a given reference frame. It takes two arguments, the left hand argument being the name of the answer frame and the right hand argument being a list. This mass point list is itself a list of lists. The component lists are made up of two entries, the mass value and the Evector position. For the body shown in

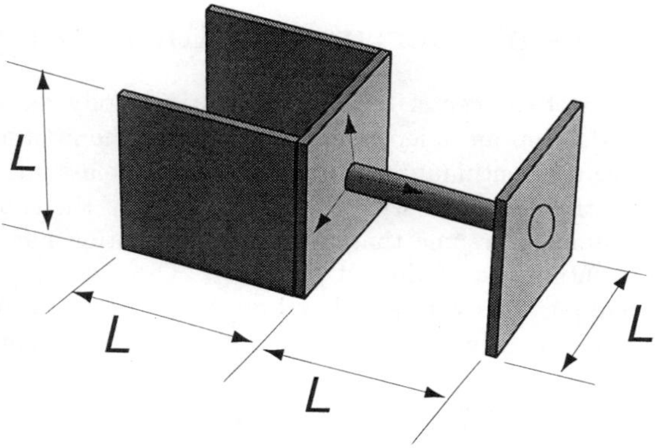

Figure 5.3: Compound Body

the figure take the reference point as located at the rod connection point to the three plate object. Also choose a frame, B, in which $\boldsymbol{b}_1$ is aligned with the connecting rod. The vectors $\boldsymbol{b}_2$ and $\boldsymbol{b}_3$ are aligned with the sides of the attached plate. With the assumptions made above the location of the mass center of each component body is evident. These are given by the following Evectors:

```
> r1 := B &ev [-L/2,0,L/2]:
> r2 := B &ev [0,0,0]:
> r3 := B &ev [-L/2,0,-L/2]:
> r4 := B &ev [L/2,0,0]:
> r5 := B &ev [L,0,0]:
```

Because of the way Sophia deals with frame transformations it is necessary to introduce at least two frames into a problem. Therefore in order to use the mass center operator use the statement:

```
> &rot [A,B,3,0]:
```

The center of mass list is formed using the fact that all the objects have equal mass, thus:

```
> rcm := B &cm [[m,r1],[m,r2],[m,r3],[m,r4],[m,r5]];
```

This gives the result:

$$[[\frac{L}{10}, 0, 0], B]$$

Sometimes it is convenient to have Sophia calculate the total mass. In such a case use the form

```
> massPointCenter( [[m,r1],[m,r2],[m,r3],[m,r4],[m,r5]],B);
```

which returns a two component list giving the mass center Evector and the total mass, i.e.

$$[[[\frac{L}{10}, 0, 0], B], 5\,m]$$

5.3 Moments of Inertia and Total Torque

Now that the motion of the center of mass of the rigid body has been dealt with, the next step is to develop an independent equation for the attitude motion of the body. Again we take the continuum point of view where we assume a distribution of mass and force. To this end start with an integral form of the equation of motion, due to Euler, where again assume that the force distribution includes the internal constraints that maintain the rigidity of the body. The picture of the body being composed of discrete particles with equal and opposite forces of constraint, which is popular in many modern texts, is replaced by the use of Euler's equation. Thus our starting point is

$$\frac{{}^A d}{dt}\int_B \boldsymbol{r} \times \rho \boldsymbol{v} d\mathcal{V} = \int_B \boldsymbol{r} \times \boldsymbol{f} d\mathcal{V} \tag{5.36}$$

where the explicit dependence of ρ and $\boldsymbol{f}$ on $\boldsymbol{\xi}$ has been suppressed. For a rigid body the velocity of any arbitrary point may be expressed in terms of the velocity of an index point and the angular velocity, i.e. that $\boldsymbol{v} = \boldsymbol{u} + \boldsymbol{\omega} \times \boldsymbol{\xi}$. Using this fact and noting that $\boldsymbol{r} = \boldsymbol{\eta} + \boldsymbol{\xi}$, equation 5.36 becomes:

$$\frac{{}^A d}{dt}\int_B \rho(\boldsymbol{\eta} + \boldsymbol{\xi}) \times (\boldsymbol{u} + \boldsymbol{\omega} \times \boldsymbol{\xi}) d\mathcal{V} = \int_B (\boldsymbol{\eta} + \boldsymbol{\xi}) \times \boldsymbol{f} d\mathcal{V}. \tag{5.37}$$

Note that $\boldsymbol{\eta}, \boldsymbol{u}$ and $\boldsymbol{\omega}$ are independent of $\boldsymbol{\xi}$. Taking advantage of this rewrite the left side of the equation as the sum of the following four terms:

$$e_1 = \frac{{}^A d}{dt}\boldsymbol{\eta} \times (\int_B \rho d\mathcal{V})\boldsymbol{u}, \tag{5.38}$$

$$e_2 = \frac{{}^A d}{dt}\boldsymbol{\eta} \times (\boldsymbol{\omega} \times \int_B \rho \boldsymbol{\xi} d\mathcal{V}), \tag{5.39}$$

$$e_3 = \frac{{}^A d}{dt}\int_B \rho \boldsymbol{\xi} d\mathcal{V} \times \boldsymbol{u}, \tag{5.40}$$

$$e_4 = \frac{{}^A d}{dt}\int_B \rho \boldsymbol{\xi} \times (\boldsymbol{\omega} \times \boldsymbol{\xi}) d\mathcal{V} \tag{5.41}$$

The right side of the equation gives two terms,

$$e_5 = \boldsymbol{\eta} \times \int_B \boldsymbol{f} d\mathcal{V}, \tag{5.42}$$

$$e_6 = \int_B \boldsymbol{\xi} \times \boldsymbol{f} d\mathcal{V}. \tag{5.43}$$

First consider term e_1, where the integral is the total mass, so that

$$e_1 = \frac{{}^A d\boldsymbol{\eta}}{dt} \times m\boldsymbol{u} + \boldsymbol{\eta} \times \frac{{}^A d m\boldsymbol{u}}{dt}. \tag{5.44}$$

The first term on the right vanishes as $\frac{{}^A d\boldsymbol{\eta}}{dt} = \boldsymbol{u}$ and we are left with the cross product of collinear vectors which is the null vector. Now use the result for the center of mass motion derived in the previous section to see that the second term on the right is same as e_5. Hence term e_1 and e_5 cancel each other out. The integrals in both terms e_2 and e_3 vanish when the index point is considered to be at the center of mass of the body. Hence only terms e_4 and e_6 remain. Recall that the *total torque* about the center of mass was defined as:

$$\boldsymbol{T} = \int_B \boldsymbol{\xi} \times \boldsymbol{f} d\mathcal{V}. \tag{5.45}$$

This quantity clearly depends on the choice of index point. As it is useful to consider various choices of index point this should be reflected in the notation. This will be done in the discussion of equivalent force systems. With this definition of the total torque the equation for the attitude motion is:

$$\tfrac{{}^A d}{dt} \int_B \rho \boldsymbol{\xi} \times (\boldsymbol{\omega} \times \boldsymbol{\xi}) d\mathcal{V} = \boldsymbol{T}. \tag{5.46}$$

Equations 5.28 and 5.46 thus appear to provide the means to separately calculate the motion of the center of mass and the attitude of the body. This is not actually the case as the total torque and total force will in general depend on the bodies position and attitude, leading to a coupling between the equations for the motion of the mass center and attitude.

The form of equation 5.46 leaves much to be desired. The angular velocity term, which depends on time, appears to be trapped under the integral in the cross product term. The terms that do depend on $\boldsymbol{\xi}$ represent a property of the body which should be determined once and for all. The key observation to make is that the expression is *linear* in $\boldsymbol{\omega}$, hence it is possible to write the integral as a dyad dotted into $\boldsymbol{\omega}$, where the dyad is only a property of the bodies geometry and mass distribution. The dyad in question represents the moments of inertia of the body. To sort this matter out it is convenient to rewrite the term under the integral in 5.46 so as to obtain a dyadic operation in some explicit form. Manipulations involving the triple product come up frequently in mechanics and it is often helpful to have available the result that for any three arbitrary vectors, $\boldsymbol{a}, \boldsymbol{b}$ and $\boldsymbol{c}$,

$$\boldsymbol{a} \times (\boldsymbol{b} \times \boldsymbol{c}) = \boldsymbol{b}(\boldsymbol{a} \cdot \boldsymbol{c}) - \boldsymbol{c}(\boldsymbol{a} \cdot \boldsymbol{b}), \tag{5.47}$$

which can also be put into the dyadic form:

$$\boldsymbol{a} \times (\boldsymbol{b} \times \boldsymbol{c}) = (\boldsymbol{bc} - \boldsymbol{cb}) \cdot \boldsymbol{a}. \tag{5.48}$$

Application of this result to the above integrand yields:

$$\rho \boldsymbol{\xi} \times (\boldsymbol{\omega} \times \boldsymbol{\xi}) = \rho(\boldsymbol{\omega}(\boldsymbol{\xi} \cdot \boldsymbol{\xi}) - \boldsymbol{\xi}(\boldsymbol{\xi} \cdot \boldsymbol{\omega})). \tag{5.49}$$

To obtain a dyadic expression recall that the unit dyad, $\boldsymbol{U}$ operating on a vector yields that vector. Use the notation that $\boldsymbol{\xi} \cdot \boldsymbol{\xi} = \xi^2$. Thus write the above as

$$\rho(\xi^2 \boldsymbol{U} - \boldsymbol{\xi}\boldsymbol{\xi}) \cdot \boldsymbol{\omega} \tag{5.50}$$

hence equation 5.46 can be cast in the form:

$$\tfrac{^A d}{dt}[(\int_B \rho(\xi^2 \boldsymbol{U} - \boldsymbol{\xi}\boldsymbol{\xi}) d\mathcal{V}) \cdot \boldsymbol{\omega}] = \boldsymbol{T}, \tag{5.51}$$

which leads to the definition of the *moment of inertia dyad* as:

$$\boldsymbol{I} = \int_B \rho(\xi^2 \boldsymbol{U} - \boldsymbol{\xi}\boldsymbol{\xi}) d\mathcal{V}. \tag{5.52}$$

The attitude equation of motion now takes the form:

$$\tfrac{^A d}{dt} \boldsymbol{I} \cdot \boldsymbol{\omega} = \boldsymbol{T}. \tag{5.53}$$

The observation regarding the total torque, i.e. that it depends on the index point used, also pertains to the moment of inertia dyad, and again it is in many instances desirable that the notation used reflect this fact.

An example of this occurs when one point of the rigid body is fixed in the inertial frame of reference. Instead of the center of mass as index point use the fixed point as the index point, in which case the velocity $\boldsymbol{u} = \mathbf{o}$, and the first equation of motion gives the relationship:

$$\tfrac{^A d}{dt}(\boldsymbol{\omega} \times m\boldsymbol{\xi}^*) = \boldsymbol{R}. \tag{5.54}$$

Now consider the various terms in the second equation of motion. The terms e_1 and e_3 clearly vanish. Use of the above result from the first equation shows that

$$e_2 = e_5 = \boldsymbol{\eta} \times \boldsymbol{R}. \tag{5.55}$$

Therefore only e_4 and e_6 remain, so that the resulting equation, with notation that indicates the involved bodies and the reference point for the moment of inertia and the torque, is

$$\tfrac{^A d}{dt} \boldsymbol{I}^{B/\mathcal{O}} \cdot \boldsymbol{\omega} = \boldsymbol{T}^{B/\mathcal{O}}. \tag{5.56}$$

This is the same form as obtained for the center of mass case, except the moment of inertia and total torque of the body B is computed with respect to the fixed point $\mathcal{O}$. The force $\boldsymbol{R}$ must be interpreted as the force exerted on the body by the constraint which maintains the fixed point in its spatial location. Thus in this special case the center of mass does not provide the most useful form for the equations of motion.

The discussion of the equation of motion of the center of mass noted that any force system with a vanishing integral over the body would not effect the bodies center of mass motion. Equation 5.53 implies that any force system with a vanishing torque about the center of mass will not influence the attitude motion. Thus the total torque of any internal force system which serves to maintain the rigid bodies shape during a motion must also vanish. This is a *consequence* of the rigid body constraint and Euler's equation of motion. Many authors derive this fact as a consequence of Newton's third law of action and reaction in which the constraint forces between the particles comprising the body are assumed to act along a line connecting them and are equal and opposite. The latter condition leads to the satisfaction of the vanishing total torque requirement. This simply provides an example of how the condition, that the total force and torque of the distributed constraint forces vanish, may be met. Euler's equations thus imply Newton's laws. The converse is not true. These considerations provide the motivation for the discussion of *equivalent force systems* given in the following section.

For later reference, the forms taken by the equations are listed in the notation introduced in the last chapter for multiple body mechanisms:

$$\frac{^A d}{dt} m^{(L)} \boldsymbol{v}^{<L} = \boldsymbol{R}^{(L)} \tag{5.57}$$

$$\frac{^A d}{dt} \boldsymbol{I}^{(L)} \cdot \boldsymbol{\omega}^{(L)} = \boldsymbol{T}^{(L)} \tag{5.58}$$

$$\boldsymbol{R}^{(L)} = \int_{(L)} \boldsymbol{f}^{(L)} d\mathcal{V}_{(L)} \tag{5.59}$$

$$\boldsymbol{T}^{(L)} = \int_{(L)} \boldsymbol{r}^{L>} \times \boldsymbol{f}^{(L)} d\mathcal{V}_{(L)} \tag{5.60}$$

$$m^{(L)} = \int_{(L)} \rho d\mathcal{V}_{(L)} \tag{5.61}$$

$$\boldsymbol{I}^{(L)} = \int_{(L)} \rho^{(L)}((\boldsymbol{r}^{L>})^2 \boldsymbol{U} - \boldsymbol{r}^{L>}\boldsymbol{r}^{L>}) d\mathcal{V}_{(L)}. \tag{5.62}$$

It is of course assumed that $\boldsymbol{r}^{<L}$ is a vector from the inertial observers origin to the center of mass of the Lth body, i.e. that:

$$\int_{(L)} \rho^{(L)} \boldsymbol{r}^{L>} d\mathcal{V}_{(L)} = \mathbf{o}. \tag{5.63}$$

Following the notation of chapter four construct the K column equivalents for these terms for a system of rigid bodies by simply dropping the L from the notation, e.g. $\boldsymbol{I}^{()}$ represents a K column of moment of inertia dyads for the K rigid bodies comprising the system. In such a treatment the particles of the last chapter can be considered as degenerate cases of the full rigid body.

•Illustration

The calculation of the moment of inertia dyad is illustrated by continuing with the three point mass example of the last two sections. Recall that the mass center was located at the position

$$\boldsymbol{\xi}^* = \frac{l}{3}(\boldsymbol{b}_1 + \boldsymbol{b}_2 + \boldsymbol{b}_3). \tag{5.64}$$

In addition to moving the index point of the body frame to the center of mass it may also be desirable to change to a new orientation of the frame's triad. In the present example one might expect that the moment of inertia dyad will be in its 'simplest' form if a symmetrical choice of origin and orientation is made. Based on this, seek a triad which has two of its vectors in the plane of the three point configuration, the first triad vector aligned with the mass at $l\boldsymbol{b}_1$ to the mass at $l\boldsymbol{b}_2$, the second aligned with a line from the center point between the first two masses to the third. The third vector can be taken as the cross product of the first two, hence it will be orthogonal to the plane of the three masses. The vector $\boldsymbol{\xi}_2 - \boldsymbol{\xi}_1 = l(\boldsymbol{b}_2 - \boldsymbol{b}_1)$ points between the first two mass points. Divide this vector by its magnitude to obtain the first of the new triad vectors:

$$\boldsymbol{b}_1{}' = \frac{1}{\sqrt{2}}(\boldsymbol{b}_2 - \boldsymbol{b}_1). \tag{5.65}$$

The vector $\boldsymbol{\xi}_3 - \boldsymbol{\xi}_1$ is also in the plane of the configuration. Subtract the projection of this vector onto the vector $\boldsymbol{b}_1{}'$ from itself and divide by the length of the result to obtain:

$$\boldsymbol{b}_2{}' = \sqrt{\frac{1}{6}}(-\boldsymbol{b}_1 - \boldsymbol{b}_2 + 2\boldsymbol{b}_3). \tag{5.66}$$

The third of the new triad vectors is then

$$\boldsymbol{b}_3{}' = \boldsymbol{b}_1{}' \times \boldsymbol{b}_2{}' = \frac{1}{\sqrt{3}}(\boldsymbol{b}_1 + \boldsymbol{b}_2 + \boldsymbol{b}_3). \tag{5.67}$$

The direction cosine matrix is therefore:

$$R_{b'b} = \begin{bmatrix} -\frac{1}{\sqrt{2}} & \frac{1}{\sqrt{2}} & 0 \\ -\sqrt{\frac{1}{6}} & -\sqrt{6} & \sqrt{\frac{2}{3}} \\ \frac{1}{\sqrt{3}} & \frac{1}{\sqrt{3}} & \frac{1}{\sqrt{3}} \end{bmatrix}. \tag{5.68}$$

The only contributions to the integral for the moment of inertia dyad will come from the mass points, therefore the dyads $\boldsymbol{\xi}'^2\boldsymbol{U} - \boldsymbol{\xi}'_i\boldsymbol{\xi}'_i$ must be evaluated in terms of the triad b', where $\boldsymbol{\xi}'_i$ are the position vectors of the mass points in the new frame. The inertia dyad will then consist of the sum of these three terms multiplied by $m_0/3$. The coordinates of the mass points in the center of mass frame with the new orientation are given by terms of the form $(\boldsymbol{\xi}_i - \boldsymbol{\xi}^*) \cdot \boldsymbol{b}_j{}'$. By direct computation using such forms or by geometrical reasoning, one finds:

$$\boldsymbol{\xi}'_1 = -\frac{1}{\sqrt{2}} l\boldsymbol{b}_1{}' - \frac{1}{\sqrt{6}} l\boldsymbol{b}_2{}', \tag{5.69}$$

$$\boldsymbol{\xi}'_2 = +\frac{1}{\sqrt{2}} l\boldsymbol{b}_1{}' - \frac{1}{\sqrt{6}} l\boldsymbol{b}_2{}' \tag{5.70}$$

$$\boldsymbol{\xi}'_3 = \frac{2}{\sqrt{6}} \boldsymbol{b}_2{}'. \tag{5.71}$$

Carrying out the indicated operations the moment of inertia dyad is then found to be

$$\boldsymbol{I} = \frac{1}{3} l^2 m_0 (\boldsymbol{b}_1{}'\boldsymbol{b}_1{}' + \boldsymbol{b}_2{}'\boldsymbol{b}_2{}' + 2\boldsymbol{b}_3{}'\boldsymbol{b}_3{}'). \tag{5.72}$$

Thus in the representation b' the moment of inertia dyad has the matrix form

$$^{b'}I = \frac{m_0 l^2}{3} \begin{bmatrix} 1 & 0 & 0 \\ 0 & 1 & 0 \\ 0 & 0 & 2 \end{bmatrix}. \tag{5.73}$$

Such diagonal representations can lead to simplified forms of equations for Newtonian motions, therefore it is worth making the effort to see when and how one can find triads which lead to diagonal forms of the inertia dyad. The problem of computing the total applied force and torque relative to the new system is left as an exercise for the reader.

5.4 Equivalent Force Systems

Some of the forces acting on a rigid body arise from internal constraints, that is from the bodies attempt to maintain its rigidity. The physical concept of a rigid body is such that motion would not be expected to arise from these internal forces. Therefore the very idea of rigidity, which of course is only an approximation to the bodies internal structure, requires that the center of mass and attitude motion be uneffected by the internal constraining forces. From the work of the last two sections it should be clear that if such an internal force distribution is given by $\boldsymbol{f}'$, then:

$$\int_B \boldsymbol{f}' d\mathcal{V} = \mathbf{o}, \tag{5.74}$$

$$\int_B \boldsymbol{\xi} \times \boldsymbol{f}' d\mathcal{V} = \mathbf{o}. \tag{5.75}$$

Any force distribution which satisfies the above conditions is called a *Null Force System*. It is taken as a basic assumption of rigid body mechanics that internal constraint forces are null force systems. External forces will also consist of constraints arising from other bodies and applied forces. These can also be null systems. As an example consider a ball with loads applied to a number of surface points, as shown in Figure 5.4. It is assumed that each force has a counterpart at the opposite pole of the ball. The center of mass is taken at the geometric center of the ball. This is clearly a null force system, and the result that it will not effect the motion of a *rigid ball* clearly accords with our intuition and expectations of how a *rigid* body will behave. Quite different results would be expected for a deformable body!

Based on these considerations for rigid body systems, any force systems which differ by a null force system are equivalent, i.e. lead to the same Newtonian motion. For this reason a special theory has been developed for force distributions that only

Figure 5.4: Null force system acting on a rigid body.

differ from one another by a null force system. The accepted terminology is that *force systems that differ by null force systems are called equipollent.* Therefore two force systems, $\boldsymbol{f}_1$ and $\boldsymbol{f}_2$ are equipollent if:

$$\int_B \boldsymbol{f}_1 d\mathcal{V} = \int_B \boldsymbol{f}_2 d\mathcal{V} \tag{5.76}$$

$$\int_B \boldsymbol{\xi} \times \boldsymbol{f}_1 d\mathcal{V} = \int_B \boldsymbol{\xi} \times \boldsymbol{f}_2 d\mathcal{V}. \tag{5.77}$$

For this definition to be useful, equipollence should not depend on the base point used in the computation of total torque. Shift the base point from $\mathcal{P}$ to $\mathcal{P}'$, so that if $\boldsymbol{\mu}$ is a displacement vector running from $\mathcal{P}$ to $\mathcal{P}'$ we have $\boldsymbol{\xi} = \boldsymbol{\xi}' + \boldsymbol{\mu}$. Therefore rewrite the condition for equipollence as:

$$\int_B (\boldsymbol{\xi}' + \boldsymbol{\mu}) \times \boldsymbol{f}_1 d\mathcal{V} = \int_B (\boldsymbol{\xi}' + \boldsymbol{\mu}) \times \boldsymbol{f}_2 d\mathcal{V} \tag{5.78}$$

so that using the fact that the shift in base point, $\boldsymbol{\mu}$ is independent of the integration variable:

$$\int_B \boldsymbol{\xi}' \times \boldsymbol{f}_1 d\mathcal{V} - \int_B \boldsymbol{\xi}' \times \boldsymbol{f}_2 d\mathcal{V} = \boldsymbol{\mu} \times \int_B (\boldsymbol{f}_2 - \boldsymbol{f}_1) d\mathcal{V}. \tag{5.79}$$

If the force systems are equipollent, the last integral must vanish, hence the concept of equipollence is independent of the base point used to compute the total torque. What is important is that the *same* base point be used for any force systems that are

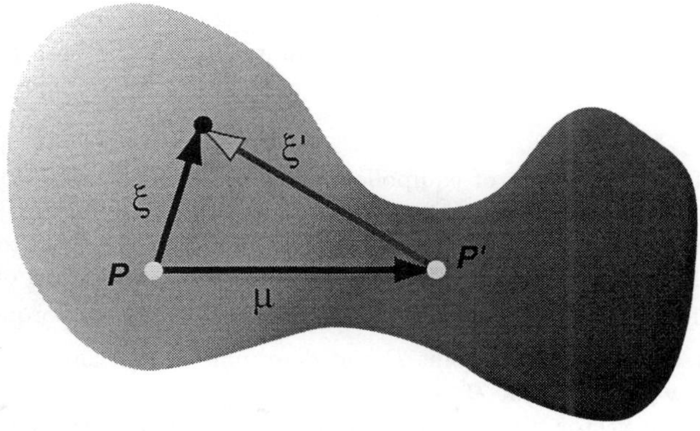

Figure 5.5: Displacement of reference point for an equipollent force distribution.

to be compared! That is the total torque of any force distribution will depend on the base point. Thus with the vector $\boldsymbol{\mu}$ still describing the shift from $\mathcal{P}$ to $\mathcal{P}'$:

$$\int_B \boldsymbol{\xi} \times \boldsymbol{f} d\mathcal{V} = \int_B \boldsymbol{\xi}' \times \boldsymbol{f} d\mathcal{V} + \boldsymbol{\mu} \times \int_B \boldsymbol{f} d\mathcal{V}, \tag{5.80}$$

which expresses the relationship between the total torque calculated at two different base points. To reflect this dependence on base point in our notation write $\boldsymbol{T}^{/\mathcal{P}}$ for the torque computed with $\mathcal{P}$ as base point. To emphasize the identity of the body involved in the calculation one writes this as $\boldsymbol{T}^{B/\mathcal{P}}$. In terms of this notation the last equation can be expressed as:

$$\boldsymbol{T}^{/\mathcal{P}} = \boldsymbol{T}^{/\mathcal{P}'} + \boldsymbol{\mu} \times \boldsymbol{R}. \tag{5.81}$$

For the purpose of recalling this result note that the displacement vector $\boldsymbol{\mu}$ runs from the point on the left side of the equation, $\mathcal{P}$ to the base point used on the right side, $\mathcal{P}'$.

There are two special cases in which the total torque is not influenced by a change of base point. The first is when the total force vanishes, i.e. when $\boldsymbol{R} = \mathbf{o}$. In this case a change in base point does not change the total torque. Such a force distribution is called a *couple* as it is equipollent with any system consisting of two opposed forces which have the same total torque as the given system. Sometimes the total torque is referred to as the torque or moment of the couple.

The other case of interest is when the displacement is parallel to the total force. As in the previous case the torque does not change. Thus any point on a line parallel to the resultant force and passing through a given base point will yield the same torque as that calculated about that base point. While the resultant force is really a free vector it is convenient when discussing equipollent systems to think of it as being attached to the base point used to evaluate the torque of the force distribution. The line with the same direction as this vector passing through the base point is called the *line of action* of the force distribution.

•Illustration

To illustrate some aspects of equipollent force systems consider the problem of calculating a particular representation of equipollent systems called a *wrench*. The idea here is to shift the base point to a location such that the total torque will be aligned with the force. Thus the term wrench which invokes the idea of a 'twisting' and 'turning' force acting on the rigid body. Start with a given torque, $\boldsymbol{T}^{/\mathcal{P}}$, and resultant $\boldsymbol{R}$ at the known base point $\mathcal{P}$ and determine a new torque $\boldsymbol{T}^{/\mathcal{P}'}$ and the location of a new base point at $\mathcal{P}'$, such that the torque is parallel to the total force. Therefore find the vector displacement, $\boldsymbol{\mu}$, to the location of the new base point which accomplishes the goal of aligning the torque and force. The condition that the new torque aligns itself with the resultant force can be stated in terms of a to be determined strength parameter λ. Therefore because of the presumed alignment:

$$\boldsymbol{T}^{/\mathcal{P}'} = \lambda \boldsymbol{R}. \tag{5.82}$$

The quantity λ is known as the *pitch* of the wrench. Using the relation between torques of a force distribution calculated in terms of different base points, and substituting the above expression for the new torque, gives

$$\boldsymbol{T}^{/\mathcal{P}} = \boldsymbol{\mu} \times \boldsymbol{R} + \lambda \boldsymbol{R}. \tag{5.83}$$

Observe that shifting the base point along the line of action of $\boldsymbol{R}$ will not change the torque. This means that all we can expect to determine is the shift in position of the base point in a plane perpendicular to $\boldsymbol{R}$. Thus we must seek $\boldsymbol{\mu}$ as a *two dimensional vector in a plane perpendicular to the total force.* Another way of expressing this is to note that $\boldsymbol{\mu} \cdot \boldsymbol{R} = \mathbf{o}$. Solve for λ by dotting the above equation for the torque shift with $\boldsymbol{R}$. The properties of the vector triple product then show that the first term on the right will vanish and we will be left with the result that:

$$\boldsymbol{T}^{/\mathcal{P}} \cdot \boldsymbol{R} = \lambda \boldsymbol{R} \cdot \boldsymbol{R}. \tag{5.84}$$

Referring to the magnitude of $\boldsymbol{R}$ as R, which is also called the *intensity* of the wrench:

$$\lambda = \frac{1}{R^2} \boldsymbol{R} \cdot \boldsymbol{T}^{/\mathcal{P}}. \tag{5.85}$$

Introduce a unit vector, $\boldsymbol{e}$ along the direction of $\boldsymbol{R}$ so that:

$$\boldsymbol{T}^{/\mathcal{P}'} = (\boldsymbol{T}^{/\mathcal{P}} \cdot \boldsymbol{e})\boldsymbol{e}, \tag{5.86}$$

that is that the torque of the wrench is simply the projection of the torque at the initial base point onto the direction of the total force. Substituting this result into the equation relating the torques shows that:

$$\boldsymbol{T}^{/\mathcal{P}} = \boldsymbol{\mu} \times \boldsymbol{R} + (\boldsymbol{T}^{/\mathcal{P}} \cdot \boldsymbol{e})\boldsymbol{e}. \tag{5.87}$$

Now as $\boldsymbol{\mu} \times \boldsymbol{R}$ is orthogonal to both $\boldsymbol{\mu}$ and $\boldsymbol{R}$, the vector $(\boldsymbol{\mu} \times \boldsymbol{R}) \times \boldsymbol{R}$ will be parallel to and opposite in direction to $\boldsymbol{\mu}$. Thus the displacement to the line of action of the wrench is

$$(\boldsymbol{\mu} \times \boldsymbol{R}) \times \boldsymbol{R} = -R^2\boldsymbol{\mu}. \tag{5.88}$$

Therefore by taking the cross product of the torque equation with $\boldsymbol{R}$ and using the above result determine that the displacement vector to the base position where the torque is parallel to the total force is given by:

$$\boldsymbol{\mu} = \frac{1}{R}\boldsymbol{e} \times \boldsymbol{T}^{/\mathcal{P}}. \tag{5.89}$$

Therefore the pitch of the wrench, λ and the position of the wrenche's line of action, $\boldsymbol{\mu}$, are obtainable by respectively dotting and crossing the vector $\boldsymbol{e}/R$ into the total torque. •

In many treatments of mechanics the idea of equipollent force systems is introduced in an apriori manner, with little or no motivation as to why such force systems are of interest. The above discussion is meant to motivate the introduction of the concept of equipollence by showing that it is a consequence of the assumption of the idea of a rigid body and the Newtonian laws of motion. Thus for a rigid body only the total force and torque of any distributed force comes into the calculation of a Newtonian motion. Any null force system, i.e. one for which the integral of the local force and torque density over the body vanish, will contribute nothing to the motion. In addition the imposition of the constraint of rigidity implies that the internal forces that maintain the form of the rigid body must be a null force system. The common derivation of rigid body equations of motion by summing over individual particles usually includes the assumption that each particle pair interacts with equal and opposite forces along the line that joins them i.e. assumes Newton's third law. Instead of this Euler's integral form of the equations of motion has been used as the basic postulate!

5.5 The Moment of Inertia Dyad

The dynamical behavior of a rigid body is governed by its mass distribution and geometry in the form of several integrated quantities. One of these, the moment of inertia dyad calls for some more detailed consideration. The determination of the moment of inertia of a rigid body is often simplified by the use of some of the derived properties of this dyad. In addition it is useful to have a clear physical understanding in order to improve ones ability to formulate rigid body dynamics problems in as forthright manner as possible. Our starting point is equation 5.52, which expresses the moment of inertia dyad as an integral over the body in question, i.e. an integration over the position vector $\boldsymbol{\xi}$ which gives the displacement between an arbitrary point in the body and the center of mass. To understand the meaning of this integral it is helpful to appreciate the significance of the dyadic part of the integrand, $\boldsymbol{\xi}\cdot\boldsymbol{\xi}\boldsymbol{U}-\boldsymbol{\xi}\boldsymbol{\xi}$. This is a symmetric dyad, as the unit dyad, $\boldsymbol{U}$ and $\boldsymbol{\xi}\boldsymbol{\xi}$ are both symmetric.Therefore the moment of inertia dyad itself is also symmetric. Also we see a dependence on both the particular body and the fact that the vector $\boldsymbol{\xi}$ is taken as the displacement from the center of mass to the integration point in the body. As it is of interest to compute the moment of inertia dyad about other points and as we will be dealing with multi-body systems, it is useful to introduce a notation that reflects this. Therefore we will write $\boldsymbol{I}^{B/\mathcal{P}^*}$ for the moment of inertia dyad of body B about point $\mathcal{P}^*$, where the '*' indicates that the point $\mathcal{P}^*$ is at the center of mass of the body. To appreciate the meaning of the dyadic integrand, let $\boldsymbol{\xi}_1$ and $\boldsymbol{\xi}_2$ denote two orthogonal unit vectors chosen so that their cross product is in the direction of the vector $\boldsymbol{\xi}$. Thus if $\boldsymbol{\xi}_3$ is a unit vector having the direction of $\boldsymbol{\xi}$, then $\boldsymbol{\xi}_1\times\boldsymbol{\xi}_2=\boldsymbol{\xi}_3$. Then using the properties of the unit dyad:

$$\boldsymbol{\xi}\cdot\boldsymbol{\xi}\boldsymbol{U}-\boldsymbol{\xi}\boldsymbol{\xi}=\xi^2(\boldsymbol{\xi}_1\boldsymbol{\xi}_1+\boldsymbol{\xi}_2\boldsymbol{\xi}_2), \tag{5.90}$$

which shows that the dyad in parentheses is a *projection operator* that projects a vector into a vector parallel to the plane orthogonal to $\boldsymbol{\xi}$. Therefore define the orthogonal projection operator:

$$\boldsymbol{\Pi}_\xi=\boldsymbol{U}-\frac{1}{\xi^2}\boldsymbol{\xi}\boldsymbol{\xi}. \tag{5.91}$$

The moment of inertia dyad can now be written as:

$$\boldsymbol{I}^{B/\mathcal{P}^*}=\int_B\rho\xi^2\boldsymbol{\Pi}_\xi d\mathcal{V}. \tag{5.92}$$

The geometric interpretation of this equation can be seen by examining Figure 5.6 This figure shows a differential element of mass located at a displacement $\boldsymbol{\xi}$ from the center of mass of the body. A vector $\boldsymbol{\omega}$, is also shown. This is a 'free' vector, however for clarity in the figure it is indicated as being placed at the mass center. A plane normal to $\boldsymbol{\xi}$ is also shown. The dyad $\boldsymbol{\Pi}_\xi$ will project the vector $\boldsymbol{\omega}$ into a vector parallel to this normal plane. This vector is denoted as $\boldsymbol{\omega}_\perp$. If $\boldsymbol{\omega}$ is the vector of angular velocity of the body B with respect to the inertial observer, $\xi\boldsymbol{\omega}_\perp$ will equal the speed of the mass element as it moves around the axis of the vector $\boldsymbol{\omega}_\perp$, and ξ times this quantity will give the moment of momentum of the element. This quantity, the moment of momentum, is known as the angular momentum of the element. Therefore $\boldsymbol{I}^{B/\mathcal{P}^*}$ dotted into $\boldsymbol{\omega}$ in effect integrates the elements of the angular momentum over

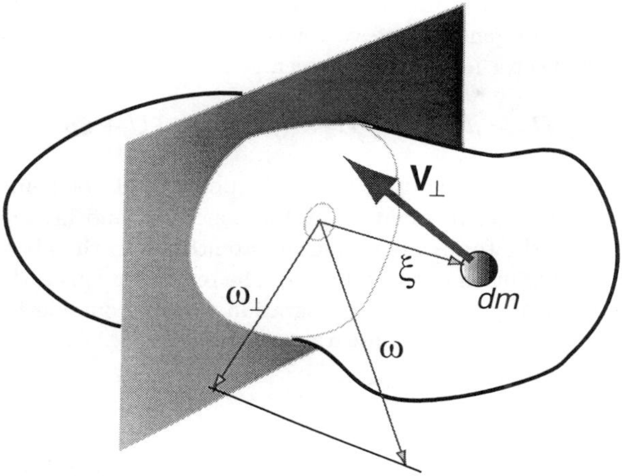

Figure 5.6: Geometric Interpretation of the moment of inertia dyad.

the body. Thus $\boldsymbol{I}^{B/\mathcal{P}^*} \cdot \boldsymbol{\omega}$ yields the total angular momentum of the body. Thus define

$$\boldsymbol{H}^{B/\mathcal{P}^*} = \int_B \rho \boldsymbol{\xi} \times \boldsymbol{v} d\mathcal{V} \tag{5.93}$$

implying that

$$\boldsymbol{H}^{B/\mathcal{P}^*} = \boldsymbol{I}^{B/\mathcal{P}^*} \cdot \omega. \tag{5.94}$$

Thus the equation of rotational motion with reference to the center of mass may be written as

$$\frac{{}^A d}{dt} \boldsymbol{H}^{B/\mathcal{P}^*} = \boldsymbol{T}^{B/\mathcal{P}^*}. \tag{5.95}$$

There are two forms of transformation of the moment of inertia dyad that are useful and important in both application and development of the theory. First consider the relation between $\boldsymbol{I}^{B/\mathcal{P}^*}$, the moment of inertia dyad computed about the mass center of the body, and $\boldsymbol{I}^{B/\mathcal{Q}}$ an arbitrary point, also *fixed* in the body.The points $\mathcal{P}^*$ and $\mathcal{Q}$ are the center of mass point and the arbitrary point. The vector, $\boldsymbol{\xi}$, is the displacement from the center of mass to the integration point, and $\boldsymbol{\mu}$ the displacement from $\mathcal{Q}$, to $\mathcal{P}^*$. Finally, $\boldsymbol{\eta}$ is the displacement from $\mathcal{Q}$ to the integration point. A slightly different approach to this problem is taken than that found in most texts, mainly for the purpose of gaining some more insight about the roll of the projection operator $\boldsymbol{\Pi}$. To this end consider the effect of a projection operator, $\boldsymbol{\Pi}_\eta$ on some arbitrary vector $\boldsymbol{w}$. Noting that $\boldsymbol{\eta}/\eta$ is a unit vector in the direction of $\boldsymbol{\eta}$, and considering the action of the vector cross product in producing vectors that are perpendicular to the vectors involved in the product, we see that:

$$\eta^2 \boldsymbol{\Pi}_\eta \cdot \boldsymbol{w} = \boldsymbol{\eta} \times (\boldsymbol{w} \times \boldsymbol{\eta}) = (\boldsymbol{\eta} \times \boldsymbol{w}) \times \boldsymbol{\eta} \tag{5.96}$$

We leave it for the reader to draw a convincing diagram to illustrate this and to see that in this case (not in general) the vector triple product is associative. Using this result and standard vector identities it is seen that:

$$\eta^2 \boldsymbol{\Pi}_\eta = \mu^2 \boldsymbol{\Pi}_\mu + \xi^2 \boldsymbol{\Pi}_\xi - (\boldsymbol{\mu}\boldsymbol{\xi} - 2\boldsymbol{\mu} \cdot \boldsymbol{\xi} \boldsymbol{U} + \boldsymbol{\xi}\boldsymbol{\mu}). \tag{5.97}$$

Therefore the moment of inertia dyad about the point $\mathcal{Q}$ can be computed by multiplying both sides of the above equation by the density ρ, and integrating the result over the body. All of the terms appearing in parentheses in the above equation will yield a vanishing contribution to this integral. The reason for this is that after moving quantities that do not depend on the integration variable outside the integral each one of these terms will contain the integral

$$\int_B \rho \boldsymbol{\xi} d\mathcal{V}$$

and as $\boldsymbol{\xi}$ is the displacement from the mass center, this integral evaluates to zero. Therefore using the fact that $\boldsymbol{\mu}$ is a constant vector, and that $\int_B \rho d\mathcal{V} = m$, the mass, we are left with the result:

$$\boldsymbol{I}^{B/\mathcal{Q}} = m\mu^2 \boldsymbol{\Pi}_\mu + \boldsymbol{I}^{B/\mathcal{P}^*}. \tag{5.98}$$

The first term on the right hand side represents the moment of inertia due to the entire mass of the body taken as a point body about the reference point $\mathcal{Q}$. If this moment of inertia dyad is indicated by the notation:

$$\boldsymbol{I}^{B*/\mathcal{Q}},$$

we can write our result, which is the dyadic form of the so called *parallel axis theorem* for the moment of inertia, as:

$$\boldsymbol{I}^{B/\mathcal{Q}} = \boldsymbol{I}^{B*/\mathcal{Q}} + \boldsymbol{I}^{B/\mathcal{P}^*}. \tag{5.99}$$

This result can be shown to apply in suitable form to linear functions of the inertia dyad. An important case is to the component representation in terms of a standard triad. If the triad is fixed in the body the components will be time independent. Thus if b represents a standard triad fixed in body B our previous work with dyads shows that the matrix of components of the inertia dyad about the center of mass will be:

$$b \cdot \boldsymbol{I}^{B/\mathcal{P}^*} \cdot b^T = \int_B \rho \xi^2 b \cdot \boldsymbol{\Pi}_\xi \cdot b^T d\mathcal{V}. \tag{5.100}$$

Now from the definition of the projection operator:

$$\xi^2 \boldsymbol{\Pi}_\xi = \xi^2 \boldsymbol{U} - \boldsymbol{\xi}\boldsymbol{\xi}. \tag{5.101}$$

The displacement vector $\boldsymbol{\xi} = \sum \xi_i \boldsymbol{b}_i$, hence we have the result that:

$$b \cdot \boldsymbol{I}^{B/\mathcal{P}^*} \cdot b^T = b \cdot \int_B \rho \sum (\xi^2 \delta_{ij} - \xi_i \xi_j) \boldsymbol{b}_i \boldsymbol{b}_j d\mathcal{V} \cdot b^T. \tag{5.102}$$

Therefore using a left superscript to indicate the standard triad used we find that the components of the moment of inertia matrix referred to it are given by:

$${}^b I_{ij}^{B/\mathcal{P}*} = \int_B \rho (\xi^2 \delta_{ij} - \xi_i \xi_j) d\mathcal{V} \tag{5.103}$$

the matrix itself being indicated by the notation:

$${}^b I^{B/\mathcal{P}*} = b \cdot \boldsymbol{I}^{B/\mathcal{P}^*} \cdot b^T. \tag{5.104}$$

Our previous work on the effect of orthonormal transformations on dyads and their matrix representations provides us with the answer to the problem of how to compute the effect of a triads orientation on the moment of inertia dyad's representation. The answer follows from our general discussion of this in chapter two, i.e. if $R_{bb'}$ is the direction cosine matrix for transformation from the triad b' to the triad b, we have that:

$${}^{b'} I^{B/\mathcal{P}} = R_{b'b} \, {}^b I^{B/\mathcal{P}} R_{bb'}. \tag{5.105}$$

The ability to transfer the moment of inertia quantities from one reference position to another in a body combined with the means to refer them to rotated standard triads are basic tools for the calculation of the moments of inertia of complex bodies from simpler ones. The additivity property of the moment of inertia is useful in calculations involving composite bodies where the individual body components have tabulated moment of inertia properties. Thus bodies composed of simple solids combined with surface, line and point components can be dealt with by use of the above results rather than by the use of direct integration. While the emphasis of this text is on theory, skill in the calculation of moments of inertia is a requirement for the solution of non trivial rigid body problems!

The moment of inertia dyad was seen to be symmetric, hence any matrix representation in terms of a standard triad must also be symmetric. It is expected that the reader is familiar with the eigenvalue problem for symmetric matrices, in particular that the eigenvalues are real and the relation of the eigenvalue problem with the geometry of quadratic surface in three dimensional space. This leads to the result that the matrix representation is diagonal when two of the triad's vectors point along the directions of minimum and maximum moment of inertia. The determination of the principal axis follows closely our discussion of orthogonal transformations. Geometric quantities having the dimensions of length can be defined by considering the moment of inertia as computed about the principal axis system. Thus if in this system the diagonal components of the inertia matrix are given by I_j then the radius of gyration, k_j associated with I_j is given by the equation $I_j = mk_j^2$. Exploitation of body symmetries may be used in setting up the body axis system so as to minimize the algebraic manipulations. In many cases it is assumed without comment that the body axes are principal axes of the moment of inertia dyad.

•Illustration

In the calculation of the moment of inertia dyad for the rigid body composed of three point masses intuitive arguments were used to obtain a diagonal representation about the mass center. The properties discussed in this section can be used to obtain this representation from an arbitrary one. Start with the original description of the body from the first illustration used in this chapter, i.e. with three masses $m_0/3$ located at the points $l\boldsymbol{b}_j$. The mass density ρ for this description was also given in the illustration. The matrix representation of the inertia dyad can be calculated from the results of this section, i.e.

$$ {}^{b}I_{11}^{B/\mathcal{Q}} = \int_B \rho(\xi_2^2 + \xi_3^2)d\mathcal{V}, \tag{5.106}$$

with similar equations for the other diagonal components. The result is that

$$ {}^{b}I_{11}^{B/\mathcal{Q}} = {}^{b}I_{22}^{B/\mathcal{Q}} = {}^{b}I_{33}^{B/\mathcal{Q}} = \frac{2}{3}m_0 l^2. \tag{5.107}$$

The off diagonal components, $i \neq j$, are given by:

$$ {}^{b}I_{ij}^{B/\mathcal{Q}} = -\int_B \rho\xi_i\xi_j d\mathcal{V} = \mathbf{o}, \tag{5.108}$$

so that

$$ {}^{b}I^{B/\mathcal{Q}} = \frac{1}{3}m_0 l^2 \begin{bmatrix} 2 & 0 & 0 \\ 0 & 2 & 0 \\ 0 & 0 & 2 \end{bmatrix}. \tag{5.109}$$

It was calculated that the displacement vector from the origin to the mass center is given by

$$ \boldsymbol{\mu} = \frac{l}{3}(\boldsymbol{b}_1 + \boldsymbol{b}_2 + \boldsymbol{b}_3). \tag{5.110}$$

The *parallel axis theorem* expressed for matrix representations of the inertia dyad, shows that:

$$ {}^{b}I^{B/\mathcal{P}^*} = {}^{b}I^{B/\mathcal{Q}} - {}^{b}I^{B^*/\mathcal{Q}}. \tag{5.111}$$

The calculation of the moment of inertia matrix about the reference point $\mathcal{Q}$, with all the mass concentrated at the center of mass $\mathcal{P}^*$, gives

$$ {}^{b}I^{B^*/\mathcal{Q}} = \frac{1}{3}m_0 l^2 \begin{bmatrix} \frac{2}{3} & -\frac{1}{3} & -\frac{1}{3} \\ -\frac{1}{3} & \frac{2}{3} & -\frac{1}{3} \\ -\frac{1}{3} & -\frac{1}{3} & \frac{2}{3} \end{bmatrix}, \tag{5.112}$$

hence carrying out the matrix subtraction it is seen that

$$ {}^{b}I^{B/\mathcal{P}^*} = \frac{1}{3}m_0 l^2 \begin{bmatrix} \frac{4}{3} & \frac{1}{3} & \frac{1}{3} \\ \frac{1}{3} & \frac{4}{3} & \frac{1}{3} \\ \frac{1}{3} & \frac{1}{3} & \frac{4}{3} \end{bmatrix}. \tag{5.113}$$

While this last result is with respect to a frame with origin at the mass center, it is not diagonal. The representation is still with respect to the triad b, and it is a natural question to ask if there exists a triad b' for which the matrix representation of the inertia dyad with respect to the center of mass is diagonal. The answer is yes, as noted in the above discussion. Though it is expected that most readers have the background in linear algebra needed to find b', a short review of the reasoning behind the calculation is now provided, both as a review and to emphasize some particular points.

The general theory of the representation of dyads under orthogonal transformations, developed in chapter two, shows that

$$ {}^{b'}I = R_{b'b}\, {}^{b}I R_{bb'}, \tag{5.114}$$

where ${}^{b}I$ or ${}^{b'}I$ now stand for representations of the Inertia dyad at the center of mass with respect to the indicated triad. The goal is to find $R_{b'b}$ such that ${}^{b'}I$ is diagonal, i.e. so that

$$ {}^{b'}I = \begin{bmatrix} I_1 & 0 & 0 \\ 0 & I_2 & 0 \\ 0 & 0 & I_3 \end{bmatrix}. \tag{5.115}$$

To proceed let

$$ R_{b'b} = \begin{bmatrix} w_{11} & w_{12} & w_{13} \\ w_{21} & w_{22} & w_{23} \\ w_{31} & w_{32} & w_{33} \end{bmatrix}, \tag{5.116}$$

or with the notation that w_j represents the *row* matrix with components w_{ij},

$$ R_{b'b} = \begin{bmatrix} w_1 \\ w_2 \\ w_3 \end{bmatrix}. \tag{5.117}$$

From the orthogonality property of direction cosine matrices it is also true that

$$ R_{bb'} = [w_1^T, w_2^T, w_3^T]. \tag{5.118}$$

Multiplying the transformation equation between representations on the left by $R_{b'b}$ the desired condition for diagonality of the inertia matrix becomes

$$ [w_1^T, w_2^T, w_3^T] \begin{bmatrix} I_1 & 0 & 0 \\ 0 & I_2 & 0 \\ 0 & 0 & I_3 \end{bmatrix} = {}^{b}I[w_1^T, w_2^T, w_3^T], \tag{5.119}$$

which is equivalent to the three equations

$$ {}^{b}w_j^T = I_j w_j^T. \tag{5.120}$$

The last result shows that if the center of mass inertia dyad can be diagonalized, the diagonal terms, I_j, must be the *eigenvalues* of this *eigenvalue* problem. The *eigencolumns,* w_j^T, provide the three columns of the direction cosine matrix $R_{bb'}$, provided they also satisfy the orthonormality conditions. As ${}^bI$ is a symmetric matrix, the theory of linear algebra tells us that these conditions can be satisfied. The procedure is to first find the *principal moments of inertia,* I_k, use the eigenvalue equation to find as many independent components of the eigencolumns as possible, and normalize the eigencolumns to unity. The symmetry of the inertia dyad insures that this is always possible and is carried out by using the following three equations:

$$det({}^bI - I_kU) = 0, \tag{5.121}$$

$$({}^bI - I_kU)w_k^T = 0, \tag{5.122}$$

$$w_k w_k^T = 1. \tag{5.123}$$

To apply these results to the three mass point rigid body it is easiest to work with the matrix

$$\begin{bmatrix} 4 & 1 & 1 \\ 1 & 4 & 1 \\ 1 & 1 & 4 \end{bmatrix} = {}^bI\frac{9}{m_0 l^2}, \tag{5.124}$$

which will give eigenvalues proportional to the principal moments of inertia. The equation derived from the determinent of this system, sometimes called the *secular equation* is

$$\begin{vmatrix} 4-\lambda & 1 & 1 \\ 1 & 4-\lambda & 1 \\ 1 & 1 & 4-\lambda \end{vmatrix} = 0, \tag{5.125}$$

where the principal moments of inertia will be given by

$$I_j = \frac{1}{9}m_0 l^2 \lambda_j, \tag{5.126}$$

the subscript j on λ indicating one of the three roots of the equation. These roots are

$$\lambda_1 = 3, \tag{5.127}$$

$$\lambda_2 = 3, \tag{5.128}$$

$$\lambda_3 = 6, \tag{5.129}$$

which gives the same principal moments of inertia as obtained by our intuitive approach. The eigenvector problem is slightly complicated by the coincidence of two of the roots of the secular equation. This is easily dealt with if one realizes that it is connected with the intrinsic symmetry of the body and finds the eigencolumns for λ_1 and λ_3. The second eigencolumn is simply the one corresponding to the cross product of the others, giving three independent and orthonormal eigencolumns. Carrying out these calculations gives the same result for $R_{b'b}$ as obtained previously!

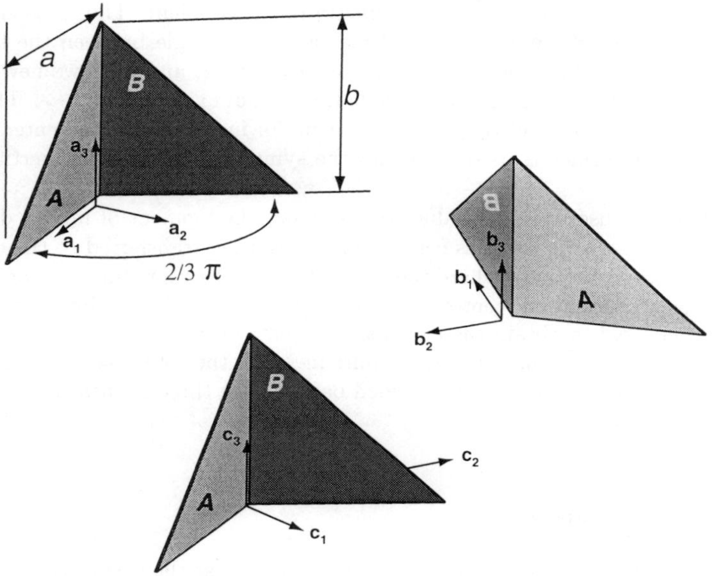

Figure 5.7: Triangular Composite Body

5.5.1 Inertia Dyads and Sophia

Even simple bodies can easily lead to tedious moment of inertia calculations so it is useful to apply computer algebra to this problem. To illustrate this we now solve the problem, taken from Kane and Levinson, associated with the figure showing a composite body made up of two identical right triangular sheets at an angle of $2\pi/3$. The task is to find the minimum moment of inertia for this object for some particular values of the ratio of the height to base width of the triangle. The moments of inertia of a right triangular laminar about its center of mass can be easily computed from first principles, or found in available reference works. To deal with the particular problem at hand we must first use the given moment of inertia components to find the moments of inertia of the total system about its center of mass. The eigenvalues of the inertia dyad referred to the center of mass can then be found. Transferring the diagonal components from the center of mass point will only increase their magnitude. Therefore the minimum eigenvalue provides the desired answer.

Three reference frames are introduced to assist in the calculation. These are indicated in the figure. The frame A aligns the $\boldsymbol{a}_1$ unit vector with the base of triangle A, while the frame B aligns the $\boldsymbol{b}_1$ unit vector with the base of triangle B. Note that the vectors $\boldsymbol{a}_2$ and $\boldsymbol{b}_2$ are oriented in the same way to each of triangles A and B. This allows us to use the same representation of the inertia dyad in the respective frames A and B fixed in the respective bodies A and B. The third frame is introduced to take advantage of the symmetry of the problem. Thus it is clear that the center of mass will lie along a line that bisects the angle between the two base lines, i.e. at a positive or negative rotation of $\pi/3$ from $\boldsymbol{a}_1$ and $\boldsymbol{b}_1$. Two new Sophia-Maple commands will be used in the calculation, `EinertiaDyad` and `&->`. The latter is used to effect the parallel transfer of moments of inertia from the center of mass values, while the former is used to set up the symmetric moment of inertia Edyad structure.

The plan is to use 'look-up' values to construct the moment of inertia dyads for each of the bodies about its mass center. These are then transferred to the center of mass of the composite body where they are added in the representation of frame C which is oriented so that the center of mass is in the $\boldsymbol{c}_1 - \boldsymbol{c}_3$ plane. Maple's eigenvalue routine is then used to obtain the principal moments of inertia.

To normalize the calculation choose units in which the total mass and the height are unity. Then define the relations needed between the three frames. In Sophia this is done with the statements:

```
> m:=1/2:
> a:='a':
> b:=1:
> &rot [A,B,3,Pi*2/3]:
> &rot [A,C,3,Pi/3]:
```

Take the point on the base where the triangles meet as the origin 'O', so that the displacement vectors from this point to the projection of mass centers of A and B onto the base plane are given by:

```
> rOAs:=Evector(a/3,0,0,A):
> rOBs:=Evector(a/3,0,0,B):
```

The distance to the projection of the system center of mass is then obtained using the previously discussed center of mass operator:

```
> rOCs:= C &cm [[m,rOAs],[m,rOBs]]:
```

The displacements from each body's mass center to the new system mass center are then given by

```
> rAsCs := rOCs &-- rOAs
> rBsCs := rOCs &-- rOBs
```

The components of the inertia dyad for each of the triangles are given by

```
> I11:=m*b^2/18:
> I33:=m*a^2/18:
> I22:=I11+I33:
> I13:=-m*a*b/36:
```

These are now used to form the Edyad data structure:

```
> IDAs:=EinertiaDyad(I11,I22,I33,0,I13,0,A):
> IDBs:=EinertiaDyad(I11,I22,I33,0,I13,0,B):
```

The general form in an obvious notation is

```
> EinertiaDyad(I11,I22,I33,I12,I13,I23,frameName).
```

The left hand argument for the parallel transfer operator consists of a list composed of the total mass and the center of mass inertia dyad. The right hand argument is the vector displacement from the mass center to the new reference point, in this case the center of mass of the composite body. Thus

```
> IDACs:= ( [m,IDAs] &-> rAsCs):
> IDBCs:= ( [m,IDBs] &-> rBsCs):
```

The moment of inertia dyad for the composite body is now obtained by adding the dyads. Sophia automatically ensures that the addition is done in the same representation. We also add the additional step of transferring the result to the frame C aligned with the symmetry axis of the new body:

```
> IDCs := C &to (IDACs &++ IDBCs);
```

which gives the result:

$$[\begin{bmatrix} \frac{a^2}{8}+1/18 & 0 & -\frac{a}{72} \\ 0 & \frac{a^2}{72}+1/18 & 0 \\ -\frac{a}{72} & 0 & \frac{5a^2}{36} \end{bmatrix}, C]$$

Two special cases, $a = 2b$ and $a = b/2$ are asked for in the book of Kane and Levinson. Maple's linear algebra package contains the function **eigenvals** which takes a matrix argument. The first case is dealt with by the statements

```
>  IDs1:=subs({a=2*b},IDCs):eigenvals(&vPart IDs1);
```

giving the result

$$1/9, \frac{7}{12}, \frac{19}{36}$$

showing that the minimum moment of inertia is $1/9$ or the minimum radius of gyration is $1/3$. For the second case:

```
> IDs2:=subs({a=b/2},IDCs):
> eigenvals(&vPart IDs2);
```

giving

$$\frac{17}{288}, \frac{35}{576}+\frac{\sqrt{241}}{576}, \frac{35}{576}-\frac{\sqrt{241}}{576}$$

•

5.6 Energy and Power

In parallel with the discussion of the equations of motion for a point mechanism, it is useful to calculate the kinetic energy and power associated with a rigid body motion. We will now carry out this task, which is greatly simplified by using the center of mass of the body as index point. In this section super and subscripts are left out, it being a given that the center of mass is the index point, that $\boldsymbol{\xi}$ is the distance from the center of mass to an arbitrary body point, moments are taken with respect to the center of mass point and that the differentiations are with respect to the body system of an inertial observer. Start with the previously derived equations of motion for the rigid body, i.e.

$$m\tfrac{d}{dt}\boldsymbol{u} = \boldsymbol{R} \tag{5.130}$$

and

$$\tfrac{d}{dt}\boldsymbol{I}\cdot\boldsymbol{\omega}=\boldsymbol{T}, \tag{5.131}$$

where $\boldsymbol{u}$ is the velocity of the mass center. As before $\boldsymbol{v}$ will be used to indicate the velocity of some arbitrary point that is fixed in the body. Recall that the right hand side of these equations result from the integration of the force density and the moment of the force density over the body. In terms of the force density define the *local power density* as the product of the force density and the velocity of the body point in question. Therefore the total power will be given by:

$$\mathcal{P}=\int \boldsymbol{f}\cdot\boldsymbol{v}d\mathcal{V}. \tag{5.132}$$

At first sight it is not clear as to how to relate this to the above equations of motion. This is because $\boldsymbol{v}$ refers to any of the points in the moving body and cannot simply be moved into and out of the integral. The answer to this problem is to consider the sum of the two terms, $\boldsymbol{R}\cdot\boldsymbol{u}$, and $\boldsymbol{T}\cdot\boldsymbol{\omega}$. Thus in terms of the definitions of T and R write:

$$\boldsymbol{R}\cdot\boldsymbol{u}+\boldsymbol{T}\cdot\boldsymbol{\omega}=\int \boldsymbol{f}d\mathcal{V}\cdot\boldsymbol{u}+\int\boldsymbol{\xi}\times\boldsymbol{f}d\mathcal{V}\cdot\boldsymbol{\omega}=\int(\boldsymbol{u}+\boldsymbol{\omega}\times\boldsymbol{\xi})\cdot\boldsymbol{f}d\mathcal{V}. \tag{5.133}$$

It should be clear that as $\boldsymbol{u}$ and $\boldsymbol{\omega}$ are independent of $\boldsymbol{\xi}$ they can be placed under the integral as indicated. Now use the result that $\boldsymbol{v}=\boldsymbol{u}+\boldsymbol{\omega}\times\boldsymbol{\xi}$, to obtain the result that the power can be expressed as

$$\mathcal{P}=\boldsymbol{R}\cdot\boldsymbol{u}+\boldsymbol{T}\cdot\boldsymbol{\omega}. \tag{5.134}$$

This is an extremely important and useful result. In effect it provides in compact form the way the resultant force and torque interact with the velocity and angular velocity of the rigid body. It could even be used as a starting point for rigid body theory, and is the basis for the application of the principle of virtual power to rigid body dynamics. The power *consumed* by the motion, given by:

$$\mathcal{P}^*=m\boldsymbol{u}\cdot\tfrac{d}{dt}\boldsymbol{u}+\boldsymbol{\omega}\cdot\tfrac{d}{dt}\boldsymbol{I}\cdot\boldsymbol{\omega} \tag{5.135}$$

must equal the applied power. If the system starts from rest and this result is integrated, an expression for the *kinetic energy* of the body can be developed. Thus it is left as an exercise to show that

$$\mathcal{K}=\int_{t_0}^{t}\mathcal{P}^*dt=\frac{1}{2}(\boldsymbol{\omega}\cdot\boldsymbol{I}\cdot\boldsymbol{\omega}+mu^2) \tag{5.136}$$

where t_0 is taken at a time when the body is at rest.

5.7 Screws

The characteristic description of a rigid body has shown that the velocity of points in the body relative to some reference frame are related by the body's angular velocity. The constraint of rigidity has also been seen to impose a relation between the applied torque computed at different body points. In this case it is the total force acting on the body that governs the relationship. It is not too hard to see that a similar relationship applies to the angular momentum, where the linear momentum determines the relationship. It is thus reasonable to ask if there is some common thread between these facts. The answer is positive and provides the foundation for the concept of a *screw.*

The motivation for this concept is evident if we consider the transformation relationships for the velocity and the torque between the points $\mathcal{P}$ and $\mathcal{Q}$ of a rigid body with angular velocity $\boldsymbol{\omega}$ and total applied force $\boldsymbol{R}$.

$$\boldsymbol{v}^{\mathcal{Q}} = \boldsymbol{v}^{\mathcal{P}} + \boldsymbol{\omega} \times \boldsymbol{r}^{\mathcal{PQ}} \tag{5.137}$$

$$\boldsymbol{T}^{/\mathcal{Q}} = \boldsymbol{T}^{/\mathcal{P}} + \boldsymbol{R} \times \boldsymbol{r}^{\mathcal{PQ}}. \tag{5.138}$$

It should be evident that the angular velocity and the total force play similar roles in these relationships. Recall also that in contrast to the velocity and the torque, the angular velocity and the total force are free vectors, i.e. they do not depend on the body point under consideration. A screw is defined as a pair consisting of a free and a bound vector, where the bound vector follows the above transformation law [1]. The free vector is called the *vector of the screw.* There are a number of notations for these six dimensional objects, here they will simply be indicated by enclosing them in parenthesis, with the vector of the screw in the second position. This means that we must also indicate the point associated with the first or bound vector. Thus the force screw is given by

$$(\boldsymbol{T}^{/\mathcal{P}}) = (\boldsymbol{T}^{/\mathcal{P}}, \boldsymbol{R}), \tag{5.139}$$

and the velocity screw by

$$(\boldsymbol{v}^{\mathcal{P}}) = (\boldsymbol{v}^{\mathcal{P}}, \boldsymbol{\omega}). \tag{5.140}$$

It is possible to show that the linear and angular momentum pair also form a screw, i.e.

$$(\boldsymbol{H}^{\mathcal{P}}) = (\boldsymbol{H}^{\mathcal{P}}, \boldsymbol{P}) \tag{5.141}$$

obeys the screw transformation law. Thus the equations of motion for a rigid body can be written as

$$\tfrac{d}{dt}(\boldsymbol{H}^{\mathcal{P}}) = (\boldsymbol{T}^{/\mathcal{P}}). \tag{5.142}$$

In a previous section it was shown that in certain cases a point could be found at which both of the vectors comprising the force screw had the same direction. At this

[1] The algebra of screws, (see Bottema and Roth) involves a number of special notions and concepts of duality which are glossed over in this text

point the relative magnitude of the vectors was called the pitch, while the object itself was called a wrench. The equivalent calculation can be carried out for the velocity screw, giving an object that is sometimes called a *twist*. Thus the position and pitch of the twist is given by:

$$\boldsymbol{\mu} = \frac{1}{\omega^2}\boldsymbol{\omega} \times \boldsymbol{v}^{/\mathcal{P}} \tag{5.143}$$

$$\lambda = \frac{1}{\omega^2}\boldsymbol{\omega} \cdot \boldsymbol{v}^{/\mathcal{P}}. \tag{5.144}$$

In addition it is possible to form a product between screws that is invariant, i.e. that takes the same value for all points of the body. For the example of the screw product between the velocity and the force screw we have:

$$(\boldsymbol{v}^{\mathcal{Q}}) \cdot (\boldsymbol{T}^{/\mathcal{Q}}) = \boldsymbol{v}^{\mathcal{Q}} \cdot \boldsymbol{R} + \boldsymbol{T}^{/\mathcal{Q}} \cdot \boldsymbol{\omega}. \tag{5.145}$$

The invariance of this is checked by substitution of the expressions for the torque and velocity at another point of the body into the above expression. Thus

$$\begin{aligned} (\boldsymbol{v}^{\mathcal{Q}}) \cdot (\boldsymbol{T}^{/\mathcal{Q}}) &= \boldsymbol{v}^{\mathcal{P}} \cdot \boldsymbol{R} + \boldsymbol{\omega} \times \boldsymbol{r}^{\mathcal{PQ}} \cdot \boldsymbol{R} + \boldsymbol{T}^{/\mathcal{P}} \cdot \boldsymbol{\omega} + \boldsymbol{R} \times \boldsymbol{r}^{\mathcal{PQ}} \cdot \boldsymbol{\omega} \\ &= (\boldsymbol{v}^{\mathcal{P}}) \cdot (\boldsymbol{T}^{/\mathcal{P}}) \end{aligned}$$

Therefore the reference to a particular body point can be ignored as long as it is realized that the same point must be used in any particular calculation, i.e.

$$(\boldsymbol{v}) \cdot (\boldsymbol{T}) = \boldsymbol{v} \cdot \boldsymbol{R} + \boldsymbol{T} \cdot \boldsymbol{\omega}. \tag{5.146}$$

This shows that the 'power' required for a motion is an invariant, independent of which point of the rigid body is used for the computation. The reader is invited to interpret some of the other possible screw products involving the rates of change angular and linear momentum.

The screw concept is used in a number of classic works in mechanics and in the theory of mechanisms and robotics. A formal development of the concept revolves around treating screws as comprising a six dimensional linear vector space. No particular use of screw formalism will be made in the present text with the exception of using some of the terminology in Sophia commands. The general state of a rigid body expressed by its position and attitude is sometimes called the body's *pose*. There are also a number, ever growing it seems, of mathematical objects used to describe rigid body motions. Among the more famous are quaternions, spinors and dual numbers, which are discussed in the text of Bottema and Roth. Many of these objects have found important uses in mathematics and modern theoretical physics, however it will be seen that the matrix arrays of vectors, which we call Kvectors, provide an adequate and convenient description of rigid body motions.

•Illustration

Several Sophia commands are available for dealing with equipollent force systems and screws in general. This illustration provides a simple example. Figure 5.8 shows a 5 bar truss in the form of a tetrahedron having equilateral triangles with side of length L as faces. The three legs rest on a smooth surface at points A, B, and C, while the segments are connected by ball and socket joints at points D,E,F,G and H. A vertical load w is applied at point F. The points G, D and E are located at a distance S from the base as measured along the legs. The problem is to determine the reaction forces at A,B and C.

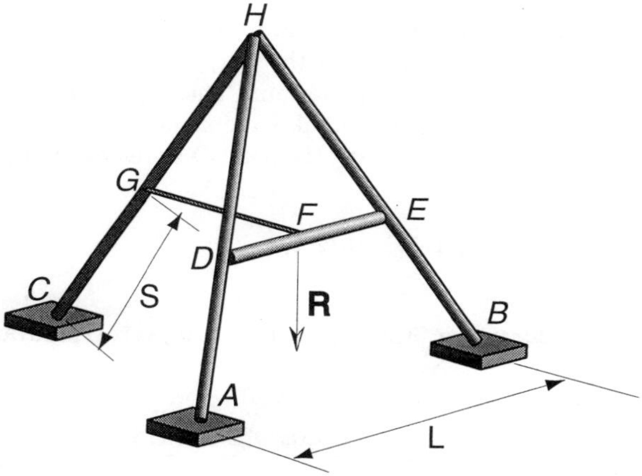

Figure 5.8: Equilateral Truss

To solve the problem it is necessary to determine a number of displacement vectors between various points. It is convenient to introduce two reference frames for this purpose. The reference frame A is arranged so that $\boldsymbol{a}_1$ and $\boldsymbol{a}_2$ are oriented along the lines OC and OA respectively, the point O bisecting the line AB. The second frame, B is chosen so that $\boldsymbol{b}_3$ lies in the face of the triangle ABH, and is in the direction OH. The following statements simply set up a data base of useful displacements:

```
> &rot [A,B,2,arcsin(sqrt(3)/2)]:
> rOH := B &ev [0,0,L]:
> rOA := A &ev [0,L/2,0]:
> rOB := A &ev [0,-L/2,0]:
> rOF := B &ev [0,0,S]:
> rAH := rOH &-- rOA:
> rBH := rOH &-- rOB:
> eAH := (1/L) &** rAH:
> eBH := (1/L) &** rBH:
> rAD := (S) &** eAH:
```

```
> rBE := (S) &** eBH:
> rFD := rOA &++ (rAD &-- rOF):
> rFE := rOB &++ (rBE &-- rOF):
```

Note that eAH and eBH are unit vectors. In addition it is helpful to have a null vector, i.e. let

```
> rzero := A &ev [0,0,0]:
```

The condition of smoothness at the supports implies that the reaction forces will all be orthogonal to the base, therefore take:

```
> RA := A &ev [0,0,RA3]:
> RB := A &ev [0,0,RB3]:
> RC := A &ev [0,0,RC3]:
```

The quantities RA3, RB3 and RC3 are to be determined. The load at point F is

```
> Rw := A &ev [0,0,-w]:
```

This forms an equipollent force system and we will now evaluate the screw associated with the external forces made up of the load and reaction. Both component vectors of the screw must vanish for equilibrium. The Sophia command we use is `A &screw [[r1,R1],[r2,R2], ...]` which takes the name of a frame and a list of displacement force pairs as arguments. Using the point D as reference the displacements needed are:

```
> rAB := A &ev [0,-L,0]:
> rAC := A &ev [(L*sin(Pi/3)),-L/2,0]:
> rAF := rAD &-- rFD:
```

The desired screw is then:

```
> TAscrew := A &screw [[rAB,RB],[rAC,RC],[rAF,Rw],[rzero,RA]]:
```

which gives:

$$[[[\frac{Lw}{2} - \frac{LRC3}{2} - LRB3, \frac{\sqrt{3}Sw}{2} - \frac{L\sqrt{3}RC3}{2}, 0], A],$$
$$[[0, 0, RA3 - w + RC3 + RB3], A]].$$

From the symmetry of the problem we expect the vector of the screw, i.e. the total force to have only a vertical component, while the torque will be parallel to the base plane. This is the case, as can be verified by examining the above expression. The equations for the reactions are then formed and solved:

```
> Tex := TAscrew[1];
> Rex := TAscrew[2];
> reactions :=
>     solve({(Tex &c 1),(Tex &c 2),(Rex &c 3)},{RA3,RB3,RC3}):
```

The result is that

$$\begin{aligned} RA3 &= \frac{1}{2}\frac{L-S}{L} \\ RB3 &= \frac{1}{2}\frac{L-S}{L} \\ RC3 &= \frac{Sw}{L}. \end{aligned}$$

Appendix 1 also lists a Sophia function for transforming the screw to other points in space. •

5.8 The Darboux Vector

The constraint principle of D'Alembert is also taken to be valid for the rigid body. It is not immediately obvious how to formulate the description of a body's configuration so as to apply the principle. The constraint forces in the general case take the form of distributions over the body. Each point of the body is restricted in its possible position by both the applied constraints and the rigidity of the body. As the body moves about, each of its points will be restricted to its own configuration hypersurface. For example, if the overall body is constrained so that it rotates about a fixed point any arbitrary point of the body will be restricted to motion in a spherical surface. If the constraint point undergoes a given oscillatory displacement the instantaneous configuration surface of the arbitrary point will still be a sphere. By D'Alembert's principle the constraint force distribution must have a vanishing power functional with tangent vectors that are in the instantaneous constraint surface for each point in the body. The general rigid body has a continuous distribution of matter, hence an infinite number of points, however as no more than six parameters are required to describe the *full configuration* it is possible to eliminate the constraints in a relatively simple and effective manner, closely related to the methods used for the multi-point mass mechanism treated in chapter four. The main reason for this is that the motion of any arbitrary point of the body can be described in terms of the motion of the index point and the attitude of the body's triad as a linear function of the points location in the body. To effectively do this it is useful to be able to describe the rate of change of the body's triad vectors in terms of any parameters on which they may depend. These include generalized coordinates and the time. Previous work has shown that the rates of change of triad vectors are best understood using the concept of angular velocity. It is now useful to introduce a generalization of the concept of angular velocity which assists in the description rates of change of triad vectors with respect to arbitrary parameters. This generalization is known as the *Darboux vector* after the 19th century French mathematician who introduced the idea in his work in mechanics and the differential geometry of surfaces.

Recall that a general standard triad is represented by a column of orthonormal vectors, thus the triad designated by b is given by:

$$b = \begin{bmatrix} \boldsymbol{b}_1 \\ \boldsymbol{b}_2 \\ \boldsymbol{b}_3 \end{bmatrix}. \tag{5.147}$$

In the discussion of angular velocity it was assumed that the triad vectors were functions of time, either explicitly or through dependence on generalized coordinates which were themselves functions of time. The angular velocity dyad was seen as the tool for obtaining the rate of change of an arbitrary vector with respect to one triad when represented in terms of another triad. It was seen that the angular velocity dyad was antisymmetric, hence that the linear action of the dyadic dotted into the vector could be replaced by a vector cross product operation. This vector was given by the operation $vect(\boldsymbol{\Omega})$, defined in chapter two. If the reader takes another look at these developments it should be evident that the variable of time could be replaced by any other parameter, and that the study of rates of change with respect to many parameters could be carried out by changing total derivatives to partial derivatives. The problem is illustrated in figure 5.9 which shows a set of frame vectors considered as a function of a curve parameter. We are interested in characterizing the change in attitude of the frame as it is transported along the curve and in particular in the rates of change of vectors represented in the moving frame.

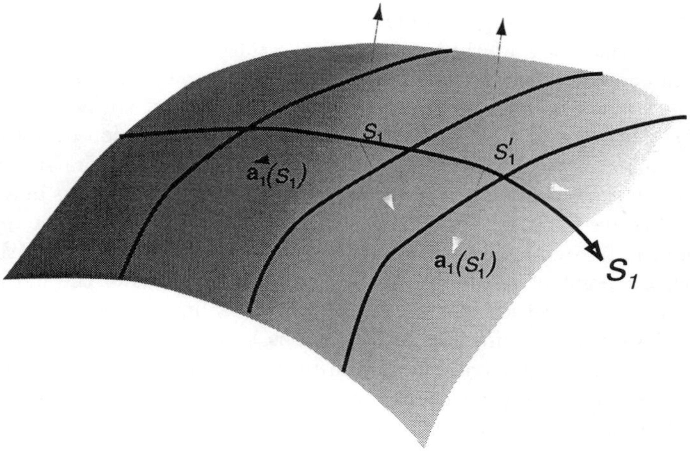

Figure 5.9: Vectors change as they are carried along curves, the change can be a function of the curve parameter.

It has in effect already been demonstrated that if b is a function of the parameters $s_1, s_2, \cdots$, which describe its attitude in terms of another triad, a, then for an arbitrary vector, $\boldsymbol{w}$, the partial derivative of $\boldsymbol{w}$ with respect to s_j relative to the triad a is related to the partial derivative with respect to s_j relative to the triad b by the equation:

$$\frac{{}^{A}\partial}{\partial s_j}\boldsymbol{w} = \frac{{}^{B}\partial}{\partial s_j}\boldsymbol{w} + b_{A/s_j} \times \boldsymbol{w}, \tag{5.148}$$

where the *vector* b_{A/s_j} is the vector associated with the antisymmetric dyad

$${}^{A}\boldsymbol{\Omega}^{B}{}_{s_j} = \frac{{}^{A}\partial \boldsymbol{b}_1}{\partial s_j}\boldsymbol{b}_1 + \frac{{}^{A}\partial \boldsymbol{b}_2}{\partial s_j}\boldsymbol{b}_2 + \frac{{}^{A}\partial \boldsymbol{b}_3}{\partial s_j}\boldsymbol{b}_3, \tag{5.149}$$

that is

$$b_{A/s_j} = \boldsymbol{vect}({}^A\boldsymbol{\Omega}^B{}_{s_j}). \tag{5.150}$$

This vector is called the *Darboux* vector for the partial rate of change of the triad with respect to the parameter s_j, and b_{A/s_j} is read as *the partial slant of b relative to A with respect to s_j*. Naturally if b depends on only one parameter, say s, the total derivative is taken. This is indicated by the notation:

$$b_{A//s} = \boldsymbol{vect}(\tfrac{{}^Ad\boldsymbol{b}_1}{ds}\boldsymbol{b}_1 + \tfrac{{}^Ad\boldsymbol{b}_1}{ds}\boldsymbol{b}_1 + \tfrac{{}^Ad\boldsymbol{b}_1}{ds}\boldsymbol{b}_1), \tag{5.151}$$

in which case:

$$\tfrac{{}^Ad}{ds}\boldsymbol{w} = \tfrac{{}^Bd}{ds}\boldsymbol{w} + b_{A//s} \times \boldsymbol{w}. \tag{5.152}$$

If the triad b is a function of parameters which are in turn functions of other parameters the chain rule of calculus can be applied to evaluate the total slant of a triad in terms of partial slants. This case is common in mechanics, where a triad may be a function of a number of generalized coordinates and the time, in which case it would possess partial slant vectors with respect to each of the generalized coordinates and time. *The total slant operator with respect to time is the angular velocity,* so that:

$$b_{A//t} = {}^A\boldsymbol{\omega}^B = \sum_{j=1}^{n} b_{A/q_j} + b_{A/t}. \tag{5.153}$$

Therefore the relation between the total time derivatives of a vector relative to two triads is:

$$\begin{aligned} \tfrac{{}^Ad}{dt}\boldsymbol{w} &= \sum_{j=1}^{n}(\tfrac{{}^B\partial}{\partial q_j} + b_{A/q_j} \times \boldsymbol{w})\dot{q}_j + \tfrac{{}^A\partial}{\partial t}\boldsymbol{w} + b_{A/t} \times \boldsymbol{w} \\ &= \tfrac{{}^Bd}{dt}\boldsymbol{w} + b_{A//t} \times \boldsymbol{w}. \end{aligned} \tag{5.154}$$

The Darboux vector provides a powerful tool for the study of curves and surfaces in three dimensional space. The Darboux dyad, as defined, provides a tool for use in spaces of higher dimension, or where it is not desirable to convert the dyadic operation to a cross product. The partial slant of a triad with respect to a generalized coordinate is called a *partial angular velocity,* in Kane's formulation of mechanics. These quantities serve a similar function to the tangent vectors of previous chapters in the task of elimination of constraint forces from the equations for Newtonian motion.

•Illustration

The theory of curves in three dimensional space is closely connected with mechanics. If one attempts to solve the problem of the motion of a bead on a wire of given shape, there is an immediate need to make use of this theory. From another point of view, the Newtonian path of a particle is a curve in three dimensional space, parameterized by the time of arrival of the particle at each point on the curve. In this example the geometric theory of curves is studied using the idea of the Darboux vector. The application of this to the path of a particle in space or on a given curve is left as an exercise.

Given a reference frame, i.e., a triad of orthonormal vectors a, a curve can be specified by providing the vectorial position of all the points on the curve as a function of a single parameter. Thus $\boldsymbol{r}(s)$ can be considered a 'mapping' of the one dimensional line into the curve $\boldsymbol{r}(s)$ specified by the three functions $\boldsymbol{a}_i \cdot \boldsymbol{r}(s)$. In mechanics the path of a particle would be specified as a function of time, and two motions along the 'same' curve would be different if the arrival times of the particle at the same locations differed. In the geometric theory it is the 'shape' of the curve that counts and except for convenience the parameterization would be irrelevant. The simplest parameterization from the point of view of a clean mathematical theory is in terms of a parameter which leads to the derivative of the displacement vector having unit magnitude, i.e. such that:

$$|\tfrac{d}{ds}\boldsymbol{r}| = 1. \tag{5.155}$$

In terms of such a parameter the *arc length* of the curve is given by:

$$\int_{\boldsymbol{r}_1}^{\boldsymbol{r}_2} |d\boldsymbol{r}| = \int_{s_1}^{s_2} ds = s_2 - s_1, \tag{5.156}$$

hence s is called the arc length. While s has theoretical advantages, frequently it is useful to use another parameter for computation. Any parameter that is monotonically related to s by a sufficiently smooth function will serve. The above equation is used to relate derivatives in the various curve formula derived using s. From the geometric point of view the main question is what kind of information is needed to specify a curve. The question of where in space a particular point on the curve is located, or the particular parameterization should be eliminated from consideration. Thus the geometrical problem is to answer the question, what is needed to specify the geometry of a curve? The theory of the Darboux vector provides an immediate and elegant answer to this question in the form of the Frenet-Serret equations. To see this, attach a standard triad to the curve, where the triad vectors alter orientation as a function of the arc length from some given point in space. Assume that at the given point the curve triad, b, coincides with a *fixed triad, a*. Take the triad vector $\boldsymbol{b}_1$ as the tangent vector to the curve. As the curve is parameterized by the arc length this will be a unit vector for all points on the curve. The second vector of the curve triad is obtained by differentiating $\boldsymbol{b}_1$ with respect to the arc length. As $\boldsymbol{b}_1 \cdot \boldsymbol{b}_1 = 1$ it is insured that:

$$\boldsymbol{b}_1 \cdot \tfrac{d}{ds}\boldsymbol{b}_1 = \mathbf{o}, \tag{5.157}$$

i.e. the new vector is orthogonal to the tangent vector. In general it will not itself be a unit vector. To obtain a unit vector define the function, $\kappa(s)$, so that if

$$\tfrac{d}{ds}\boldsymbol{b}_1 = \kappa(s)\boldsymbol{b}_2, \tag{5.158}$$

then

$$|\boldsymbol{b}_2| = 1. \tag{5.159}$$

The third unit vector in the curve triad is then defined as

$$\boldsymbol{b}_3 = \boldsymbol{b}_1 \times \boldsymbol{b}_2. \tag{5.160}$$

Now suppose that the Darboux vector of the triad has the representation

$$b_{A//s} = \gamma_1(s)\boldsymbol{b}_1 + \gamma_2(s)\boldsymbol{b}_2 + \gamma 3(s)\boldsymbol{b}_3, \tag{5.161}$$

where it has been emphasized that the expansion coefficients are functions of the arc length. The Darboux vector can now be used to calculate the three vectors

$$\tfrac{d}{ds}\boldsymbol{b}_j = b_{A//s} \times \boldsymbol{b}j. \tag{5.162}$$

Thus

$$\tfrac{d}{ds}\boldsymbol{b}1 = \gamma_3\boldsymbol{b}_2 - \gamma_2\boldsymbol{b}_3 = \kappa\boldsymbol{b}_2. \tag{5.163}$$

The last term in the above comes from the definition of the triad vector $\boldsymbol{b}_2$. This shows that

$$\gamma_2 = \mathbf{o}, \tag{5.164}$$

$$\gamma_3 = \kappa. \tag{5.165}$$

In the same manner the derivative of the rest of the triad vectors and the use of the Darboux vector shows that:

$$\tfrac{d}{ds}\boldsymbol{b}_2 = -\kappa\boldsymbol{b}_1 + \gamma_1\boldsymbol{b}_3, \tag{5.166}$$

$$\tfrac{d}{ds}\boldsymbol{b}_3 = -\gamma_1\boldsymbol{b}_2. \tag{5.167}$$

The function $\kappa(s)$ is called the 'curvature', while the common symbol for γ_1 is τ, called the 'torsion'. In most texts on differential geometry, what has been called the tangent vector, $\boldsymbol{b}_1$, is denoted by $\boldsymbol{t}$. The vector $\boldsymbol{b}_2$ is called the 'normal' and is denoted by $\boldsymbol{n}$, while $\boldsymbol{b}_3$ is called the 'binormal' and is written as $\boldsymbol{b}$. To avoid confusion with the use of these symbols in the rest of this work it is only in this illustration of the Darboux theory that this geometric notation is used. With this notation the relations derived with the Darboux vector take the form:

$$\tfrac{d}{ds}\boldsymbol{t} = \kappa\boldsymbol{n}, \tag{5.168}$$

$$\tfrac{d}{ds}\boldsymbol{n} = -\kappa\boldsymbol{t} + \tau\boldsymbol{b}, \tag{5.169}$$

$$\tfrac{d}{ds}\boldsymbol{b} = -\tau\boldsymbol{n}, \tag{5.170}$$

which are called the Serret-Frenet equations. The theory of ordinary differential equations insures a unique solution of these equations, given 'reasonable' behavior of the functions for the curvature and the torsion. The particular placement and orientation of the solution curve depends on the initial values, however the 'form' only depends on the functions $\kappa(s)$ and $\tau(s)$. Thus a geometric curve is defined by these functions. The study of individual cases shows that the curvature and torsion have meaningful geometric interpretations. One example which shows this is the case of constant curvature and torsion, which defines a helix of radius $1/\kappa$, and pitch $1/\tau$.

•

5.9 The Constrained Rigid Body

To apply D'Alembert's principle of constraint for the elimination of the constraints it is necessary to have a set of independent vectors, tangent to the configuration manifold of each point in the body that is subjected to a constraint force. Thus a *field* of tangent vectors, dependent on the position of each point of body, $\boldsymbol{\xi}$, must be found. The position of each point, relative to an inertial observer, is given by the vector field, $\boldsymbol{r} = \boldsymbol{\eta} + \boldsymbol{\xi}$. This may be written as:

$$\boldsymbol{r}(q,t) = \boldsymbol{\eta}(q,t) + \xi_1 \boldsymbol{b}_1(q,t) + \xi_2 \boldsymbol{b}_2(q,t) + \xi_3 \boldsymbol{b}_3(q,t), \tag{5.171}$$

where q represents the set of n generalized coordinates and b is a triad fixed in the moving body. The matrix containing vectors notation of chapter two can be used to write this as:

$$\boldsymbol{r} = \boldsymbol{\eta} + b^T \, {}^b\xi. \tag{5.172}$$

In a test motion each of the q is given as some function of a parameter, say s, and the time t is fixed. This defines a curve for each fixed value of the body position vector $\boldsymbol{\xi}$. Note that the body triad will in general be a function of q, hence will vary with the test parameter s. Thus there will be an infinite family of curves for a given test motion, one curve for each body position. The tangent vectors, which by D'Alembert's principle will have a null power functional with the constraint forces, are the tangents to these curves. To find these vectors simply use the chain rule to differentiate the position vector as a function of the test parameter, thus:

$$\tfrac{d}{ds}\boldsymbol{r} = \sum_{j=1}^{n} \left(\tfrac{{}^A\partial}{\partial q_j}\boldsymbol{\eta} + \tfrac{{}^A\partial}{\partial q_j} b^T \, {}^b\xi\right) \tfrac{d}{ds} q_j. \tag{5.173}$$

The second term in this summation can be evaluated in terms of the Darboux vectors for the rate of change of the body triad with respect to each of the generalized coordinates, so that

$$\tfrac{d}{ds}\boldsymbol{r} = \sum_{j=1}^{n} \left(\tfrac{{}^A\partial}{\partial q_j}\boldsymbol{\eta} + b_{A/q_j} \times \boldsymbol{\xi}\right) \tfrac{d}{ds} q_j. \tag{5.174}$$

As the generalized coordinates are independent, this gives n independent tangent vectors, each tangent vector being composed of a sum of two parts, one relating to the motion of the index point the other to the variation in body attitude. Therefore the coordinate tangent vectors, for any fixed point in the body, are given by the n equations:

$$\boldsymbol{\tau}_j = \tfrac{{}^A\partial}{\partial q_j}\boldsymbol{\eta} + b_{A/q_j} \times \boldsymbol{\xi}. \tag{5.175}$$

The force distribution function over the body can be written in the form:

$$\boldsymbol{f}(\boldsymbol{\xi},t) = \boldsymbol{f}_c(\boldsymbol{\xi},t) + \boldsymbol{f}_a(\boldsymbol{\xi},t), \tag{5.176}$$

where the subscripts c and a indicate the unknown force due to constraints and the applied force respectively. Assuming the validity of D'Alembert's principle:

$$\int_B \boldsymbol{f}_c \cdot \boldsymbol{\tau}_j d\mathcal{V} = 0. \tag{5.177}$$

To make use of this result, integrate the dot product of the full force distribution into a tangent vector over the body, so that:

$$\begin{aligned} \int_B \boldsymbol{f} \cdot \boldsymbol{\tau}_j d\mathcal{V} &= \int_B (\boldsymbol{f} \cdot \frac{{}^A\partial \boldsymbol{\eta}}{\partial q_j} + \boldsymbol{f} \cdot b_{A/q_j} \times \boldsymbol{\xi}) d\mathcal{V} \\ &= \int_B \boldsymbol{f}_a d\mathcal{V} \cdot \frac{{}^A\partial \boldsymbol{\eta}}{\partial q_j} + b_{A/q_j} \cdot \int_B \boldsymbol{\xi} \times \boldsymbol{f}_a d\mathcal{V} \\ &= \boldsymbol{R}_a \cdot \boldsymbol{\tau}_j^* + \boldsymbol{T}_a \cdot b_{A/q_j}, \end{aligned} \tag{5.178}$$

where quantities independent of the body position $\boldsymbol{\xi}$ have been moved out of the integrals, use has been made of the properties of the triple dot-cross product and a superscript '*' has been used to indicate a tangent vector at the body index point. The quantities $\boldsymbol{R}_a$ and $\boldsymbol{T}_a$ are the resultant applied force and applied torque, the latter being evaluated about the index point. The constraint generated component is eliminated by the scalar product with the tangent vector.

To obtain equations for a Newtonian motion which are independent of the constraint forces dot the equation for the center of mass motion into the vector $\boldsymbol{\tau}_j^*$, the equation for the attitude motion into b_{A/q_j} and sum the results to obtain:

$$F_j^* + F_j = 0, \tag{5.179}$$

where the generalized inertial force is

$$F_j^* = -\frac{{}^A dm\boldsymbol{u}}{dt} \cdot \boldsymbol{\tau}_{q_j}^* - \frac{{}^A d\boldsymbol{H}}{dt} \cdot b_{A/q_j}, \tag{5.180}$$

and the generalized applied force is

$$F_j = \boldsymbol{R}_a \cdot \boldsymbol{\tau}_j^* + \boldsymbol{T}_a \cdot b_{A/q_j}. \tag{5.181}$$

It should be understood that the angular momentum, $\boldsymbol{H}$, and the applied torque, $\boldsymbol{T}_a$, are to be taken about the mass center, unless the index point is fixed. The above result gives the n Kane equations for the motion of a rigid body. As with the mass point system it is possible to use arbitrary linear combinations of the coordinate tangent vectors. The full local tangent vectors, $\boldsymbol{\tau}_j^* + b_{A/q_j} \times \boldsymbol{\xi}$, provide the individual members of the combinations, hence the same transformation laws apply individually to the center of mass component and the attitude component. Thus all the relations developed in chapter four apply with out change. The attitude component tangent vector derived as a coordinate Darboux vector b_{A/q_j}, will be denoted as ω_j, and following Kane will be called a *partial angular velocity.* Also from now on, when dealing with rigid bodies, the notation $\boldsymbol{\tau}_j$ indicates the tangent vector, $\frac{{}^A\partial \boldsymbol{\eta}}{\partial q_j}$. The greek letters β and γ are used to indicate general vectors which are linear combinations of $\boldsymbol{\tau}_j$ and $\boldsymbol{\omega}_j$ respectively.

5.9.1 Kvector Notation

The Kvector notation, introduced in chapter 3, will now be modified to include rigid body systems. We will also assume the general notational convention that an over dot on a symbol indicates differentiation with respect to a suitable inertial frame or reference system, which will frequently be designated as frame N. The momentum Kvector is defined by a forming a column matrix of vectors. Each point particle in the system is, as before, represented by one vector in the matrix, while each rigid body is represented by two vectors, the body's linear and angular momentum. Thus for a system consisting of a point mass body, designated as body 1 and a rigid body, designated as body 2, the momentum Kvector has the form

$$\boldsymbol{P}^{<} = \begin{bmatrix} \boldsymbol{P}^{<1} \\ \boldsymbol{P}^{<2} \\ \boldsymbol{H}^{<2} \end{bmatrix}, \tag{5.182}$$

where $\boldsymbol{P}^{<j}$ is the linear momentum of the body of its index point, usually but not always the center of mass of a rigid body, and $\boldsymbol{H}^{<j}$ is the angular momentum of the rigid body, again usually but not always taken with respect to the center of mass. The linear and rotational inertia forces are the time derivatives of the linear and angular momentum, therefore the *inertia force K-Vector* is given by

$$\dot{\boldsymbol{P}}^{<} = \begin{bmatrix} \dot{\boldsymbol{P}}^{<1} \\ \dot{\boldsymbol{P}}^{<2} \\ \dot{\boldsymbol{H}}^{<2} \end{bmatrix}. \tag{5.183}$$

This convention, of taking the two vectors associated with a rigid body for linear and rotational motion in turn as vector components of the systems overall Kvector, is also applied to the force components. For the above case where the first body is a particle we only consider a resultant force $\boldsymbol{R}^{<1}$, while the second body requires the specification of both a resultant force and torque, $\boldsymbol{R}^{<2}$ and $\boldsymbol{T}^{<2}$. It is understood that the torque must be taken about the index point, which is most commonly taken as the center of mass. Thus the total force Kvector has the form

$$\boldsymbol{R}^{<} = \begin{bmatrix} \boldsymbol{R}^{<1} \\ \boldsymbol{R}^{<2} \\ \boldsymbol{T}^{<2} \end{bmatrix}. \tag{5.184}$$

The j *th* tangent Kvector for the system, using this notation, will have the form

$$\boldsymbol{\tau}_j^{<} = \begin{bmatrix} \boldsymbol{\tau}_j^{<1} \\ \boldsymbol{\tau}_j^{<2} \\ b_{N/j} \end{bmatrix}. \tag{5.185}$$

Here it is assumed that the tangent vector is formed from components which are coordinate tangent vectors. The general case of tangent vectors which are formed as linear combinations of coordinate tangent vectors follows from the same reasoning given in chapter 3. The rotational component, which is usually found by inspection of angular velocity expressions given in terms of generalized speeds, will often be designated as

$$\boldsymbol{\omega}^{(L)}{}_j = {}^N\boldsymbol{\omega}^{B_j}. \tag{5.186}$$

To see how this is accomplished consider the velocity Kvector for the system, which in the case of our example is given by

$$\boldsymbol{v}^< = \begin{bmatrix} v^{<1} \\ v^{<2} \\ \boldsymbol{\omega}^{(2)} \end{bmatrix}. \tag{5.187}$$

To be explicit suppose that the system under consideration has two degrees of freedom, hence two independent generalized speeds, u_j. In the next chapter we will consider situations in which the generalized speeds are not independent, but for the moment let us ignore this complication. Following the reasoning of chapter 3, the velocity Kvector can always be expanded as a linear form in the generalized speeds, thus

$$\boldsymbol{v}^< = \boldsymbol{\beta}_1^< u_1 + \boldsymbol{\beta}_2^< u_2 + \boldsymbol{\beta}_t^<. \tag{5.188}$$

In practice each component of the tangent vector is obtained from the corresponding linear and angular velocity of the relevant body.

With this notation the Kane equations for the system has an identical form to that for the individual body or the system of point mass particles. Thus using the subscript a to indicate applied forces, the constraint free equation for the j *th* body can be written as

$$\dot{\boldsymbol{P}}^< \bullet \boldsymbol{\beta}_j^< = \boldsymbol{R}_a^< \bullet \boldsymbol{\beta}_j^<. \tag{5.189}$$

5.9.2 The Compound Pendulum

To gain some familiarity with this formulation we consider an example from basic mechanics, the motion of a compound pendulum. The geometric configuration is shown in figure 5.9, the pendulum being suspended from the point $\mathcal{P}$. To specify the problem we use the moments of the mass distribution about the point $\mathcal{P}$, which provides us with the position of the center of mass and the appropriate components of the moment of inertia dyad. The body fixed reference system b is chosen so that the vector $\boldsymbol{b}_1$ is parallel to the displacement from the suspension point to the center of mass point, the coordinate q_1 being chosen so that $\boldsymbol{b}_1 \cdot \boldsymbol{n}_1 = \cos(q_1)$. The inertial based triad is picked so that $\boldsymbol{n}_1$ aligns with the direction of the uniform gravitational acceleration. The compound pendulum is designated as body B, so that in our notation the moment of inertia dyad of B around the point $\mathcal{P}$ has the form:

$$\boldsymbol{I}^{B/\mathcal{P}} = b^T \begin{bmatrix} m_{22} & -m_{12} & 0 \\ -m_{12} & m_{11} & 0 \\ 0 & 0 & I \end{bmatrix} b, \tag{5.190}$$

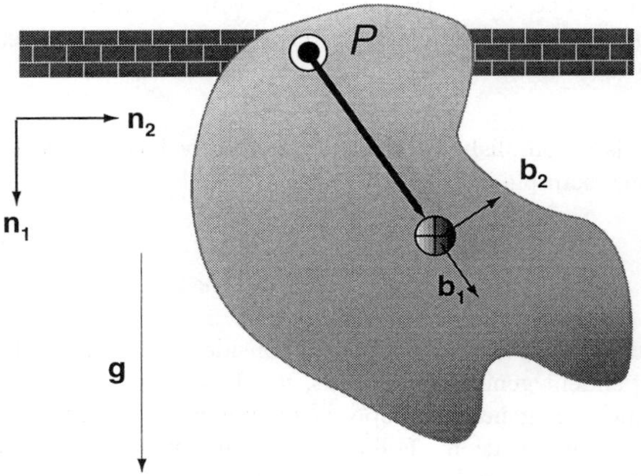

Figure 5.10: Compound Pendulum.

where we have set $I = m_{11} + m_{22}$. This simple system is holonomic and has but one degree of freedom. The obvious choice of a generalized speed is to take

$$\dot{q}_1 = u_1, \tag{5.191}$$

so that the angular velocity of the body B in the inertial frame N is given by

$${}^N\boldsymbol{\omega}^B = u_1 \boldsymbol{b}_3. \tag{5.192}$$

As the suspension point $\mathcal{P}$ is fixed, its velocity in the inertial frame is zero. Therefore

$$\boldsymbol{v}^< = \begin{bmatrix} 0 \\ u_1 \boldsymbol{b}_3 \end{bmatrix} \tag{5.193}$$

so that by inspection the single tangent Kvector is seen to be

$$\boldsymbol{\tau}_1^< = \begin{bmatrix} 0 \\ \boldsymbol{b}_3 \end{bmatrix}. \tag{5.194}$$

If the displacement vector from the suspension point to the center of mass is given by

$$\boldsymbol{\xi}^* = \mu \boldsymbol{b}_1, \tag{5.195}$$

then from our formulation of the equations of motion the linear momentum of the body is given by

$$\boldsymbol{P}^{<1} = m{}^N\boldsymbol{\omega}^B \times \boldsymbol{\xi}^* = m\mu u_1 \boldsymbol{b}_2, \tag{5.196}$$

while the angular momentum is given by

$$\boldsymbol{H}^{<1} = \boldsymbol{I}^{B/\mathcal{P}} \cdot {}^{N}\boldsymbol{\omega}^{B} = u_1 I \boldsymbol{b}_3. \tag{5.197}$$

The applied force and torque about the suspension point are given by

$$\boldsymbol{R}^{<1} = mg\boldsymbol{n}_1 \tag{5.198}$$

and

$$\boldsymbol{T}^{<1} = \mu \boldsymbol{b}_1 \times (mg\boldsymbol{n}_1). \tag{5.199}$$

The derivation of the equation of motion now proceeds from the inertial and applied force Kvectors projections in the direction of the single tangent Kvector. Thus we have

$$\dot{\boldsymbol{P}}^{<} = \begin{bmatrix} m\mu\dot{u}_1\boldsymbol{b}_2 - m\mu u_1^2\boldsymbol{b}_1 \\ I\dot{u}_1\boldsymbol{b}_3 \end{bmatrix}, \tag{5.200}$$

$$\boldsymbol{R}^{<} = \begin{bmatrix} mg\boldsymbol{n}_1 \\ \mu\boldsymbol{b}_1 \times (mg\boldsymbol{n}_1) \end{bmatrix}, \tag{5.201}$$

and

$$\boldsymbol{\tau}_1^{<} = \begin{bmatrix} 0 \\ \boldsymbol{b}_3 \end{bmatrix}. \tag{5.202}$$

The single dynamic or 'Kane' equation is thus given by

$$\dot{\boldsymbol{P}}^{<} \bullet \boldsymbol{\tau}_1^{<} = \boldsymbol{R}^{<} \bullet \boldsymbol{\tau}_1^{<}, \tag{5.203}$$

which reduces to

$$I\dot{u}_1 + \mu mg \sin q_1 = 0 \tag{5.204}$$

the kinematic equation in this case simply being the above relation that $\dot{q}_1 = u_1$.

To complete this simple problem we now use the notion of generalized Kvectors to determine the constraint force. By D'Alembert's Principle and the identification of tangent and cotangent vectors developed in chapter three the constraint force lies in the orthogonal complement to the tangent space, or in other words is a linear sum of the K-vectors that are orthogonal to the tangent K-vector. In the present case this implies that the constraint force must be given in the form:

$$\boldsymbol{R}_c^{<} = c_1\boldsymbol{\gamma}_1^{<} + c_2\boldsymbol{\gamma}_2^{<}, \tag{5.205}$$

where the K-vectors $\boldsymbol{\gamma}_1^{<}$ and $\boldsymbol{\gamma}_2^{<}$ are given by

$$\boldsymbol{\gamma}_1^{<} = \begin{bmatrix} \boldsymbol{b}_1 \\ 0 \end{bmatrix} \tag{5.206}$$

and

$$\boldsymbol{\gamma}_2^{<} = \begin{bmatrix} \boldsymbol{b}_1 \\ 0 \end{bmatrix}. \tag{5.207}$$

The constants c_1 and c_2 are then found from the equations

$$\dot{\boldsymbol{P}}^{<} \bullet \gamma_j = (\boldsymbol{R}^{<} + \boldsymbol{R}_c{}^{<}) \bullet \gamma_j. \tag{5.208}$$

The result is that

$$\frac{\boldsymbol{R}_c{}^{<1}}{mg} = -(\cos(q_1) + \frac{\mu}{g}u_1^2)\boldsymbol{b}_1 + (\frac{\mu}{g}\dot{u}_1 + \sin(q_1)\boldsymbol{b}_2). \tag{5.209}$$

It is instructive to carry out the solution of this problem using the center of mass as the index point in the body.

5.10 Computer Algebra Derivation of Equations of Motion

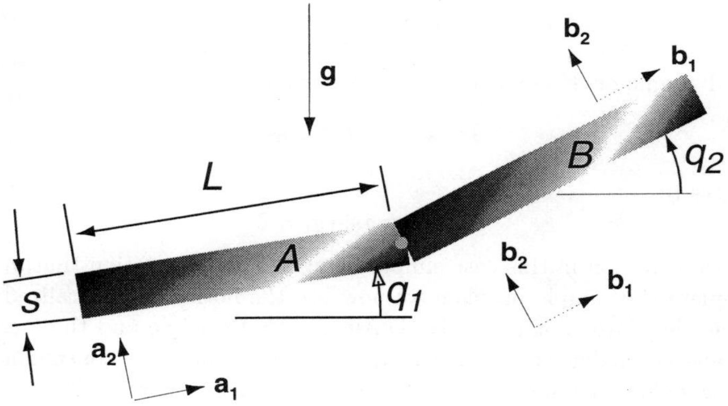

Figure 5.11: Hinged Bars

The Sophia routines have been designed to derive equations of motion for rigid body systems while retaining a reasonable connection with the way one would work when solving a mechanics problem with pen a paper. In this section we will discuss each step in detail. All the steps are put together at the end of this section so that they can serve as a model. We consider a relatively simple example, the planer motion of two identical bars of mass m, length L and thickness s that are connected by a torsion spring of strength k and placed in a uniform gravitational field g as shown in figure 5.11. By choosing a suitable reference frame we could eliminate the gravitational force, however for demonstration purposes it will be retained.

To deal with the configuration geometry we note that the mechanism has 4 degrees of freedom. The angles made by each bar with the vector $\boldsymbol{n}_1$ of an inertial reference frame serve as the generalized coordinates q_1 and q_2. The position of the hinge with respect to the inertial frame is given by the coordinates q_3 and q_4 which are distances measures along the $\boldsymbol{n}_1$ and $\boldsymbol{n}_2$ directions. Following the scaling ideas discussed in chapter 4 we choose natural units based on the mass, length and torsion. This is noted as the first steps in our input file. These could be changed if we wished to choose other sets of units as natural or desired our equations in dimensional form.

```
> m := 1:
> L := 1:
> k := 1:
```

The standard operator **&kde** is used to set up $q_{1..4}$ and$u_{1..4}$ as generalized coordinates and speeds dependent on time with the simple form of kinematic differential equations in which generalized speeds are the generalized velocities or time derivatives of the generalized coordinates.

```
> &kde 4:
```

The relationships between the inertial frame N and the frames A and B for the two bars are expressed as:

```
> &rot [N,A,3,q1]:
> &rot [N,B,3,q2]:
```

The moment of inertia dyad will be in diagonal form because of the symmetry of the bodies in the chosen reference frames. Also because the rods are two dimensional laminar the 3-3 component will be the sum of the other two diagonal components. Thus the inertia tensor will have the general form:

```
> IA:= EinertiaDyad(I1,I2,J,0,0,0,A):
> IB:= EinertiaDyad(I1,I2,J,0,0,0,B):
```

The computer algebra will be 'cleaner' if for most of the work we retain this general form. At the end of our calculations we might wish to insert the values for I1,I2 and J in terms of the body geometry. Therefore we set up a substitution set for this purpose, thus:

```
>  inertias:={I1=m*s^2/12,I2=m*(L^2)/12,J=I1+I2}:
```

The next step is to enumerate the active forces on the system, which includes the couple from the torsion spring and the gravitational force. In the following we have made use of the result that $\boldsymbol{a}_3 = \boldsymbol{b}_3 = \boldsymbol{n}_3$:

```
> Ta:= N &ev [0,0,k*(q2-q1)]:
> Tb:= N &ev [0,0,k*(q1-q2)]:
> Ra:= N &ev [0,-g,0]:
> Rb:= N &ev [0,-g,0]:
```

The configuration geometry is specified by giving the positions of the hinge and the mass centers of the two bars. The latter are expressed in terms of the relative positions of the mass centers to the hinge.

```
> rhinge:= N &ev [q3,q4,0]:
> rha:= A &ev [-1/2,0,0]:
> rhb:= B &ev  [1/2,0,0]:
```

```
> r1:= rhinge &++ rha:
> r2:= rhinge &++ rhb:
```

To obtain expressions for the linear momentum and to find tangent vectors we need the velocities of the individual mass centers. One of several ways of doing this is to simply take the time derivative of the mass center positions with respect to the inertial frame, thus:

```
> v1:=&simp subs(kde,N &fdt r1):
> v2:=&simp subs(kde,N &fdt r2):
```

In this problem the angular velocities of the frames A and B in N can be deduced by inspection as $u_1\boldsymbol{n}_3$ and $u_2\boldsymbol{n}_3$, though they can also be simply computed using the Sophia angular velocity operator `&aV`. In any case we now write:

```
> w1:= N &ev [0,0,u1]:
> w2:= N &ev [0,0,u2]:
```

Using the extended Kvector notation discussed above we form the velocity Kvector as

```
> vK:= &KM [v1,w1,v2,w2]:
```

There is now enough information to form the linear and angular momentum vectors for the two bodies, thus:

```
> p1 := m &** v1:
> p2 := m &** v2:
```

for the linear momentum and

```
> h1 := IA &o w1:
> h2 := IB &o w2:
```

for the angular momentum. These are then used to set up the extended momentum Kvector as:

```
> pK := &KM [p1,h1,p2,h2]:
```

Depending on problem complexity and the computer resources available it may be wise to first differentiate the expressions for the linear and angular momentum and to work to simplify them before forming the momentum derivative. From the viewpoint of clarity it is nicer to directly deal with the momentum Kvector. There is a special Sophia derivative operator `&Kfdt` for taking frame based time derivatives of Kvectors. Using this we have:

```
> pKt := &Ksimp subs(kde,N &Kfdt pK):
```

where we have combined the differentiation operation with a substitution and simplification command. The substitution is to obtain our equations in standard form by replacing terms such as q1t with u1. The operator `&Ksimp` is used for simplifing Kvectors in contrast to `&simp` which simplifies Evectors.

The basic data base for the problem is completed by gathering the applied force Evectors into a Kvector:

```
> RK := &KM [Ra,Ta,Rb,Tb]:
```

We can now proceed to the problem of the derivation of the equations of motion. The first step is to find a set of four independent tangent Kvectors to the configuration. The Sophia command `KMtangents` takes four arguments, the velocity Kvector, the name used for the generalized speeds and the number of degrees of freedom. It outputs a list of independent tangent Kvectors, what we have called an SKvector. We store this as the symbol tau, thus:

```
> tau := KMtangents(vK,u,4):
```

The Sophia operator **&kane** produces a list of fat dot products between the members of a list of Kvectors given as the left argument and a single Kvector given as the right argument. We store the list obtained from using tau and RK as the symbol 'gaf', for generalized active forces (a list of what Kane calls F_j). Thus the list of generalized active forces is given by:

```
> gaf := tau &kane RK:
```

This operator can also be used to obtain a list of the generalized inertia forces, or using the symbol 'mgif' to denote the list of negative generalized inertia forces (Kane's $-F_j^*$)

```
> mgif := tau &kane pKt:
```

The Kane form of the dynamic equations is then formed from the components of these two lists. Here we use some other Maple functions to produce a 'nice' result. Thus collect gathers terms as coefficients of the list given as its second argument. Sort, does just that in the order of the list given as its second argument. Finally the seq function is used to put all the equations on a single list called 'eqns':

```
> for i from 1 to 4 do
> eq.i := sort(collect(-mgif[i] + gaf[i],[u1t,u2t,u3t,u4t]),
>         [u1t,u2t,u3t,u4t])
> od:
> eqns := [seq(eq.i=0,i=1..4)]:
```

These four dynamic or Kane's equations together with the kinematic differential equations provide the eight first order equations required for the problem.

For the readers reference some specific results are now given, thus Kane's equations are:

- $$(-J - 1/4)\, u_{1t} - \frac{\sin(q_1) u_{3t}}{2} + \frac{\cos(q_1) u_{4t}}{2} + \frac{\cos(q_1) g}{2} + q_2 - q_1 = 0$$

- $$(-J - 1/4)\, u_{2t} + \frac{\sin(q_2) u_{3t}}{2} - \frac{\cos(q_2) u_{4t}}{2} - \frac{\cos(q_2) g}{2} + q_1 - q_2 = 0$$

- $$\frac{\sin(q_2) u_{2t}}{2} - \frac{\sin(q_1) u_{1t}}{2} - 2\, u_{3t} + \frac{\cos(q_2) u_2^2}{2} - \frac{\cos(q_1) u_1^2}{2} = 0$$

- $$\frac{\cos(q_1) u_{1t}}{2} - \frac{\cos(q_2) u_{2t}}{2} - 2\, u_{4t} - \frac{\sin(q_1) u_1^2}{2} + \frac{\sin(q_2) u_2^2}{2} - 2\, g = 0$$

If desired the inertia component substitution list could be used with subs to replace J in terms of body parameters. Further work on this example is suggested in the problem section. We close this section with a list of the entire set of steps as they might be input into the Sophia-Maple system:

```
> m := 1:
> L := 1:
> k := 1:
> &kde 4:
> &rot [N,A,3,q1];
> &rot [N,B,3,q2];
> IA:= EinertiaDyad(I1,I2,J,0,0,0,A);
> IB:= EinertiaDyad(I1,I2,J,0,0,0,B);
> inertias:={I1=m*s^2/12,I2=m*(L^2)/12,J=I1+I2};
> Ta:= N &ev [0,0,k*(q2-q1)];
> Tb:= N &ev [0,0,k*(q1-q2)];
> Ra:= N &ev [0,-g,0];
> Rb:= N &ev [0,-g,0];
> rhinge:= N &ev [q3,q4,0];
> rha:= A &ev [-1/2,0,0];
> rhb:= B &ev  [1/2,0,0];
> r1:= rhinge &++ rha;
> r2:= rhinge &++ rhb;
> v1:=&simp subs(kde,N &fdt r1);
> v2:=&simp subs(kde,N &fdt r2);
> w1:= N &ev [0,0,u1];
> w2:= N &ev [0,0,u2];
> vK:= &KM [v1,w1,v2,w2];
> ##########################################
> p1 := m &** v1;
> p2 := m &** v2;
> h1 := IA &o w1;
> h2 := IB &o w2;
> pK := &KM [p1,h1,p2,h2];
> pKt := &Ksimp subs(kde,N &Kfdt pK);
> RK := &KM [Ra,Ta,Rb,Tb];
> ##########################################
> tau := KMtangents(vK,u,4);
> ###########################################
> ### generalized active forces
> gaf := tau &kane RK;
> ### - generalized inertial forces
> mgif := tau &kane pKt;
> ##### Equations of Motion
> for i from 1 to 4 do
> eq.i := sort(collect(-mgif[i] + gaf[i],[u1t,u2t,u3t,u4t]),
>          [u1t,u2t,u3t,u4t])
> od;
> eqns := [seq(eq.i=0,i=1..4)];
```

The reader who has mastered the material of the last five chapters is equipped to solve a large class of multibody mechanics problems. The computer algebra tools should permit the solution of systems which can produce quite complex equations of motion. Even a few rigid bodies can lead to great complexity which will defeat the abilities of most computer algebra systems. To solve such problems may require the use of purely numerical procedures for the elimination of constraints or careful rethinking of the basic model in order to match the capabilities of computer algebra. The remaining chapters are intended to provide tools for the treatment of some of the typical complications that arise. These include closed mechanical loops in the system, nonholonomic constraints, impact and dry friction.

5.11 Problems

•Problem 5.1

The flywheel of mass m, height s and radius L rotates about an axis oriented in the $\boldsymbol{n}_3$ direction. A pin joint is fixed to the flywheel at half the radial distance from the center. The light rod connects a sphere of diameter s to the wheel. The distance between the pin joint and the spheres center is L. The rod is pinned so that it is constrained to move in a plane orthogonal to the wheel that also passes through its axis of rotation. Derive a set of first order equations for the description of the motion of this mechanism. Carry out some numerical simulations using your equations.

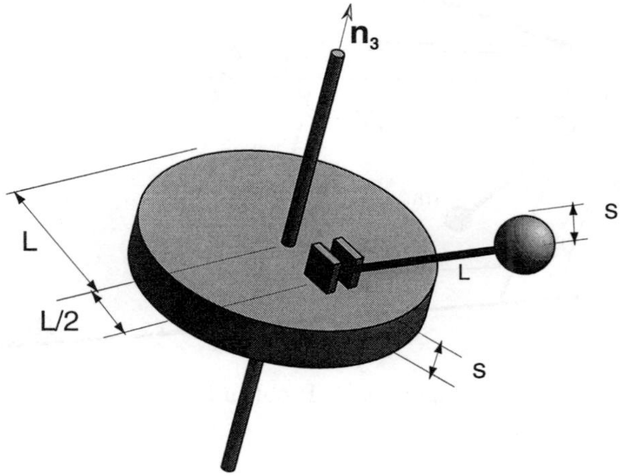

Figure 5.12: Problem 5.1

•Problem 5.2

In the 1970's the physicist G.K. O'Niel suggested the possibility of placing artificial habitats at the so called L5 points in the Earth-Moon gravitational system. One of the major reasons was the easy availability of solar power for these 'space-colonies'. O'Niel's model for such a colony consisted of two tethered cylinders, each a kilometer long rotating about their common axis and spinning about their individual axis to replicate earth gravity. The colony axis would point toward the sun and three large mirrors would be formed by longitudinal sections which could open up to a desired angle. The above sketch is intended to suggest a simplified model for the study of the effect on the main cylinder of opening or changing the angle of the mirror system. Another problem would be the effect of small differences in the mirror angles on the stability of the colony. The two small masses are attached by movable rods to the base of the mechanism as indicated to represent a simplified mirror system. Assume that the colony cylinder rotates about its own axis with variable angular velocity $\boldsymbol{\Omega}_c$ and the whole system rotates about the line of symmetry between the tethered colonies with fixed angular velocity $\boldsymbol{\Omega}_s$. Derive a set of equations to study this problem based on the suggested model. The colony cylinder is to be treated as a thin shell. Also treat the attached masses as spheres. Assume that they move in the plane passing through the axis of the colony cylinder. Choose parameters and constants at what you believe are reasonable values.

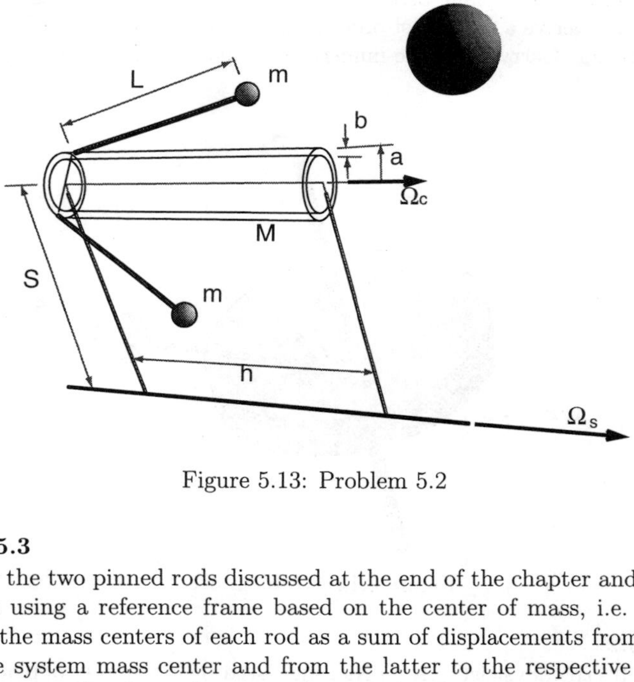

Figure 5.13: Problem 5.2

•Problem 5.3

Consider the two pinned rods discussed at the end of the chapter and reformulate the problem using a reference frame based on the center of mass, i.e. express the positions of the mass centers of each rod as a sum of displacements from the inertial frame to the system mass center and from the latter to the respective rod centers. Pick suitable generalized coordinates and speeds for this representation.

•Problem 5.4

Develop the equations of motion and expressions for the constraint forces for the three mass point system used as an example in section 5.1.

•Problem 5.5

Find the moment of inertia dyad about the center of mass of the compound body shown in figure 5.14. The body consists of a right circular cone attached to a main cylinder. Four identical smaller cylinders are attached by four thin square plates. Assume that the density of the cones and cylinders is ρ and the area density of the plates is σ.

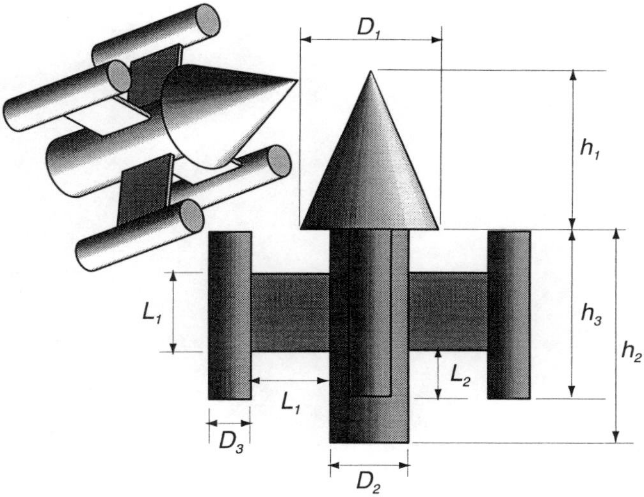

Figure 5.14: Problem 5.5

•Problem 5.6

Prove the expression for the kinetic energy of a rigid body.

Chapter 6

Redundant Variables and Nonholonomic Systems

The first part of this chapter explores the idea of introducing extra coordinates into a problem. This has several advantages, the most obvious of which is a simplification in the specification of constraint conditions. The technique also has a natural extension to the treatment of what are known as *nonholonomic* problems. A class of these problems involve velocity constraints. This is a situation that frequently arises in problems with rolling conditions. It will soon be evident that even positional constraints can be replaced by velocity constraints. This is not a two way path! Velocity constraints cannot always be converted to positional constraints. Therefore what starts out as a convenient reformulation becomes a necessity. The distinction and solution is very easily seen in Kane's approach to mechanics, which is not surprising as the approach was first taken to confront this very problem.

In the first part of this discussion we will examine the extra coordinate or redundant variable technique. This will be followed by a treatment of closed loop systems for which the technique has major advantages. It will then be seen that a large class of nonholonomic problems can be resolved by the same technique.

6.1 A Complex Solution of The Simple Pendulum Problem

The simple pendulum provides a familiar mechanism for the introduction of the idea of redundant coordinates. General purpose numerical programs for the treatment of complex mechanical systems require explicit techniques for the specification of a mechanism's geometry. A popular way of doing this is somewhat misguidedly referred to as *absolute coordinates.* Absolute coordinates are simply the use of cartesian coordinates in a common, usually inertial, frame. All the critical parts of the mechanism are assigned cartesian coordinates in that frame. For example a bar constrained to move in two dimensions normally requires three generalized coordinates to define its position and attitude. Absolute coordinates might be taken as the cartesian coordinates of two points fixed in the bar, e.g. its center of mass and an end point. The configuration is specified by four rather than three coordinates. The coordinates are redundant, because they can not be chosen independently of each other. In the case of the rigid bar the distance between the two chosen coordinate points is invariant. This constraint relation must play a role in the derivation of appropriate equations of motion in such absolute coordinates. This is illustrated by the problem of the simple pendulum problem as shown in figure 6.1. The origin of the *absolute* coordinate

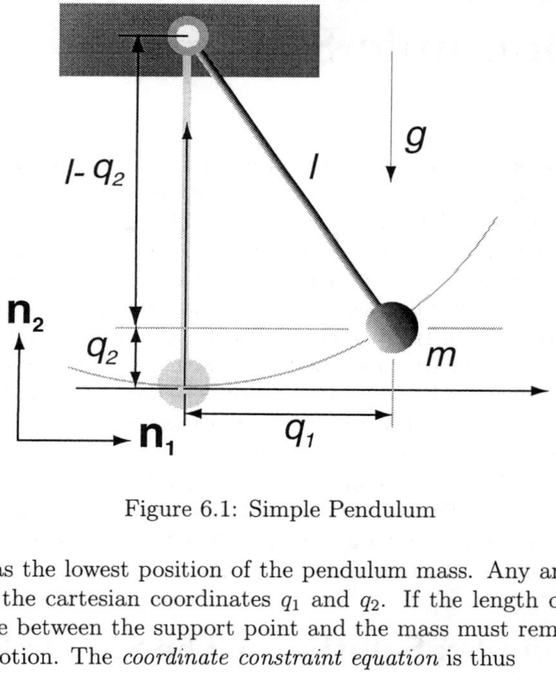

Figure 6.1: Simple Pendulum

frame is taken as the lowest position of the pendulum mass. Any arbitrary position is described by the cartesian coordinates q_1 and q_2. If the length of the pendulum is l, the distance between the support point and the mass must remain l during the course of any motion. The *coordinate constraint equation* is thus

$$(l - q_2)^2 + {q_1}^2 = l^2. \tag{6.1}$$

An obvious way of proceeding would be to solve this equation for q_1 and derive equations in terms of this one coordinate. There are several reasons for not using this idea. The nonlinearity of the constraint condition implies that several solutions exist, hence we would have to be careful about the possibility of jumping from one solution branch to another. Also it must be recognized that in a general situation it is not likely that it will be possible to explicitly solve the constraint conditions so as to eliminate the redundant coordinates. Finally it may be that the cartesian description is the most useful one and should be retained as part of the solution procedure.

The key to this problem is the realization that the constraint relation implies *linear relations* between the rates of change of the cartesian coordinates. It might be expected that such linear relations will be more convenient to manipulate. Thus obtain a *velocity constraint equation* by simply differentiating the positional constraint as:

$$q_1\dot{q}_1 - (l - q_2)\dot{q}_2 = 0. \tag{6.2}$$

To emphasize the linear structure of this relation, introduce generalized speeds u_j defined by the simple kinematic differential equations,

$$\dot{q}_1 = u_1 \tag{6.3}$$
$$\dot{q}_2 = u_2. \tag{6.4}$$

The differentiated constraint relation is seen to provide a linear equation between the generalized speeds, i.e.

$$q_1 u_1 - (l - q_2)u_2 = 0. \tag{6.5}$$

This equation could now be used to replace u_2 by u_1 in all velocity expressions. An equivalent but somewhat more systematic way of proceeding is to define a *generalized speed transformation* to a new set of generalized speed parameters, i.e.

$$w_1 = u_1 \tag{6.6}$$
$$w_2 = q_1 u_1 - (l - q_2)u_2. \tag{6.7}$$

The velocity constraint equation for the new generalized speed variables is then given by

$$w_2 = 0. \tag{6.8}$$

Since we will need to replace occurences of the original generalized speeds, u_j by w_j it is useful to obtain the inverse of the generalized speed transformation. For the present problem the *inverse generalized speed transformation* equations are:

$$u_1 = w_1 \tag{6.9}$$
$$u_2 = \frac{q_1}{l - q_2}w_1 - \frac{1}{l - q_2}w_2 \tag{6.10}$$

An immediate application of the inverse generalized speed transformation is to derive the *transformed kinematic differential equations.* In general the kinematic differential equations equate the derivatives of the generalized coordinates to functions of the generalized coordinates and speeds. In the simple case, where generalized speeds are generalized velocities, replace the left hand side of the inverse generalized speed transformation equations with the coordinate derivatives. Thus the transformed kinematic differential equations for the pendulum problem are:

$$\dot{q}_1 = w_1 \tag{6.11}$$
$$\dot{q}_2 = \frac{q_1}{l - q_2}w_1 - \frac{1}{l - q_2}w_2 \tag{6.12}$$

The application of the *velocity constraint equation* to the transformed kinematic differential equation and the inverse generalized speed transformation provides the basic relations needed to resolve the problem. The choice of the new generalized speeds w_j was made to make this task as trivial as possible, i.e. we simply set $w_2 \to 0$. The *reduced inverse generalized speed transformation* is thus

$$u_1 = w_1 \tag{6.13}$$
$$u_2 = \frac{q_1}{l - q_2} w_1. \tag{6.14}$$

The *reduced transformed kinematic differential equations* are

$$\dot{q}_1 = w_1 \tag{6.15}$$
$$\dot{q}_2 = \frac{q_1}{l - q_2} w_1.$$

Equipped with these results it is possible to proceed as in the non-redundant coordinate case. Thus in terms of the original generalized speeds the velocity Kvector (in this case an ordinary vector) is given by:

$$\boldsymbol{v}^< = u_1 \boldsymbol{n}_1 + u_2 \boldsymbol{n}_2. \tag{6.16}$$

The coefficients of the generalized speeds are not independent tangent vectors because of the velocity constraint equation. To obtain a set (in this case one) of independent tangent vectors substitute for the generalized speeds using the reduced inverse generalized speed transformation. Thus the *reduced tangent vector set* is found from the reduced velocity expression

$$\boldsymbol{v}^< = w_1(\boldsymbol{n}_1 + \frac{q_1}{l - q_2} \boldsymbol{n}_2). \tag{6.17}$$

The single independent reduced tangent vector for the problem at hand is

$$\boldsymbol{\beta}_1^< = \boldsymbol{n}_1 + \frac{q_1}{l - q_2} \boldsymbol{n}_2. \tag{6.18}$$

Kane's equations are now derived in the usual manner. Form the momentum and applied force vectors and take their dot products with the single independent reduced tangent vector. Thus

$$p^< = m w_1 \boldsymbol{\beta}_1^<, \tag{6.19}$$
$$\dot{p}^< = m \dot{w}_1 \boldsymbol{\beta}_1^< + \frac{m w_1^2}{l - q_2}(1 + \frac{{q_1}^2}{(l - q_2)^2}) \boldsymbol{n}_2. \tag{6.20}$$

The applied force is

$$\boldsymbol{R}^< = -mg \boldsymbol{n}_2. \tag{6.21}$$

The force balance is then given by

$$\dot{p}^{<} \cdot \beta_1^{<} = \tag{6.22}$$

$$= (1 + \frac{{q_1}^2}{(l-q_2)^2})(m\dot{w}_1 + \frac{m w_1^2 q_1}{(l-q_2)^2} \tag{6.23}$$

$$= \boldsymbol{R}^{<} \cdot \beta_1^{<} \tag{6.24}$$

$$= -\frac{m g q_1}{l - q_2}. \tag{6.25}$$

The coordinate constraint equation can be used to put the result in simpler form. The *reduced projected force balance equation* is thus:

$$\dot{w}_1 = -\frac{q_1 w_1^2}{(l-q_2)^2} - \frac{g q_1 (l - q_2)}{l^2}. \tag{6.26}$$

This equation, together with the reduced transformed kinematic differential equations 6.15, form a set of three first order equations in the variables q_1, q_2 and w_1. In solving these equations it is extremely important to enforce a *consistency* requirement between the initial values of q_1, q_2, that is they must satisfy the coordinate constraint equation that the distance between the fixation point and the pendulum mass is l.

6.2 The Reduction Algorithm

Following the example of the pendulum represented in redundant coordinates it useful to set forth a general procedure for the derivation of independent equations of motion. To state the procedure as an algorithm it helps to use abbreviations for the various equations that arise. These abbreviations are also useful as meaningful variable names in computer algebra work. The original set of kinematic differential equations are denoted by *kde*, the coordinate constraint equations by *cce*. The velocity constraint equations, *vce*, are derived by differentiation of the *cce* with respect to time. It will *always be assumed* that both the *cce* and the *vce* are expressed so that the left side equates numerically to a zero right hand side. The generalized speeds, *gs,* are transformed, as discussed in chapter 3, by an affine transformation. This transformation is denoted as the generalized speed transformation *gst*. The inverse general speed transformation *igst* provides expressions for the original *gs* in terms of the transformed generalized speeds, *tgs*. The application of the velocity constraint equations, *vce*, are applied to the inverse generalized speeds transformation, *igst*, to obtain the reduced inverse generalized speed transformation, *rigst*. The kinematic differential equations become the transformed kinematic differential equations, *tkde*. Application of the transformed velocity constraint equation,*tvce*, provides the reduced transformed kinematic differential equations, *rtkde*. These abbreviations are translated by noting that the letters always stand for the same word combinations. Thus cc → coordinate constraint, d → differential, e → equation, g → generalized, i →

inverse, k → kinematic, r → reduced, s → speeds and t → transformed. The symbol n_g represents an integer equal to the assumed number of coordinates. The number of coordinate constraint equations, *cce*, is represented by n_c, and the number of independent coordinates by n_i. The relation

$$n_i + n_c = n_g, \tag{6.27}$$

is expected to hold. The number n_i represents the *minimum number of coordinates* needed to give a unique *local* specification of the mechanisms configuration. Any more coordinates are redundant. The qualification *local* is in recognition of the fact that in general it is not possible to find a single coordinate system which describes all possible configurations. The convention is that one starts with a set of generalized speeds denoted by u and transforms to a new set denoted by w. The original *kde* equate the time derivatives of the coordinates q, that is $\dot{q}$ to functions of the generalized speeds u and coordinates q. The transformed kinematic differential equations, *tkde,* equate $\dot{w}$ to functions of w and q.

The *gst* from u to w is chosen so that $w_{n_i+1}, w_{n_i+2} \ldots w_{n_g}$ equal the n_g, *vce.* Therefore the transformed velocity constraint equations are:

$$w_j = 0, \tag{6.28}$$

where $j = n_i \ldots n_g$. Note that the ordering of these relations is arbitrary. The transformed generalized speeds $w_{n_1} \ldots w_{n_i}$ can be chosen as any independent subset of n_i of the original n_g generalized speeds.

The algorithm is thus:

1. cce → vce
2. gst(gs → tgs) or u → w
3. subset of n_i independent u → $w_1 \ldots w_{n_i}$
4. The remaining n_c w are set equal to the vce forming the tvce
5. invert the gst → igst
6. substitute the igst into the kde → tkde
7. substitute the tvce into the igst → rigst
8. substitute the tvce into the tkde → rtkde
9. substitute rtkde and rigst as needed → reduced Kvector velocity.
10. proceed as in the non-redundant case to obtain the reduced set of tangent Kvectors and the equations of motion

11. note that the rtkde should be used after time differentiations to obtain equations in w and q.

The end result will be n_i dynamic or Kane equations and n_g kinematic differential equations in the n_g coordinates and the n_i generalized speeds. If non-independent choices of the n_i u are made, problems will arise with finding the needed inverse relations. This can be corrected by making new choices.

•**Illustration** Figure 6.2 shows a simple mechanical system composed of three mass

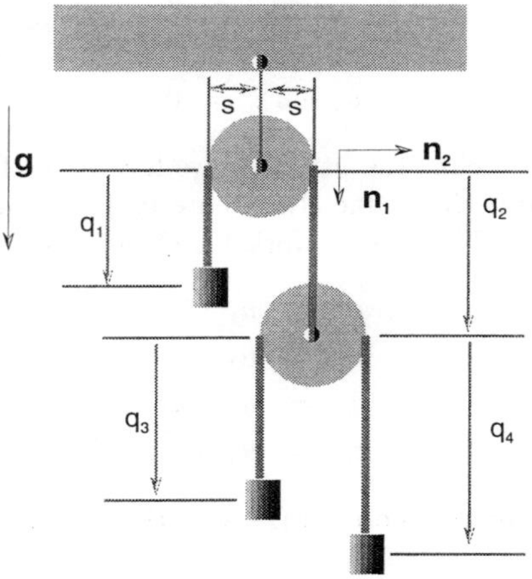

Figure 6.2: Pulley System

points suspended from two massless pulleys. The coordinate constraint is provided by the constant length of the suspension cords. Let l_1 and l_2 denote the exposed length, that is the actual lengths minus the half circumference of the pulleys. The constant gravitational acceleration is in the direction of the $\boldsymbol{n}_1$ unit vector, which is fixed in an inertial frame. The pulleys have radius s and are of equal mass. The positions of the mass points are:

$$\boldsymbol{r}^{<1} = q_1\boldsymbol{n}_1 - s\boldsymbol{n}_2, \tag{6.29}$$

$$\boldsymbol{r}^{<2} = (q_2 + q_3)\boldsymbol{n}_1, \tag{6.30}$$

$$\boldsymbol{r}^{<3} = (q_2 + q_4)\boldsymbol{n}_1 + 2s\boldsymbol{n}_2. \tag{6.31}$$

The kinematic differential equations, *kde* are chosen so that the generalized speeds are generalized velocities for the chosen coordinates, thus $\dot{q}_j = u_j$, $j = 1 \ldots 3$. Therefore the velocity Kvector is:

$$\boldsymbol{v}^{<} = \begin{bmatrix} u_1 \boldsymbol{n}_1 \\ (u_2 + u_3)\boldsymbol{n}_1 \\ (u_2 + u_4)\boldsymbol{n}_1 \end{bmatrix}. \tag{6.32}$$

Now we apply the algorithm to obtain the *rigst* and the *rtkde.* First express the *cce* which shall be differentiated to obtain the *vce*:

$$\begin{aligned} q_1 + q_2 &= l_1 \\ q_3 + q_4 &= l_2. \end{aligned} \tag{6.33}$$

After differentiation and substitution of the base generalized speeds for the coordinate derivatives one arrives at the vce's:

$$\begin{aligned} u_1 + u_2 &= 0 \\ u_3 + u_4 &= 0. \end{aligned} \tag{6.34}$$

Step two, three and four define the *gst.* Note we have some freedom with step three, however it should be clear that some choices are inappropriate. For example u_1 and u_2 are clearly dependent. Therefore a workable choice is given by:

$$w_1 = u_1 \tag{6.35}$$

$$w_2 = u_3 \tag{6.36}$$

$$w_3 = u_1 + u_2 \tag{6.37}$$

$$w_4 = u_3 + u_4. \tag{6.38}$$

This system is easily inverted to obtain the *igst*, thus

$$\begin{aligned} u_1 &= w_1 \\ u_2 &= w_3 - w_1 \\ u_3 &= w_2 \\ u_4 &= w_4 - w_2. \end{aligned} \tag{6.39}$$

Substitution of this result into the *kde* gives the *tkde*, which are

$$\begin{aligned} \dot{q}_1 &= w_1 \\ \dot{q}_2 &= w_3 - w_1 \\ \dot{q}_3 &= w_2 \\ \dot{q}_4 &= w_4 - w_2. \end{aligned} \tag{6.40}$$

The transformed velocity constraint equations, *tvce*, are

$$\begin{aligned} w_3 &= 0 \\ w_4 &= 0. \end{aligned} \tag{6.41}$$

Steps 7 and 8 provide us with the *rigst* and *rtkde*, hence

$$\begin{aligned} u_1 &= w_1 \\ u_2 &= -w_1 \\ u_3 &= w_2 \\ u_4 &= -w_2, \end{aligned} \tag{6.42}$$

and

$$\begin{aligned} \dot{q}_1 &= w_1 \\ \dot{q}_2 &= -w_1 \\ \dot{q}_3 &= w_2 \\ \dot{q}_4 &= -w_2. \end{aligned} \tag{6.43}$$

The *rigst* is now applied to the velocity Kvector to obtain:

$$\boldsymbol{v}^< = \begin{bmatrix} w_1\boldsymbol{n}_1 \\ (-w_1 + w_2)\boldsymbol{n}_1 \\ (-w_1 - w_2)\boldsymbol{n}_1) \end{bmatrix}. \tag{6.44}$$

The independent tangent vectors are picked off by inspection of this result as:

$$\boldsymbol{\beta}_1^< = \begin{bmatrix} \boldsymbol{n}_1 \\ -\boldsymbol{n}_1 \\ -\boldsymbol{n}_1 \end{bmatrix} \tag{6.45}$$

$$\boldsymbol{\beta}_2^< = \begin{bmatrix} 0 \\ \boldsymbol{n}_1 \\ -\boldsymbol{n}_1 \end{bmatrix}. \tag{6.46}$$

We now proceed as in the non-redundant coordinate case, using the *rigst* and the *rtkde* as needed to eliminate any u_j from the equations. Thus the momentum Kvector is:

$$\boldsymbol{p}^< = mw_1\boldsymbol{\beta}_1^< + mw_2\boldsymbol{\beta}_2^<. \tag{6.47}$$

The rate of change of the momentum with respect to the inertial frame N is;

$$\dot{\boldsymbol{p}}^< = m\dot{w}_1\boldsymbol{\beta}_1^< + m\dot{w}_2\boldsymbol{\beta}_2^<. \tag{6.48}$$

The applied force Kvector, in this case a result of the gravitational or weight terms, is

$$\boldsymbol{R}^< = mg\begin{bmatrix} \boldsymbol{n}_1 \\ \boldsymbol{n}_1 \\ \boldsymbol{n}_1 \end{bmatrix}. \tag{6.49}$$

The two dynamic equations are obtained in the usual manner, thus

$$\begin{aligned} \dot{\boldsymbol{p}}^{<} \bullet \boldsymbol{\beta}_1^{\lessdot} &= 3m\dot{w}_1 \\ &= \boldsymbol{R}^{<} \bullet \boldsymbol{\beta}_1^{\lessdot} \\ &= -mg, \end{aligned} \tag{6.50}$$

and

$$\begin{aligned} \dot{\boldsymbol{p}}^{<} \bullet \boldsymbol{\beta}_2^{\lessdot} &= 2m\dot{w}_2 \\ &= \boldsymbol{R}^{<} \bullet \boldsymbol{\beta}_2^{\lessdot} \\ &= 0. \end{aligned} \tag{6.51}$$

These and the *rtkde* provide six first order equations for the w and q. In this case they are easy to integrate, thus

$$\begin{aligned} q_1 &= -\frac{1}{6}gt^2 + w_1(0)t + q_1(0) \\ q_2 &= \frac{1}{6}gt^2 - w_1(0)t + q_2(0) \\ q_3 &= w_2(0)t + q_3(0) \\ q_4 &= -w_2(0)t + q_4(0). \end{aligned} \tag{6.52}$$

. Note that the initial conditions must satisfy the constraint relations, thus $q_2(0) = l_1 - q_1(0)$, with similar relations for the other terms. •

6.3 Computer Algebra Techniques For Redundant Systems

If there is any aspect of the formulation of dynamic equations of motion that involves tedious symbol manipulation, redundant coordinate techniques are a prime example. In this section we examine a simple closed loop mechanism, the crank-slider device of figure 6.3, by use of redundant coordinates. Thus three coordinates are used to describe this one degree of freedom system. The notations introduced in the last section to denote various equation sets and transformations provides meaningful variable names, e.g. the set of reduced transformed kinematic differential equations is stored in the variable *rtkde*. The work is presented in a format very close to a Maple session with comments. A new Sophia function, *ReducedSpeeds* is introduced to implement the laborious aspects of the reduction algorithm.

The crank-slider consists of a rotating wheel with an attached connecting rod and piston. The piston is constrained to move in a straight line which passes through the wheel center. Thus the mechanism consists of two revolute joints and three rigid bodies. The assumption is made that we can attack this as a planar problem, hence certain simplifications are made in specifying the inertia dyads and angular velocities that would not suffice for a three dimensional mechanism. The system represents a degenerate case of a four bar mechanism, such as was treated in chapter two, where one of the grounded links is moved to infinite distance. As a notational matter q1, q2 etc. are used for q_1, q_2, q_3 and other similar quantities. This is a convenient representation for Maple. In a similar manner derivatives of declared variables are indicated by the short hand notation of attaching strings of the letter 't' to a variable, e.g. q1t for $\dot{q}_1$.

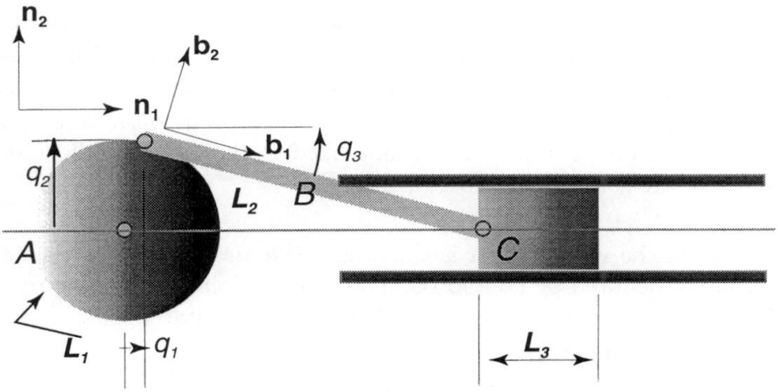

Figure 6.3: Crank-Slider

The first step is to set the size of the rotating wheel and the length of the connecting rod. The latter is used as the characteristic dimension of the problem, hence following the discussion of chapter 4, it is taken as unity. A case is studied in which the wheel has a radius that is 1/10 of this dimension. It is also convenient to define a zero valued Evector.

```
> L1:= 1/10: L2:= 1: zeroK := N &ev [0,0,0]:wA3:='wA3':
```

The kinematic differential equations are defined in the simplest form and the time dependence of generalized coordinates and velocities are defined by the following statement.

```
> &kde 3;
q1   declared
q2   declared
q3   declared
u1   declared
u2   declared
u3   declared
```

Generalized Velocity Form of Kinematic d.e.

Saved as the GLOBAL Varable kde

$$\{ q1t = u1 , q2t = u2 , q3t = u3 \}$$

The redundant coordinate technique requires a transformation to a new set of generalized speeds, which will be designated by w1 through w3. The time dependence of these, as well as a possible torque applied to the wheel, is declared.

```
> dependsTime(w1,w2,w3,TA3):
w1   declared
w2   declared
w3   declared
TA3   declared
```

The coordinates q1 and q2 describe the position of the revolute joint connecting the rod to the wheel in a coordinate system with origin at the wheel's center. The coordinate q3 denotes the angle of the rod with the line between the wheel center and the piston center. Note the convention that this angle is positive in the counter clockwise direction from this line. The frame fixed in the rod, as indicated in figure 6.3, is defined by the rotation statement, thus

```
> &rot [N,B,3,q3]:
```

The position of the wheel center is the origin, the mass centers of the connecting rod and piston are denoted by rB and rC. The revolute joint connecting rod to wheel is at r1. These points are defined by the following Sophia statements:

```
> r1 := N &ev [q1,q2,0]:
> rB := (B &ev [L2/2,0,0]) &++ r1:
> rC := (B &ev [L2,0,0]) &++ r1 &++ (N &ev [L3/2,0,0]):
```

The objects are considered as rigid bodies, hence in general we must specify their moment of inertia dyads. In two dimensional problems only the out of plane dyad components need be specified. As all three bodies possess the same out of frame triad vector as the inertial frame N, all the inertia dyads are defined so as only to take one component into account and are given in the inertial frame. Note that in a true spatial problem this would lead to incorrect results, i.e. it is only a valid simplification for planar problems. Thus:

```
> IA := EinertiaDyad(0,0,JA,0,0,0,N):
> IB := EinertiaDyad(0,0,JB,0,0,0,N):
> IC := EinertiaDyad(0,0,JC,0,0,0,N):
```

The Kvector velocity requires the computation of the center of mass velocities and angular velocities of the involved bodies, some of which are the zero vector. The linear velocities are calculated for bodies A, B and C and for the revolute joint 1:

```
> vA := zeroK:
> vB := N &fdt rB;
```

$$vB := \left[\left[-\frac{1}{2} \sin(q3)\, q3t + q1t \; \frac{1}{2} \cos(q3)\, q3t + q2t \; 0 \right] , N \right]$$

```
> vC := N &fdt rC;
```

$$vC := [[-\sin(q3)\,q3t + q1t\,\cos(q3)\,q3t + q2t\,0], N]$$

```
> v1 := N &fdt r1;
```

$$v1 := [[q1t\;q2t\;0], N]$$

The specification of the angular velocity of body A is done by using the velocity of the point 1. That is we compute the velocity of this point using the standard angular velocity equation for two points fixed in the same rigid body. First define the angular velocity vector in terms of its component, which is to be found.

```
> wA := N &ev [0,0,wA3];
```

$$wA := [[0\;0\;wA3], N]$$

The velocity of point 1, calculated from the angular velocity is:

```
> v1Q := wA &xx r1;
```

$$v1Q := [[-wA3\;q2\;wA3\;q1\;0], N]$$

The result, designated as v1Q must be the same as v1, hence we appear to have two equations for the single component wA3. The constraint conditions on the redundant coordinates make it necessary to only consider one of the equations, the other being a velocity constraint condition. We will derive the velocity constraint conditions from coordinate constraint conditions. Thus we only use the equation obtained from equating the first components. The angular velocities of body B is found using the Sophia angular velocity operator. The result for body C is clearly the zero vector. Therefore:

```
> eqw := {(v1 &c 1) = (v1Q &c 1)};
```

$$eqw := \{\, q1t = -wA3\; q2 \,\}$$

```
> wA3 := rhs(op(solve(eqw,wA3)));
```

$$wA3 := -\frac{q1t}{q2}$$

```
> wA := N &ev [0,0,wA3];
```

$$wA := \left[\left[0\;0\; -\frac{q1t}{q2}\right], N\right]$$

```
> wB := N &aV B; wC := zeroK;
```

$$wB := [[0\;0\;q3t], B]$$

$$wC := [[0\;0\;0], N]$$

The coordinate constraint equations are found by using the redundant coordinates to specify the radius of the wheel and the length of the connecting rod. It should be evident that the present system's configuration can be specified by only one coordinate, thus with three coordinates we expect two independent constraint conditions. Two possible candidates are:

```
> cce1 := q1^2+q2^2-L1^2=0;
```

$$cce1 := q1^2 + q2^2 - \frac{1}{100} = 0$$

```
> cce2 := L2 * sin(q3) + q2 = 0;
```

$$cce2 := \sin(\, q3\,) + q2 = 0$$

It is convenient to place the constraint equations into a Maple set structure, hence define cce as:

```
> cce := {cce1,cce2}:
```

The velocity constraint equations are defined by taking the scalar time derivatives of the cce, giving us the set vce:

```
> vce := &dt cce;
```

$$vce := \{\, 2\, q1\; q1t + 2\, q2\; q2t = 0, \cos(\, q3\,)\; q3t + q2t = 0\,\}$$

The next step is to define and obtain the transformation to a reduced set of transformed kinematic differential equations, what we designate as rtkde. Sophia contains a function which assists in this task. It is called *ReducedSpeeds(gs,kde,vce,u,w)*. The arguments are as follows: gs is a list (not a set) of the generalized speed which will be retained to describe the problem. This list can only contain a subset of the maximum number of generalized speeds which are independent. In the present case only one generalized speed can be used, for example u1, but u2 or u3 could be used instead. The term kde is the set of original kinematic differential equations, while vce is the set of velocity constraint equations. The last arguments specify the names used for both the original and the transformed generalized speeds. In the present case these are the symbols u and w. The output of ReducedSpeeds is a sequence of two sets, the reduced transformed kinematic differential equations and the reduced inverse generalized speed transformation. Both of these are useful in various approaches to deriving the equations of motion. Choosing u1 as the independent generalized speed in the present problem we have:

```
> gs := [u1]:
> rtkderigst:= ReducedSpeeds(gs,kde,vce,u,w);
```

$$rtkderigst := \left\{ u2 = -\frac{q1\; w1}{q2}, u3 = \frac{q1\; w1}{q2\,\cos(\, q3\,)}, u1 = w1 \right\},$$
$$\left\{ q1t = w1, q2t = -\frac{q1\; w1}{q2}, q3t = \frac{q1\; w1}{q2\,\cos(\, q3\,)} \right\}$$

```
> rtkde := rtkderigst[2]: rigst := rtkderigst[1]:
```

For simplicity we have stored each of the sets rtkde and rigst under symbols of those names. It is now possible to proceed in the same manner as used for problems where only independent coordinates were used, though the results will be slightly different. The velocity Kvector is defined and expressed in the new generalized speeds. This is accomplished simply by substitution of the rtkde into the Kvector:

```
> vK := subs(rtkde,&KM [vA,wA,vB,wB,vC,wC]);
```

$$vK := \left[[\,[0\,0\,0], N\,], \left[\left[0\,0\, -\frac{w1}{q2}\right], N\right], \right.$$
$$\left[\left[-\frac{1}{2}\frac{\sin(q3)\,q1\,w1}{q2\cos(q3)} + w1\, -\frac{1}{2}\frac{q1\,w1}{q2}\, 0\right], N\right],$$
$$\left[\left[0\,0\,\frac{q1\,w1}{q2\cos(q3)}\right], B\right], \left[\left[-\frac{\sin(q3)\,q1\,w1}{q2\cos(q3)} + w1\, 0\, 0\right], N\right],$$
$$\left. [\,[0\,0\,0], N\,], 6 \right]$$

The single independent tangent Kvector, designated by betaK[1], is found by using the KMtangents function. The list of output vectors will only contain one Kvector. Thus:

```
> betaK := KMtangents(vK,w,1);
```

$$[\%1, \%1, \%1, [\,[0\,0\,0], B\,], \%1, \%1, 6\,]$$
$$\%1 := [\,[0\,0\,0], N\,]$$

$$betaK := \left[\left[[\,[0\,0\,0], N\,], \left[\left[0\,0\, -\frac{1}{q2}\right], N\right], \right.\right.$$
$$\left[\left[-\frac{1}{2}\frac{\sin(q3)\,q1 - 2\,q2\cos(q3)}{q2\cos(q3)}\, -\frac{1}{2}\frac{q1}{q2}\, 0\right], N\right],$$
$$\left[\left[0\,0\,\frac{q1}{q2\cos(q3)}\right], B\right], \left[\left[-\frac{\sin(q3)\,q1 - q2\cos(q3)}{q2\cos(q3)}\, 0\, 0\right], N\right],$$
$$\left.\left. [\,[0\,0\,0], N\,], 6 \right]\right]$$

The next step is to obtain the inertial force Kvector. The following steps, which set up the momentum and angular momentum vectors and gather them into a Kvector, should be clear from previous discussions. It is important to note the use of rtkde in substitutions to insure that the final result is expressed in terms of the new generalized speed, w1 and that the constraint relations are correctly applied:

```
> pA := mA &** vA; hA := IA &o wA;pB := mB &** vB; hB := IB &o wB;
```

$$pA := [\,[0\,0\,0], N\,]$$

$$hA := \left[\left[0\,0\, -\frac{JA\,q1t}{q2}\right], N\right]$$

$$pB := \left[\left[mB\left(-\frac{1}{2}\sin(\,q3\,)\,q3t + q1t\right)\,mB\left(\frac{1}{2}\cos(\,q3\,)\,q3t + q2t\right)\,0\right], N\right]$$

$$hB := [\,[\,0\,0\,JB\,q3t\,], N\,]$$

```
> pC := mC &** vC; hC := IC &o wC;
```

$$pC := [\,[\,mC\,(\,-\sin(\,q3\,)\,q3t + q1t\,)\,mC\,(\,\cos(\,q3\,)\,q3t + q2t\,)\,0\,], N\,]$$

$$hC := [\,[\,0\,0\,0\,], N\,]$$

```
> pK := subs(rtkde,&KM [pA,hA,pB,hB,pC,hC]);
```

$$pK := \Bigg[[\,[\,0\,0\,0\,], N\,], \left[\left[0\,0 - \frac{JA\,w1}{q2}\right], N\right],$$
$$\left[\left[mB\left(-\frac{1}{2}\,\frac{\sin(\,q3\,)\,q1\,w1}{q2\cos(\,q3\,)} + w1\right) - \frac{1}{2}\,\frac{mB\,q1\,w1}{q2}\,0\right], N\right],$$
$$\left[\left[0\,0\,\frac{JB\,q1\,w1}{q2\cos(\,q3\,)}\right], N\right], \left[\left[mC\left(-\frac{\sin(\,q3\,)\,q1\,w1}{q2\cos(\,q3\,)} + w1\right)\,0\,0\right], N\right],$$
$$[\,[\,0\,0\,0\,], N\,], 6\Bigg]$$

```
> pKt := subs(rtkde,N &Kfdt pK);
```

$$pKt := \Bigg[[\,[\,0\,0\,0\,], N\,], \left[\left[0\,0 - \frac{JA\,w1t}{q2} - \frac{JA\,w1^2\,q1}{q2^3}\right], N\right], \Bigg[\Bigg[mB\Bigg($$
$$-\frac{1}{2}\,\frac{q1^2\,w1^2}{q2^2\cos(\,q3\,)} - \frac{1}{2}\,\frac{\sin(\,q3\,)\,w1^2}{q2\cos(\,q3\,)} - \frac{1}{2}\,\frac{\sin(\,q3\,)\,q1\,w1t}{q2\cos(\,q3\,)} - \frac{1}{2}\,\frac{\sin(\,q3\,)\,q1^2\,w1^2}{q2^3\cos(\,q3\,)}$$
$$-\frac{1}{2}\,\frac{\sin(\,q3\,)^2\,q1^2\,w1^2}{q2^2\cos(\,q3\,)^3} + w1t\Bigg) - \frac{1}{2}\,\frac{mB\,w1^2}{q2} - \frac{1}{2}\,\frac{mB\,q1\,w1t}{q2} - \frac{1}{2}\,\frac{mB\,q1^2\,w1^2}{q2^3}$$
$$0\Bigg], N\Bigg], \Bigg[$$
$$\left[0\,0\,\frac{JB\,w1^2}{q2\cos(\,q3\,)} + \frac{JB\,q1\,w1t}{q2\cos(\,q3\,)} + \frac{JB\,q1^2\,w1^2}{q2^3\cos(\,q3\,)} + \frac{JB\,q1^2\,w1^2\sin(\,q3\,)}{q2^2\cos(\,q3\,)^3}\right]$$
$$, N\Bigg], \Bigg[\Bigg[mC\Bigg(-\frac{q1^2\,w1^2}{q2^2\cos(\,q3\,)} - \frac{\sin(\,q3\,)\,w1^2}{q2\cos(\,q3\,)} - \frac{\sin(\,q3\,)\,q1\,w1t}{q2\cos(\,q3\,)}$$
$$-\frac{\sin(\,q3\,)\,q1^2\,w1^2}{q2^3\cos(\,q3\,)} - \frac{\sin(\,q3\,)^2\,q1^2\,w1^2}{q2^2\cos(\,q3\,)^3} + w1t\Bigg)\,0\,0\Bigg], N\Bigg],$$
$$[\,[\,0\,0\,0\,], N\,], 6\Bigg]$$

```
> MGIF := betaK &kane pKt;
```

$$MGIF := \Big[\frac{1}{4}(-4\,mC\ q1\ w1^2\ q2^2\cos(\ q3\ )^4 + 4\,mC\ q1\ w1^2\ q2^2\cos(\ q3\ )^2$$
$$- 4\,mC\ q1^2\ w1t\ q2^2\cos(\ q3\ )^4 + 4\,mC\ q1^2\ w1t\ q2^2\cos(\ q3\ )^2$$
$$+ 4\,q1\ JB\ w1^2\ q2^2\cos(\ q3\ )^2 + 4\,q1^2\ JB\ w1t\ q2^2\cos(\ q3\ )^2$$
$$+ 4\,q1^3\ JB\ w1^2\cos(\ q3\ )^2 + 4\,q1^3\ JB\ w1^2\sin(\ q3\ )\ q2$$
$$+ 4\,mC\ q1^3\ w1^2\ q2\sin(\ q3\ ) - 4\,mC\ q1^3\ w1^2\cos(\ q3\ )^4$$
$$+ 4\,mC\ q1^3\ w1^2\cos(\ q3\ )^2 - 4\,mC\ q2^2\cos(\ q3\ )\ q1^2\ w1^2$$
$$+ mB\ q1\ w1^2\ q2^2\cos(\ q3\ )^2 + mB\ q1^3\ w1^2\ q2\sin(\ q3\ )$$
$$+ mB\ q1^3\ w1^2\cos(\ q3\ )^2 + mB\ q1^2\ w1t\ q2^2\cos(\ q3\ )^2$$
$$- 8\,mC\sin(\ q3\ )\ q1\ w1t\ q2^3\cos(\ q3\ )^3 - 4\,mC\ q2^3\cos(\ q3\ )^3\sin(\ q3\ )\ w1^2$$
$$- 4\,mC\ q2\cos(\ q3\ )^3\sin(\ q3\ )\ q1^2\ w1^2 + 4\,mC\ q2^4\cos(\ q3\ )^4\ w1t$$
$$+ 4\,JA\cos(\ q3\ )^4\ w1^2\ q1 - 4\,mB\sin(\ q3\ )\ q1\ w1t\ q2^3\cos(\ q3\ )^3$$
$$- 2\,mB\ q2^3\cos(\ q3\ )^3\sin(\ q3\ )\ w1^2 - 2\,mB\ q2\cos(\ q3\ )^3\sin(\ q3\ )\ q1^2\ w1^2$$
$$+ 4\,mB\ q2^4\cos(\ q3\ )^4\ w1t + 4\,JA\cos(\ q3\ )^4\ w1t\ q2^2$$
$$- 2\,mB\ q2^2\cos(\ q3\ )\ q1^2\ w1^2)\Big/(\ q2^4\cos(\ q3\ )^4\ )\Big]$$

The following steps set up the applied force Kvector. In this case we consider the gravitational force on the bodies and an assumed, possibly time dependent, torque applied to the wheel:

```
> RA := N &ev [0,-mA*g,0]: TA := N &ev [0,0,TA3]:RB:= N &ev [0,-mB*g,0]:
> TB:= zeroK: RC:= N &ev [0,-mC*g,0]:TC:=zeroK:
> RK := &KM [RA,TA,RB,TB,RC,TC];
```

$$RK := [[[0\ -mA\,g\,0], N], [[0\,0\ TA3\,], N], [[0\ -mB\,g\,0], N],$$
$$[[0\,0\,0], N], [[0\ -mC\,g\,0], N], [[0\,0\,0], N], 6]$$

```
> GAF := betaK &kane RK;
```

$$GAF := \left[\frac{1}{2}\,\frac{-2\,TA3 + q1\ mB\ g}{q2}\right]$$

The single dynamic or Kane equation for the problem is thus:

```
> KaneEqn := MGIF[1] = GAF[1];
```

$$KaneEqn := \frac{1}{4}(-4\,mC\ q1\ w1^2\ q2^2\cos(\ q3\ )^4 + 4\,mC\ q1\ w1^2\ q2^2\cos(\ q3\ )^2$$
$$- 4\,mC\ q1^2\ w1t\ q2^2\cos(\ q3\ )^4 + 4\,mC\ q1^2\ w1t\ q2^2\cos(\ q3\ )^2$$
$$+ 4\,q1\ JB\ w1^2\ q2^2\cos(\ q3\ )^2 + 4\,q1^2\ JB\ w1t\ q2^2\cos(\ q3\ )^2$$
$$+ 4\,q1^3\ JB\ w1^2\cos(\ q3\ )^2 + 4\,q1^3\ JB\ w1^2\sin(\ q3\ )\ q2$$
$$+ 4\,mC\ q1^3\ w1^2\ q2\sin(\ q3\ ) - 4\,mC\ q1^3\ w1^2\cos(\ q3\ )^4$$

$$+4\,mC\,q1^3\,w1^2\cos(\,q3\,)^2 - 4\,mC\,q2^2\cos(\,q3\,)\,q1^2\,w1^2$$
$$+ mB\,q1\,w1^2\,q2^2\cos(\,q3\,)^2 + mB\,q1^3\,w1^2\,q2\sin(\,q3\,)$$
$$+ mB\,q1^3\,w1^2\cos(\,q3\,)^2 + mB\,q1^2\,w1t\,q2^2\cos(\,q3\,)^2$$
$$- 8\,mC\sin(\,q3\,)\,q1\,w1t\,q2^3\cos(\,q3\,)^3 - 4\,mC\,q2^3\cos(\,q3\,)^3\sin(\,q3\,)\,w1^2$$
$$- 4\,mC\,q2\cos(\,q3\,)^3\sin(\,q3\,)\,q1^2\,w1^2 + 4\,mC\,q2^4\cos(\,q3\,)^4\,w1t$$
$$+ 4\,JA\cos(\,q3\,)^4\,w1^2\,q1 - 4\,mB\sin(\,q3\,)\,q1\,w1t\,q2^3\cos(\,q3\,)^3$$
$$- 2\,mB\,q2^3\cos(\,q3\,)^3\sin(\,q3\,)\,w1^2 - 2\,mB\,q2\cos(\,q3\,)^3\sin(\,q3\,)\,q1^2\,w1^2$$
$$+ 4\,mB\,q2^4\cos(\,q3\,)^4\,w1t + 4\,JA\cos(\,q3\,)^4\,w1t\,q2^2$$
$$- 2\,mB\,q2^2\cos(\,q3\,)\,q1^2\,w1^2)\Big/(\,q2^4\cos(\,q3\,)^4\,) = \frac{1}{2}\,\frac{-2\,TA3 + q1\,mB\,g}{q2}$$

This, together with the rtkde, provides a set of 4 equations for w1,q1,q2 and q3. In solving a problem it is important to recall that the cce must be used to insure that the values assigned to q1, q2 and q3 are consistent with the constraints! The dynamic equation obtained above is quite lengthy and it is important to realize that one can use the Maple system itself to obtain useful approximations. In particular if the connecting rod is long compared to the wheel radius the values of the angle and the coordinates q1 and q2 will be small (recall we normalized all lengths to the connecting rod). Also note that the system is in a singular position when the angle q3 is zero. For example the application of a force parallel to the connecting line on the piston will result in an unstable equilibrium of the mechanism. A slight perturbation in the vertical direction can make the wheel turn in either the clockwise or counterclockwise direction. This shows itself in the singular appearance of q2 in the equations of motion. Simplified equations are obtained by using the Maple multiple taylor series function, which must first be read in from the Maple library file, thus:

```
> readlib(mtaylor);
proc() ... end
```

We now apply this function to obtain an expansion up to terms of order 2 in the variables q1 and q3:

```
> MGIFa := simplify( mtaylor(MGIF[1],[q1,q3],2));
```

$$MGIFa := \frac{1}{4}(4\,mC\,q2^4\,w1t + 4\,JA\,w1t\,q2^2 + 4\,mB\,q2^4\,w1t + 4\,q1\,JB\,w1^2\,q2^2$$
$$+ mB\,q1\,w1^2\,q2^2 + 4\,JA\,w1^2\,q1 - 2\,mB\,q2^3\,q3\,w1^2 - 4\,mC\,q2^3\,q3\,w1^2)\Big/$$
$$q2^4$$

The result can be put in standard form to obtain an approximate equation for w1t:

```
> dynStateEqn := simplify(solve({MGIFa=GAF[1]},{w1t}));
```

$$dynStateEqn := \Big\{w1t = -\frac{1}{4}(4\,q1\,JB\,w1^2\,q2^2 + mB\,q1\,w1^2\,q2^2 + 4\,JA\,w1^2\,q1$$
$$- 2\,mB\,q2^3\,q3\,w1^2 - 4\,mC\,q2^3\,q3\,w1^2 + 4\,TA3\,q2^3 - 2\,q1\,mB\,g\,q2^3)\Big/(\,q2^2$$

$$(\, mC\; q2^2 + JA + mB\; q2^2\,))\Big\}$$

The subject of approximate equations is discussed more fully in chapter 7.

6.4 Nonholonomic Systems

The addition of a dependent coordinate to the simple pendulum problem has mainly been for the purpose of illustrating what has been called the redundant coordinate method. The best motivation for the approach at this stage is that if numerical methods are to be used for producing the solution it is possible to reduce the numerical work by directly calculating quantities of mechanical interest as part of the integration process. A more pressing reason that may lead to a similar situation as that arising in the redundant coordinate approach arises when the problem of interest is directly formulated in terms of velocity rather than position constraint relations. The pendulum constraint has the form

$$f(x_1, x_2) = x_1^2 + (1 - x_2)^2 - 1 = 0. \tag{6.53}$$

A velocity constraint form was obtained by direct differentiation, which in more general terms may be expressed as

$$\tfrac{df}{dt} = \tfrac{\partial f}{\partial x_1}\tfrac{\partial x_1}{\partial t} + \tfrac{\partial f}{\partial x_2}\tfrac{\partial x_2}{\partial t} = 0. \tag{6.54}$$

If the kinematic differential equations have the simple form $\dot{x_j} = u_j$, this implies that $u_2 = \lambda u_1$, with

$$\lambda = -\frac{\frac{\partial f}{\partial x_1}}{\frac{\partial f}{\partial x_2}}. \tag{6.55}$$

Note that this process of differentiating a position constraint to obtain a velocity constraint leads to linear relations among the generalized speeds. The quantity λ has a very definite relation to the positional constraint relation.

Now turn the process around and ask the question: *given an arbitrary linear relation between the generalized speeds in a problem, does there exist a corresponding positional constraint equation which is independent of any particular solution?* If such is the case the velocity constraint relation is called *integrable.* If the constraint is integrable, the resulting relation can always be used to reduce the number of generalized coordinates, that is the degree of freedom of the system. The relation obtained for the velocity constraint equation in the problem of the simple pendulum is clearly integrable, reducing the problem to that of one degree of freedom. In a more general situation we could have a relation of the form:

$$\alpha_1(x_1, x_2)\tfrac{dx_1}{dt} + \alpha_2(x_1, x_2)\tfrac{dx_2}{dt} = 0. \tag{6.56}$$

If this relation was the result of differentiation of a positional constraint such as given above, it would imply that:

$$\alpha_j = \tfrac{\partial f}{\partial x_j}. \tag{6.57}$$

Modest assumptions about the continuity of the function f, such that imply that the two possible orders of mixed second partial derivatives are equal, provide the necessary condition for integrability

$$\frac{\partial \alpha_1}{\partial x_2} = \frac{\partial \alpha_2}{\partial x_1}. \tag{6.58}$$

In fact this is no restriction of any importance for the case of a time independent velocity constraint and two coordinates. The reason for this is that it is always possible to multiply the constraint relation by a function which will insure that the integrability condition is met.

The question of the existance of an integration factor changes as soon as we add the possibility of time dependence. The reader is asked to show that a condition of the form

$$\alpha_0 + \alpha_1 u_1 + \alpha_2 u_2 = 0, \tag{6.59}$$

where α_j is a function of x_1, x_2 and t, is not generally integrable. This demonstrates that there is not necessarily a function of the form $f(x_1, x_2, t)$ that will have a total time derivative of the required form!

A system with linear velocity constraints which are not integrable, that is for which equivalent positional constraints do not exist, is called *a linear nonholonomic system*, the set of constraints being called *linear nonholonomic constraints.* If all the constraints are equivalent to positional constraints which can be expressed as equality conditions the system is termed holonomic. Thus a constraint of the form $x_1 > 0$ is not holonomic! Opinions about this terminology are not universal, for example Kane considers any problem formulated in terms of velocity constraints as nonholonomic and calls such constraint relations nonholonomic constraints. It is of course also possible to formulate nonlinear relations among generalized speeds, giving nonholonomic relations which are not linear. It was seen above that coordinate constraints lead to linear velocity constraints. In practice linear velocity constraints are the rule for most practical nonholonomic systems.

6.4.1 A Simple Nonholonomic System

One of the simplest examples of a nonholonomic system is a particle which is constrained to track another particle, that is its velocity is always required to be in the direction of the tracked particle. In general this will be a true nonholonomic problem, that is it will not be possible to replace the velocity constraint by a position constraint. A two dimensional version of this situation is depicted in the figure. The particle of mass m and coordinates q_1 and q_2 is required to track the particle with the given time dependent position coordinates $x(t), y(t)$. Choose the generalized speeds to satisfy the kinematic differential equations

$$\dot{q}_j = u_j. \tag{6.60}$$

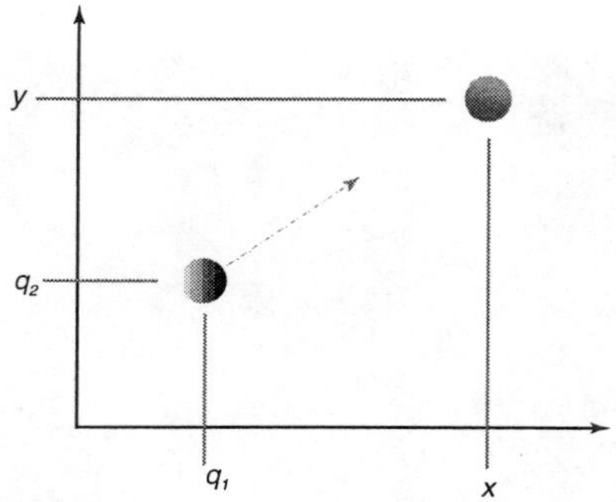

Figure 6.4: Tracking and Pursuit Problem

The tracking requirement is equivalent to the velocity constraint relation

$$u_2 = \lambda(q_1, q_2, t)u_1, \tag{6.61}$$

where

$$\lambda = \frac{y(t) - q_2}{x(t) - q_1}. \tag{6.62}$$

The particle velocity is

$$\boldsymbol{v} = u_1\boldsymbol{n}_1 + u_2\boldsymbol{n}_2 = u_1(\boldsymbol{n}_1 + \lambda\boldsymbol{n}_2) \tag{6.63}$$

which of course implies that the single independent tangent vector is given by

$$\boldsymbol{n}_1 + \lambda\boldsymbol{n}_2. \tag{6.64}$$

With no active applied force Kane's equation is simply obtained by projection of the rate of change of the particles momentum onto the tangent vector and equating the result to zero, thus

$$\dot{u}_1 + \frac{\lambda\dot{\lambda}}{1+\lambda^2}u_1 = 0. \tag{6.65}$$

The use of the velocity constraint equation in the kinematic differential equation for q_2 gives

$$\dot{q}_2 = \lambda u_1. \tag{6.66}$$

If numerical integration is used it is convenient to work in terms of the quantity λ which can be differentiated numerically as part of the solution process. This is done in the example where

$$x(t) = 1 + t \tag{6.67}$$

$$y(t) = \frac{1}{1+t} + \sin 5t. \tag{6.68}$$

Initial conditions are chosen so that the particle starts at the origin, $q_1 = q_2 = 0$, and $u_1(0) = 2$. Figure 6.5 shows a plot of the distance between the moving particle and the target, from which it is seen that the target evades the particle. It is left

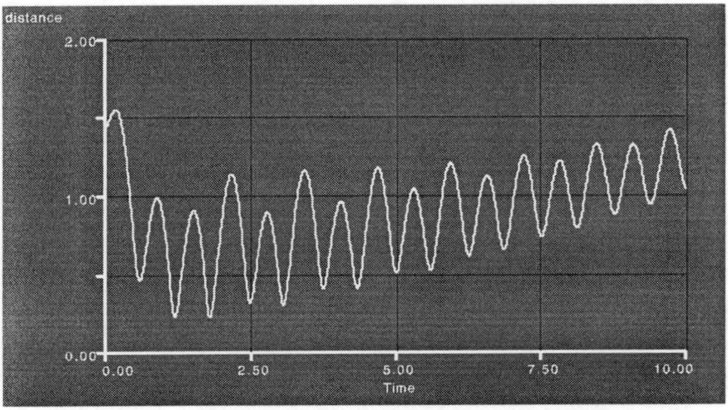

Figure 6.5: Distance between target and tracker.

as an exercise to obtain conditions under which the particle captures the target. The constraint force that acts on the particle is of considerable importance as it is the control force needed to carry out the pursuit strategy. The determination of a procedure for the calculation of this force is also left as an exercise.

6.5 Typical Velocity Constraint Problems

Problems involving velocity constraints frequently arise when non slip or rolling conditions are imposed on a mechanism. Most of the calculations are identical for truly nonholonomic problems and problems where velocity constraints are obtained by the differentiation of coordinate constraints. The main difference is in the imposition of initial conditions. In holonomic problems treated by redundant coordinates it is important to insure that initial coordinate conditions are compatible. Initial velocity conditions must also be compatible, but as for almost all problems of technical interest these conditions are linear in generalized speed variables they are relatively simple to deal with. The Sophia syntax will be used in the two problems treated. Even without using a computer the syntax should be helpful in understanding the solution algorithm.

6.5.1 The Knife Edged Pendulum

Figure 6.6 shows a typical problem in which a nonholonomic condition arises out of a non slip requirement. The mechanism consists of two masses, one of which is constrained to move in a straight line. Another point mass is allowed to slide along a guide rod which is attached to the first mass by a revolute joint. The second mass

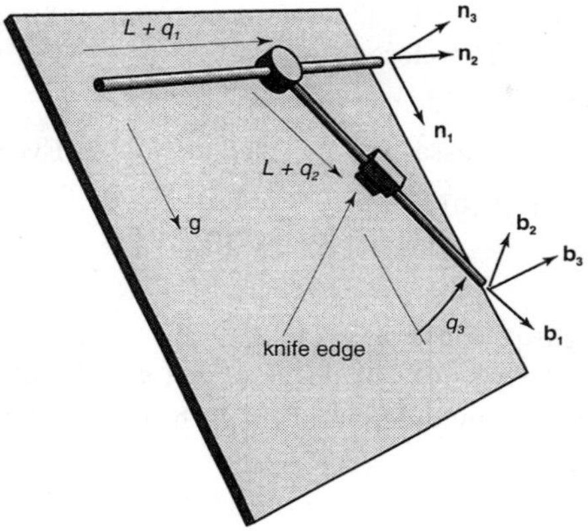

Figure 6.6: Pendulum with knife edge

is part of a 'knife' edge. The knife edge is interpreted as a constraint which only allows movement in the orientation direction of the edge. Thus no slip is allowed in the direction orthogonal to the orientation direction. It is an idealization of the kind of condition that would be expected on a skate or ski. For simplicity the masses are taken as equal. In addition the springs which connect one mass to a reference point and the other which connects the two masses are assumed to have the same natural lengths and spring constants, L and K. The Sophia statement for forming the equations of motion start with a convenient specification of the problem constants, just in case we wish to change them in another derivation of the equations. Natural units are assumed in which m=L=g=1, hence the only parameter left will be the spring constant as measured in the natural units.

```
> g:= 1: m:= 1: L:= 1:
```

The simple form of kde is assumed for the original generalized speeds. This is set up by the Sophia statement:

```
> &kde 3:
q1   declared
q2   declared
q3   declared
u1   declared
u2   declared
u3   declared
```

Generalized Velocity Form of Kinematic d.e.

Saved as the GLOBAL Varable kde

$$\{\, q1t = u1\,, q2t = u2, q3t = u3\,\}$$

The coordinates are shown in figure 6.6. Note that q1 and q2 are chosen with the natural spring lengths as reference. The reference frame of body B, the sliding mass, is defined by the angle q3. Therefore

```
> &rot [N,B,3,q3]:
```

The position of the two masses are specified as Sophia Evectors, thus

```
> r1 := N &ev [L+q1,0,0];
```

$$r1 := [\,[\,1 + q1\; 0\; 0\,], N\,]$$

```
> r2 := (B &ev [L+q2,0,0]) &++ r1;
```

$$r2 := [\,[\cos(\,q3\,) + \cos(\,q3\,)\,q2 + 1 + q1\; \sin(\,q3\,) + \sin(\,q3\,)\,q2\; 0], N\,]$$

The velocities are obtained by differentiation with respect to the inertial frame:

```
> v1 := N &fdt r1;
```

$$v1 := [\,[\,q1t\; 0\; 0\,], N\,]$$

```
> v2 := N &fdt r2;
```

$$v2 := [[-\sin(\,q3\,)\,q3t - \sin(\,q3\,)\,q3t\,q2 + \cos(\,q3\,)\,q2t + q1t \quad \cos(\,q3\,)\,q3t + \cos(\,q3\,)\,q3t\,q2 + \sin(\,q3\,)\,q2t\; 0], N]$$

The velocity constraint condition, that the knife edge does not slip, is imposed by requiring that the dot product of the knife edge's velocity with the direction orthogonal to the edge vanishes, therefore

```
> vce := {v2 &o (B&>2) = 0};
```

$$vce := \{\, q3t + q3t\,q2 - \sin(\,q3\,)\,q1t = 0\,\}$$

The reduced transformed kde are now derived using the Sophia ReducedSpeeds function, where we choose to use ul= q1t and u2 = q2t as the retained generalized speeds. We also must declare the time dependence of the new generalized speeds, which will be denoted by w, thus:

```
> dependsTime(w1,w2,w3):
w1   declared
w2   declared
w3   declared
```

```
> res := ReducedSpeeds([u1,u2],kde,vce,u,w);
```

$$res := \left\{ u1 = w1\,, u3 = \frac{\sin(\,q3\,)\,w1}{1 + q2}, u2 = w2 \right\},$$
$$\left\{ q1t = w1\,, q2t = w2, q3t = \frac{\sin(\,q3\,)\,w1}{1 + q2} \right\}$$

Using our abbreviated notation the results are stored as:

```
> rigst := res[1]; rtkde := res[2];
```

$$rigst := \left\{ u1 = w1,\, u3 = \frac{\sin(\,q3\,)\,w1}{1+q2},\, u2 = w2 \right\}$$

$$rtkde := \left\{ q1t = w1,\, q2t = w2,\, q3t = \frac{\sin(\,q3\,)\,w1}{1+q2} \right\}$$

The Kvector velocity in terms of the reduced speeds is:

```
> vK := &Ksimp subs(rtkde,&KM [v1,v2]);
```

$$vK := [[[\,w1\,0\,0\,], N\,], [[\,w1\cos(\,q3\,)^2 + \cos(\,q3\,)\,w2\,\cos(\,q3\,)\sin(\,q3\,)\,w1 + \sin(\,q3\,)\,w2\,0\,], N\,], 2]$$

Examination of this provide two independent tangent Kvectors,

```
> betaK := KMtangents(vK,w,2);
```

$$[[[\,0\,0\,0\,], N\,], [[\,0\,0\,0\,], N\,], 2\,]$$

$$betaK := [[[[\,1\,0\,0\,], N\,], [[\,\cos(\,q3\,)^2 \cos(\,q3\,)\sin(\,q3\,)\,0\,], N\,], 2\,], [[[\,0\,0\,0\,], N\,], [[\,\cos(\,q3\,)\sin(\,q3\,)\,0\,], N\,], 2]]$$

The momenta of the particles is identical to their velocities if natural coordinates with $m = 1$. For completeness, and in case we wish to alter the natural unit system, we continue to proceed formally, hence:

```
> p1 := m &** v1: p2 := m &** v2: pK := &Ksimp subs(rtkde,&KM [p1,p2]):
```

Therefore the momentum rate of change Kvector is:

```
> pKt :=  &Ksimp subs(rtkde,N &Kfdt pK);
```

$$pKt := [[[\,w1t\,0\,0\,], N\,], [[(w1t\cos(\,q3\,)^2 + w1t\cos(\,q3\,)^2\,q2 - 2\,w1^2\cos(\,q3\,) + 2\,w1^2\cos(\,q3\,)^3 - w1\,w2 + w1\,w2\cos(\,q3\,)^2 + \cos(\,q3\,)\,w2t + \cos(\,q3\,)\,w2t\,q2)\,/(\,1+q2\,)\sin(\,q3\,)(-w1^2 + 2\cos(\,q3\,)^2\,w1^2 + \cos(\,q3\,)\,w1t + \cos(\,q3\,)\,w1t\,q2 + \cos(\,q3\,)\,w1\,w2 + w2t + w2t\,q2)\,/(\,1+q2\,)0], N], 2]$$

Note that rtkde was used to express the result in the chosen transformed generalized speeds. Before obtaining the dynamic equations of motion we need the applied force Kvector, the applied forces being due to gravity and spring deformation.

```
> R1 := (B &ev [K * q2,0,0]) &++ (N &ev [m*g,-K * q1,0]):
> R2 := (B &ev [-K * q2,0,0]) &++ (N &ev [m*g,0,0]):
> RK := &Ksimp (&KM [R1,R2]);
```

$$RK := [[[\,\cos(\,q3\,)\,K\,q2 + 1\sin(\,q3\,)\,K\,q2 - K\,q1\,0\,], N\,], [[\,-\cos(\,q3\,)\,K\,q2 + 1 - \sin(\,q3\,)\,K\,q2\,0\,], N\,], 2]$$

The list of generalized applied forces, GAF, and the negative of the generalized inertia forces, MGIF, are now obtained in the usual manner. These are used to write down the two dynamic or Kane equations for the problem:

```
> GAF := betaK &kane RK:
> MGIF := betaK &kane pKt:
> KaneEqns := seq(simplify(MGIF[j])=simplify(GAF[j]),j=1..2);
```

$$KaneEqns := (w1t + w1t\, q2 + w1^2 \cos(q3)^3 - w1^2 \cos(q3) + w1t \cos(q3)^2 + w1t \cos(q3)^2 q2 + \cos(q3)\, w2t + \cos(q3)\, w2t\, q2) \big/ (1 + q2) = 1 + \cos(q3)^2, (\cos(q3)^2 w1^2 - w1^2 + \cos(q3)\, w1t + \cos(q3)\, w1t\, q2 + w2t + w2t\, q2) \big/ (1 + q2) = \cos(q3) - K\, q2$$

The integration of these equations, together with the rtkde is assisted by putting them into state space form:

```
> dynStateEqns := solve({KaneEqns},{w1t,w2t});
```

$$dynStateEqns := \{w1t = \cos(q3)\, K\, q2 + 1, w2t = -(\cos(q3)^2 K\, q2 + \cos(q3)^2 K\, q2^2 + \cos(q3)^2 w1^2 - w1^2 + K\, q2 + K\, q2^2) \big/ (1 + q2)\}$$

The rtkde are already in state space form, hence we have 5 equations for q1,q2,q3,w1 and w2. It is also useful to have an expression for the position of the knife edge. This is easily obtained, thus:

```
> xPos := r2 &c 1; yPos := r2 &c 2;
```

$$xPos := \cos(q3) + \cos(q3)\, q2 + 1 + q1$$

$$yPos := \sin(q3) + \sin(q3)\, q2$$

In the remainder of this discussion an illustration is given of how the solution may be obtained and plotted using Maple's numeric facilities. First the differential equations must be put into time functional form. Also it is useful to gather all the applicable equations into a set. Thus:

```
> stateEqns := subs(toTimeFunction,dynStateEqns union rtkde);
```

As a numeric solution is to be obtained all parameters must be specified. In this case the only parameter is K, which we take as 0.2. Therefore

```
> stateEqnsK := subs(K=0.2,stateEqns):
```

It is also necessary to specify the initial conditions:

```
> ic := {q1(0)=0.3,q2(0)=-0.3,q3(0)=0.4,w1(0)=0,w2(0)=0}:
```

The numeric solutions interpolation function is obtained with Maple's dsolve command as

```
> state := dsolve(stateEqnsK union ic,{q1(t),q2(t),q3(t),w1(t),w2(t)},
> type=numeric, output=procedurelist);
state := proc(rkf45_x) ... end
```

```
> state(0.1);
```

$$[11 = .1, q1(11) = .3036206951748513, q2(11) = -.2972311816950761,\\ q3(11) = .4020150538051340, w1(11) = .07246062939511792,\\ w2(11) = .05530324142115989]$$

The next few steps are made for the purpose of plotting the position of the knife edge.

```
> xPost := subs(toTimeFunction,xPos): yPost := subs(toTimeFunction,yPos):
> xK := proc(t) evalf(subs(state(t),xPost)) end:
> yK := proc(t) evalf(subs(state(t),yPost)) end:
> PLOT(CURVES([seq([evalf(xK(j/10.0)),evalf(yK(j/10.0))],j=1..75)]));
```

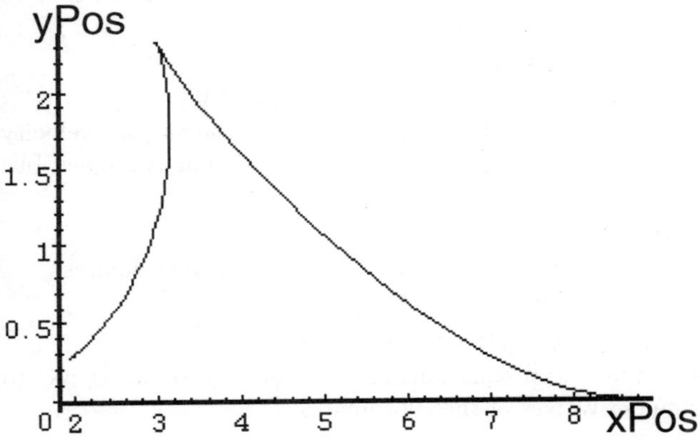

Figure 6.7: Position trace of knife edge

Figure 6.7 shows the result of this plot command. The cusp shape is typical of problems with rolling constraints. To understand it think of the tracks your car make in the snow when you back out of your driveway!

6.5.2 The Rolling Coin

The problem of the rolling coin is one of the most popular textbook examples of a nonholonomic problem. Some of the kinematic aspects of the problem have already been referred to in chapter 3. Figure 3.15 shows a set of 5 generalized coordinates that can be used to describe the configuration of a rolling coin. The Sophia frame fixed velocity function was used to describe two velocity constraint equations for the case where the disk is not allowed to slip. Therefore it is expected that the problem will involve 8 first order equations, 5 describing the kinematics and 3 the dynamics. For completeness the steps for deriving the constraint relations is repeated here without explanation. Natural units are chosen in which the mass of the disk, the disk radius and the gravitational acceleration are unity. As the kde will not be chosen in the simplest manner a separate time declaration statement is made for the generalized speeds and coordinates used.

```
> m:=1: R:=1: g:=1:
> dependsTime(seq(q.j,j=1..5),seq(u.j,j=1..5),seq(w.j,j=1..5)):
> chainSimpRot([[N,H,3,q1],[H,B,1,Pi/2-q2],[B,C,3,q3]]):
> rQ := N &ev [q4,q5,0]:
> rC := rQ &++ (R &** (B&>2)):
> vNCQ := N &to ([N,C] &ffv [rC,rQ]);
```

$$vNCQ := [\,[\cos(\,q1\,)\,q3t + q4t\;q5t + \sin(\,q1\,)\,q3t\;0\,], N\,]$$

The base plane is fixed in the inertial frame N, therefore setting the last result to zero gives the velocity constraint equations:

```
> vce:= {(vNCQ &c 1) = 0, (vNCQ &c 2) = 0};
```

$$vce := \{\,q5t + \sin(\,q1\,)\,q3t = 0, \cos(\,q1\,)\,q3t + q4t = 0\,\}$$

Three of the generalized speeds are chosen as the components of the angular velocity vector of the disk when represented in the intermediate frame B that is aligned but not rotating with its axial motion. Thus:

```
> wC := B &ev [u1,u2,u3]:
```

The next statement computes this quantity as a function of the coordinates:

```
> wCq := &simp (B &to (N &aV C));
```

$$wCq := [\,[\,-q2t\,\cos(\,q2\,)\,q1t\;q3t + q1t\,\sin(\,q2\,)\,], B\,]$$

Three of the five kinematic differential equations are obtained by equating wC to wCq and solving for the time derivatives of the coordinates:

```
> kde := solve({seq((wCq &c j) = (wC &c j), j=1..3)},{seq(q.j.t,j=1..3)});
```

$$kde := \left\{ q1t = \frac{u2}{\cos(\,q2\,)}, q3t = -\frac{u2\,\sin(\,q2\,) - u3\,\cos(\,q2\,)}{\cos(\,q2\,)}, q2t = -u1 \right\}$$

The remaining equations are obtained by choosing them as the derivatives of the coordinates of the contact point in the inertial frame N, hence the full set of kde is:

```
> kde := kde union {q4t = u4, q5t = u5};
```

$$kde := \left\{ q5t = u5, q4t = u4, q1t = \frac{u2}{\cos(\,q2\,)}, q3t = -\frac{u2\,\sin(\,q2\,) - u3\,\cos(\,q2\,)}{\cos(\,q2\,)},\right.$$
$$\left. q2t = -u1 \right\}$$

The reduced transformed kde are chosen by picking u1,u2 and u3 as independent generalized speeds, therefore using Sophia's ReducedSpeeds function we have:

```
> rigstrtkde := ReducedSpeeds([u1,u2,u3],kde,vce,u,w):
> rigst := rigstrtkde[1];
```

$$rigst := \left\{ u5 = \frac{-\sin(\,q1\,)\,w3\,\cos(\,q2\,) + \sin(\,q1\,)\,\sin(\,q2\,)\,w2}{\cos(\,q2\,)},\right.$$

$$u4 = \frac{-\cos(\,q1\,)\,w3\cos(\,q2\,) + \cos(\,q1\,)\sin(\,q2\,)\,w2}{\cos(\,q2\,)},\ u2 = w2,\ u3 = w3,$$
$$u1 = w1\Big\}$$

```
> rtkde := rigstrtkde[2];
```

$$rtkde := \Big\{q5t = \frac{-\sin(\,q1\,)\,w3\cos(\,q2\,) + \sin(\,q1\,)\sin(\,q2\,)\,w2}{\cos(\,q2\,)},$$
$$q4t = \frac{-\cos(\,q1\,)\,w3\cos(\,q2\,) + \cos(\,q1\,)\sin(\,q2\,)\,w2}{\cos(\,q2\,)},\ q1t = \frac{w2}{\cos(\,q2\,)},$$
$$q3t = -\frac{w2\sin(\,q2\,) - w3\cos(\,q2\,)}{\cos(\,q2\,)},\ q2t = -w1\Big\}$$

Note the happy circumstance that the equations for the angular speeds q1t, q2t and q3t are independent of q4 and q5, which makes the numerical integration of the final system simpler.

We can now proceed in the usual manner, but first define the inertia properties of the body:

```
> J := m * R^2 / 4: IC := EinertiaDyad(J,J,2*J,0,0,0,C);
```

$$IC := \left[\begin{bmatrix} \frac{1}{4} & 0 & 0 \\ 0 & \frac{1}{4} & 0 \\ 0 & 0 & \frac{1}{2} \end{bmatrix}, C\right]$$

The velocity of the mass center of the disk is found from the expression for its position, the rtkde being used to put the result in the reduced generalized speeds:

```
> vC := &simp subs(rtkde,(N &fdt rC));
```

$$vC := [\,[\,-w3\;0\;w1\,], B\,]$$

The angular velocity wC is also expressed in the new generalized speeds and the result used to obtain the velocity Kvector and three independent tangent Kvectors:

```
> wC :=  subs(rigst,wC);
```

$$wC := [\,[\,w1\;w2\;w3\,], B\,]$$

```
> vK := &KM [vC,wC]:
> betaK := KMtangents(vK,w,3);
```

$$[\,[\,[\,0\,0\,0\,], B\,], [\,[\,0\,0\,0\,], B\,], 2\,]$$

$$betaK := [[\,[\,[\,0\,0\,1\,], B\,], [\,[\,1\,0\,0\,], B\,], 2\,],$$
$$[\,[\,[\,0\,0\,0\,], B\,], [\,[\,0\,1\,0\,], B\,], 2\,],$$
$$[\,[\,[\,-1\,0\,0\,], B\,], [\,[\,0\,0\,1\,], B\,], 2\,]]$$

We now compute the momentum and angular momentum vectors to obtain pK the momentum Kvector. The transformation of the angular momentum to the frame B is carried out because it is expected that the result will be simplest for this frame. It is not necessary for the solution process.

```
> pC := m &** vC: hC := &simp (B &to (IC &o wC)):
> pK := &KM [pC,hC];
```

$$pK := \left[[[-w3\ 0\ w1\ ], B\ ], \left[\left[\frac{1}{4}\ w1\ \frac{1}{4}\ w2\ \frac{1}{2}\ w3\right], B\right], 2\right]$$

The inertial force Kvector follows by differentiation and substitution. The substitution step insures we have the result in the chosen variables and is not strictly needed. In the present case it also leads to an elegant form of the equations.

```
> pKt := &Ksimp subs(rtkde,(N &Kfdt pK));
```

$$pKt := \left[\left[\left[-w3t + w2\ w1\ - \frac{\sin(\ q2\ )\ w2\ w3 + w1^2 \cos(\ q2\ )}{\cos(\ q2\ )}\ w2\ w3 + w1t\right], B\right]\right.$$
$$, \left[\left[\frac{1}{4}\ \frac{w1t \cos(\ q2\ ) - w2^2 \sin(\ q2\ ) + 2\ w2\ w3 \cos(\ q2\ )}{\cos(\ q2\ )}\right.\right.$$
$$\left.\left.\frac{1}{4}\ \frac{\sin(\ q2\ )\ w2\ w1\ - 2\ w3\ w1 \cos(\ q2\ ) + \cos(\ q2\ )\ w2t}{\cos(\ q2\ )}\ \frac{1}{2}\ w3t\right], B\right], 2\right]$$

The only applied force is the weight of the disk, therefore if we assume that no torque is applied we have:

```
> RC := N &ev [0,0,-m*g]: TC := N &ev [0,0,0]:
> RK := &KM [RC,TC];
```

$$RK := [[[0\,0\ -1], N], [[0\,0\,0], N], 2]$$

We can now form the dynamic equations for the disk motion:

```
> MGIF := betaK &kane pKt:
> GAF := betaK &kane RK:
> dynEqns := seq(MGIF[j]=GAF[j],j=1..3);
```

$$dynEqns := \frac{1}{4}\ \frac{6\ w2\ w3 \cos(\ q2\ ) + 5\ w1t \cos(\ q2\ ) - w2^2 \sin(\ q2\ )}{\cos(\ q2\ )} = -\sin(\ q2\ ),$$
$$\frac{1}{4}\ \frac{\sin(\ q2\ )\ w2\ w1\ - 2\ w3\ w1 \cos(\ q2\ ) + \cos(\ q2\ )\ w2t}{\cos(\ q2\ )} = 0,$$
$$\frac{3}{2}\ w3t - w2\ w1 = 0$$

```
> dynStateEqns := solve({dynEqns},{w1t,w2t,w3t});
```

$$dynStateEqns := \left\{w3t = \frac{2}{3}\ w2\ w1\ , w2t = \frac{w1\ (-w2 \sin(\ q2\ ) + 2\ w3 \cos(\ q2\ ))}{\cos(\ q2\ )},\right.$$
$$\left. w1t = -\frac{1}{5}\ \frac{6\ w2\ w3 \cos(\ q2\ ) - w2^2 \sin(\ q2\ ) + 4 \sin(\ q2\ ) \cos(\ q2\ )}{\cos(\ q2\ )}\right\}$$

The final set of equations includes both the dynamic state equations and the reduced transformed kinematic differential equations:

$$q1t = \frac{w2}{\cos(\, q2\,)}$$

$$q2t = -w1$$

$$q3t = -\frac{w2\sin(\, q2\,) - w3\cos(\, q2\,)}{\cos(\, q2\,)}$$

$$q4t = \frac{-\cos(\, q1\,)\, w3\cos(\, q2\,) + \cos(\, q1\,)\sin(\, q2\,)\, w2}{\cos(\, q2\,)}$$

$$q5t = \frac{-\sin(\, q1\,)\, w3\cos(\, q2\,) + \sin(\, q1\,)\sin(\, q2\,)\, w2}{\cos(\, q2\,)}$$

$$w1t = -\frac{1}{5}\,\frac{6\, w2\, w3\cos(\, q2\,) - w2^2\sin(\, q2\,) + 4\sin(\, q2\,)\cos(\, q2\,)}{\cos(\, q2\,)}$$

$$w2t = \frac{w1\,(-w2\sin(\, q2\,) + 2\, w3\cos(\, q2\,))}{\cos(\, q2\,)}$$

$$w3t = \frac{2}{3}\, w2\, w1$$

6.6 Problems

•**Problem 6.1** The planar mechanism shown in figure 6.8 consists of three bars of length L and width a. The central bar is fixed to ground by a revolute joint at its center. The outer bars are attached to the central bar by revolute joints at the bar's edges. Their centers are fitted with pin into slots at distance S/2 from the central revolute joint as shown. The bars have equal mass, m and are in a uniform gravitational field of acceleration g. Find equations of motion for the problem using the indicated coordinate $q_1 \ldots q_6$ by the redundant coordinate technique.

•**Problem 6.2** Consider the problem of the knife edged pendulum, treated in section

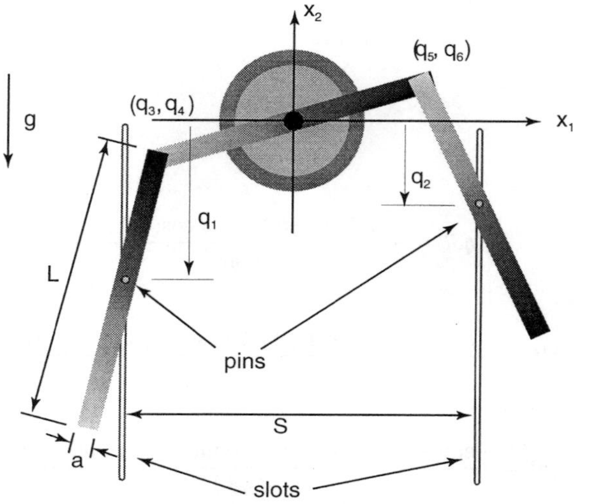

Figure 6.8: Slotted Pendulum Mechanism

6.3, but with the distance along the pendulum rod to the second mass fixed, i.e. with the coordinate q_2 eliminated. Determine if this is a holonomic or nonholonomic problem and set up the equations of motion by the redundant coordinate technique. If you decide the problem is holonomic determine integral forms of the equations of constraint by integration of the appropriate velocity constraint equation.

•**Problem 6.3** Carry out the solution of the rolling coin problem when the coin is placed on a rotating table. Assume the rotation speed is a constant.

•**Problem 6.4** Complete the calculations for the control force in the pursuit problem discussed in section 6.4.

•**Problem 6.5** The mechanism shown in figure 6.9 is intended to test a castor wheel system. A light rod is attached to the central shaft S. The shaft can rotate freely about its axis. An L shaped bracket, B is attached to the shaft by means of a sliding ring A. The fixture allows rotation along the long axis of the bracket. Finally a wheel C is attached to the short end of the bracket. The wheel is free to rotate about the attachment axis and rolls without slip on the plane N. Assume that a linear spring damper is placed between A and S and that the bracket B is attached to the slider A by a torsion spring. Develop equations of motion for this system and find some solutions which appear 'interesting'.

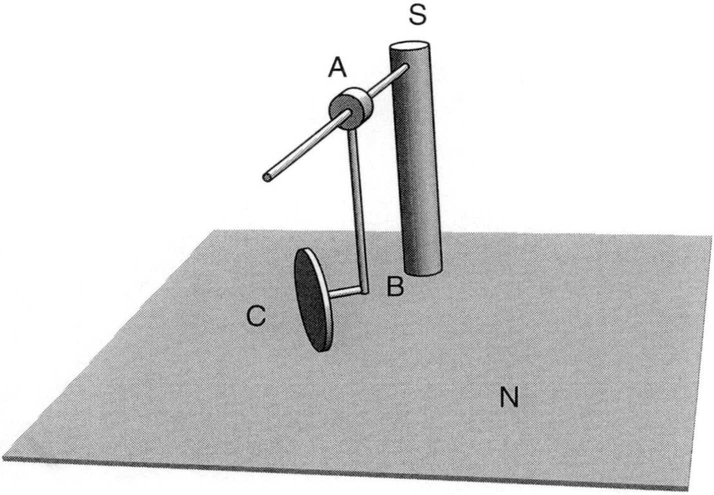

Figure 6.9: Castor testing mechanism

Chapter 7

Approximate Methods

The understanding of mechanical systems can be aided by the study of approximate systems of equations. This has been well known since the formulation of the laws of mechanics, Newton himself having made use of perturbation methods to deal with the problems of a many body solar system. The *simple pendulum* is made even simpler by studying an approximation where the motion is confined to a neighborhood of its equilibrium position. Until the development of electronic computers the only means available for the mathematical study of all but a few simple mechanical systems were based on perturbation techniques. Of course even the simplest of problems at some point require the tabulation of numerical results, for example one needs to know the values of the trigonometric functions, which must be calculated. So called exact solutions, while valuable are in reality no more than identifications of problems with other problems where the work of numerical tabulation and analytical investigation is more complete. In another sense all the theories of physical science are approximate. Implicitly some effects, represented by particular parameters taking on extreme values are ignored or emphasized. Thus geometrical optics can be considered as physical optics in the small wavelength limit, classical mechanics the limit of relativistic mechanics when velocities are small compared to light and rigid body mechanics as the mechanics of deformable media where the time scales are long compared to elastic wave traversal times over the body. Thus the study of approximate methods can involve deep issues of the foundation of science as well as complex questions of technique. In this chapter only a modest introduction of two generally useful ideas is attempted. The first is expansion in a small parameter. The simple pendulum provides a model of this, where the assumed parameter might be a small maximum displacement angle. This idea also provides an entry into the study of the stability of equilibrium, small or linear vibrations and the local theory of dynamic systems. The other situation that is examined is the operation of sudden large forces that act on a system, such as in the collision of bodies. The idea is that for large forces of small time duration, compared to the other time scales of the problem, a body's velocity can be changed while its configuration is approximately unaltered. These short duration high amplitude forces are called *impulsive forces.* The integral of such forces over their period of duration is called the *impulse*.

Instead of trying to set down all possible cases, several instructive examples are studied. These are used to develop some useful principles. The *art* of these methods involves using available physical information as well as intuition. Because of the unreliability of such information or intuition it is always a good idea to have some supporting experiments. Computer algebra not only saves labor but assists in the investigation of alternative approaches. Several new Sophia functions for carrying out expansion approximations are presented.

7.1 Direct Expansion Methods

The basic assumption behind expansion methods is that a solution to a problem near to the one we are interested in is known. The closeness is usually measured by some parameter, and it will be assumed that the smaller this parameter is the closer the problem is to the known case. In dynamics problems a frequent situation is that the known case is one of static equilibrium, for example a pendulum which is suspended and not in motion at either its uppermost position or its lowest position. The assumption is made that the pendulum is given a small displacement. In one case the motion will be confined to the region of the equilibrium position, in the other it will grow. One is *stable* the other is *unstable*. The small parameter expansion provides the information as to which situation holds. The pendulum problem will be used as an example to introduce these ideas and how they can be implemented as part of the algorithms that have been introduced in this text.

First advantage is taken of the simplicity of the pendulum problem to see what might be expected in more general situations. In natural units, where the mass, length and gravitational acceleration are unity, the equation for the pendulum angle takes the form

$$\ddot{\theta} + \sin\theta = 0. \tag{7.1}$$

There are two equilibrium solutions in which the pendulum would remain at rest, $\theta = 0$ and $\theta = \pi$. The first corresponds to the pendulum being at the bottom of its path, the other to it being at the top. Ones expectation is that the latter position is unstable, that is the motion will not remain confined to the vicinity of the equilibrium point. Assume that the maximal displacement is measured by a parameter ϵ, for example it could be the initial displacement angle with the initial angular velocity taken as zero. Transform the equation to a new variable which is unity when the displacement angle is ϵ. For notational brevity we can retain the same symbol θ for the new variable, thus make the transformation $\theta \to \epsilon\theta$. The equation of motion becomes

$$\epsilon\ddot{\theta} + \sin\epsilon\theta = 0. \tag{7.2}$$

In the case where $\theta = \pi$ we make the transformation $\theta \to \pi + \epsilon\theta$ giving the equation

$$\epsilon\ddot{\theta} + \sin(\pi + \epsilon\theta) = 0. \tag{7.3}$$

Therefore if ϵ is assumed small it is reasonable to expect that the motion in the two cases will be approximated by the equations:

$$\ddot{\theta} + \theta = 0, \tag{7.4}$$
$$\ddot{\theta} - \theta = 0. \tag{7.5}$$

In the first case solutions are linear combinations of $e^{\pm it}$, while in the second they are combinations of $e^{\pm t}$. Thus in the second case, corresponding to an upper equilibrium point, there are solutions which grow in an unbounded fashion. Hence this point appears to be unstable. In the first case note that $\dot{\theta} = \pm i\theta$. Therefore $\dot{\theta}^2 + \theta^2 = 0$. This is the equation of a circle in the $\theta - \dot{\theta}$ or phase plane. Therefore the motion follows a bounded circular path near the bottom equilibrium point in the phase plane. On the other hand the solutions near the top equilibrium point show that $\dot{\theta}^2 - \theta^2 = 0$. This is the product of the factors $\dot{\theta} \pm \theta$ showing that the path in the phase plane is along two straight lines at an angle of $\pi/4$ with the axis. Slight displacement along one of the lines results in a motion back towards the origin, however in the other case it leads to motion away from the origin, that is an unbounded motion. Multiplication of the initial equation by $\dot{\theta}$ or equivalently the direct application of the principle of energy conservation leads to the equation

$$\frac{1}{2}\dot{\theta}^2 - \cos\theta = \frac{1}{2}\dot{\theta}(t_0)^2 - \cos\theta(t_0) \tag{7.6}$$

where the right hand side is evaluated at the initial time, t_0. It is left as an exercise to show that the results obtained by small parameter expansion methods also follows from this equation. The pendulum equation is unusual in its simplicity. In more complex systems it will not always be possible to obtain explicit forms of the equations of motion, even with computer algebra. It is therefore worth some effort to develop methods for the direct determination of small parameter expansion equations. As an example a small parameter expansion is developed directly for the pendulum problem using Kane's algorithm.

The pendulum angular velocity is chosen as the generalized speed. With the angular coordinate taken as q_1 the angular velocity is $\dot{q_1} = u_1$, which is then the generalized speed. The Kvector notation is used for purposes of easy generalization. The velocity Kvector $\boldsymbol{v}^< = u_1\boldsymbol{b}_2$ where the B reference frame is fixed in the pendulum so that $\boldsymbol{b}_1$ is aligned in the direction from the support hinge to the mass point. The inertial frame N is chosen so that the vector $\boldsymbol{n}_1$ is in the direction of the gravitational acceleration. Therefore, by inspection, the single independent tangent Kvector is given by $\boldsymbol{\tau}_1^< = \boldsymbol{b}_2$. The momentum Kvector is $\boldsymbol{P}^< = u_1\boldsymbol{b}_2$ (recall $m = g = L = 1$). The only applied force is gravity so $\boldsymbol{R}^< = \boldsymbol{n}_1$. Note also that $\boldsymbol{b}_1 = \cos q_1\boldsymbol{n}_1 + \sin q_1\boldsymbol{n}_2$ and $\boldsymbol{b}_2 = -\sin q_1\boldsymbol{n}_1 + \cos q_1\boldsymbol{n}_2$. The constraint free dynamic equation of motion is:

$$\dot{\boldsymbol{P}}^< \bullet \boldsymbol{\tau}_1^< = \boldsymbol{R}^< \bullet \boldsymbol{\tau}_1^<. \tag{7.7}$$

The small parameter expansion proceeds by first rescaling the terms, i.e. by transforming to variables which are expected to remain bounded during the motion. Thus if the initial angular displacement is $q_1 = \epsilon$ and the initial angular velocity $u_1 = 0$ it is reasonable to expect the new variable $\tilde{q}_1 = q_1/\epsilon$ to be less than or equal to 1. One hopes that if there is a contradiction to this it will exhibit itself in the development of the calculations.

Instead of introducing new symbols it is possible to apply a scaling transformation such as $q_1 \to Q_1 + \epsilon q_1$, where Q_1 is the base solution. For the present example $Q_1 = 0$. All terms are then expanded up to a given power of ϵ. The convention is adopted that if the expansion is called an *order n* expansion, it includes all terms that are smaller than ϵ^n, as $\epsilon \to 0$. A *power series* expansion of order 2 would include all terms up to ϵ^1. This case is frequently called *linearizing the equations of motion.* A common technique in linearization is to set ϵ to 1 in the final result. Going further one may never use a formal parameter, keeping informal accounting of the order of terms. Since the use of computer algebra requires more formality, we will always proceed by scaling the equations with the relevant parameter. Both independent and dependent variables can be scaled, which leads to the theories of boundary layers, matched asymptotic expansions and multiple time scaling. This is beyond the scope of the present work and the reader unfamiliar with these techniques may consult the references for more information.

It is very important to note that our technique for finding tangent vectors by the inspection of velocity terms requires an *exact* expression for the velocity! Therefore *tangent vectors must be found before scaling and expansion.* The single *exact* tangent vector for this problem may then be scaled and expanded:

$$\boldsymbol{\tau}_1^< = \boldsymbol{b}_2 \to -\boldsymbol{n}_1 \sin \epsilon q_1 + \boldsymbol{n}_2 \cos \epsilon q_1 \to \boldsymbol{n}_2 - \boldsymbol{n}_1 \epsilon q_1 + O(\epsilon^2), \tag{7.8}$$

where the symbol $O(\epsilon^2)$ indicates this is an expansion that includes terms that asymptotically dominate ϵ^2 as $\epsilon \to 0$. The reader should be able to confirm that the expansions for the dynamic force $\dot{\boldsymbol{P}}^<$, and the applied force $\boldsymbol{R}^<$ are:

$$\begin{aligned} \dot{\boldsymbol{P}}^< &= \dot{u}_1 \boldsymbol{b}_2 - u_1^2 \boldsymbol{b}_1 \to \epsilon \dot{u}_1 \boldsymbol{n}_2 + O(\epsilon^2) & (7.9) \\ \boldsymbol{R}^< &= \boldsymbol{n}_1 \to \boldsymbol{n}_1. & (7.10) \end{aligned}$$

Therefore

$$\begin{aligned} \dot{\boldsymbol{P}}^< \bullet \boldsymbol{\tau}_1^< &= \epsilon \dot{u}_1 + O(\epsilon^2) & (7.11) \\ \boldsymbol{R}^< \bullet \boldsymbol{\tau}_1^< &= -\epsilon q_1 + O(\epsilon^2). & (7.12) \end{aligned}$$

Therefore the linearized dynamic force or Kane equation is:

$$\dot{u}_1 + q_1 = 0, \tag{7.13}$$

which together with the linearized kinematic differential equation gives the expected simple harmonic oscillator equation. To understand why it is necessary to start with exact velocity expressiont it is a worthwhile exercise to see what happens if one first expands the velocity and uses the tangent vector obtained by the inspection of the generalized speed coefficient.

7.2 Partial Series Expansions

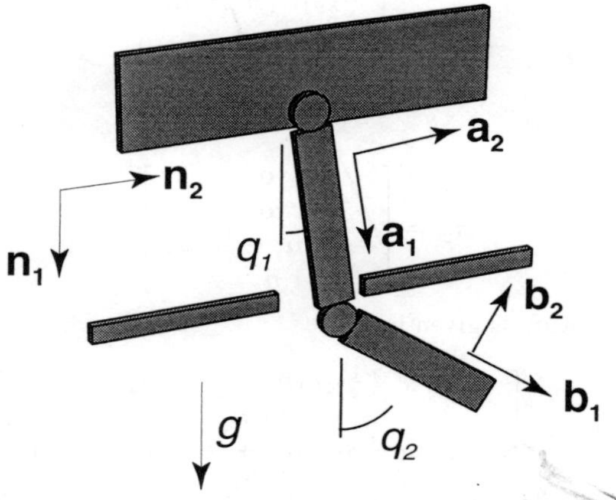

Figure 7.1: Double physical pendulum with stop

The purpose of the next example, shown in figure 7.1, is to emphasize the expansion procedure in a multibody case with several degrees of freedom. Specifically it is possible to consider situations in which the unperturbed state is itself nonlinear. The mechanism is a double physical pendulum in which one arm is restricted in its motion. The small parameter is the maximum angle that can be reached by the upper arm, which will be taken as ϵ. The bottom link is free to rotate in an unrestricted manner. The present section will not include the *problem of impact* when the upper arm contacts the body which blocks its motion. The problem of deriving impact *jump* conditions is treated in section 7.3 as part of the general topic of *impulsive forces.* The links have the same geometry and physical properties. Units are chosen in which the lengths, masses and gravitational acceleration are unity. The out of plane moments of inertia about the center of mass are designated by J and it is assumed that the mass centers are at the geometric center of the bodies. The upper link is designated as body A, the bottom as B. The angles q_1, q_2 are chosen as generalized coordinates. The time derivatives of these, i.e. the generalized velocities, are taken as the generalized speeds. For compactness the notation s_j, c_j, will be used for the trigonometric functions. From the concept of *simple angular velocity* it is clear that the generalized speeds are the angular velocities of the bodies in the Newtonian frame N. Therefore the Kvector velocity is given by:

$$\boldsymbol{v}^< = \begin{bmatrix} -\frac{1}{2}s_1u_1\boldsymbol{n}_1 + \frac{1}{2}c_1u_1\boldsymbol{n}_2 \\ u_1\boldsymbol{n}_3 \\ (-\frac{1}{2}s_2u_2 - s_1u_1)\boldsymbol{n}_1 + (\frac{1}{2}c_2u_2 + c_1u_1)\boldsymbol{n}_2 \\ u_2\boldsymbol{n}_3 \end{bmatrix} \tag{7.14}$$

from which, by inspection of the coefficients of the generalized speeds, the tangent Kvectors are seen to be:

$$\boldsymbol{\tau}_1^< = \begin{bmatrix} -\frac{1}{2}s_1\boldsymbol{n}_1 + \frac{1}{2}c_1\boldsymbol{n}_2 \\ \boldsymbol{n}_3 \\ -s_1\boldsymbol{n}_1 + c_1\boldsymbol{n}_2 \\ \mathbf{o} \end{bmatrix} \tag{7.15}$$

and

$$\boldsymbol{\tau}_2^< = \begin{bmatrix} \mathbf{o} \\ \mathbf{o} \\ -\frac{1}{2}s_2\boldsymbol{n}_1 + \frac{1}{2}c_2\boldsymbol{n}_2 \\ \boldsymbol{n}_3 \end{bmatrix}. \tag{7.16}$$

The Kvector momentum is given by:

$$\boldsymbol{P}^< = \begin{bmatrix} -\frac{1}{2}s_1u_1\boldsymbol{n}_1 + \frac{1}{2}c_1u_1\boldsymbol{n}_2 \\ Ju_1\boldsymbol{n}_3 \\ (-\frac{1}{2}s_2u_2 - s_1u_1)\boldsymbol{n}_1 + (\frac{1}{2}c_2u_2 + c_1u_1)\boldsymbol{n}_2 \\ Ju_2\boldsymbol{n}_3 \end{bmatrix} \tag{7.17}$$

and as there is no applied torque about the link mass centers the applied force Kvector is:

$$\boldsymbol{R}^< = \begin{bmatrix} \boldsymbol{n}_1 \\ \mathbf{o} \\ \boldsymbol{n}_1 \\ 0 \end{bmatrix}. \tag{7.18}$$

The approximate equations are based on the angle q_1 being bounded by the stop. The angle q_2 is not restricted, hence this suggests the scaling:

$$q_1 \rightarrow \epsilon q_1, \tag{7.19}$$
$$u_1 \rightarrow \epsilon u_1, \tag{7.20}$$
$$q_2 \rightarrow q_2 \tag{7.21}$$
$$u_2 \rightarrow u_2. \tag{7.22}$$

Thus the base solution has the upper pendulum in the bottom state of rest. After scaling the linearization is obtained in exactly the same manner as with the simple pendulum. The simplification is that it is much easier to evaluate the dynamic force terms from the partially linearized momentum Kvector. This gives the result:

$$\dot{\boldsymbol{P}}^< = \begin{bmatrix} \frac{1}{2}\epsilon\dot{u}_1\boldsymbol{n}_2 \\ \epsilon J\dot{u}_1\boldsymbol{n}_3 \\ -(\frac{1}{2}s_2\dot{u}_2 + \frac{1}{2}u_2^2c_2)\boldsymbol{n}_1 + (\frac{1}{2}s_2\dot{u}_2 - \frac{1}{2}u_2^2s_2 + \epsilon\dot{u}_1)\boldsymbol{n}_2 \\ J\dot{u}_2\boldsymbol{n}_3 \end{bmatrix} + O(\epsilon^2). \tag{7.23}$$

Notice that letting $\epsilon \rightarrow 0$ results in the dynamic force Kvector appropriate to a single pendulum with base point fixed at the bottom position of the upper link. The final results follow from using the *up to* $O(\epsilon^2)$ form of the two independent tangent vectors. The terms $\dot{\boldsymbol{P}}^< \bullet \boldsymbol{\tau}_j^<$ and $\boldsymbol{R}^< \bullet \boldsymbol{\tau}_j^<$ are formed to $O(\epsilon^2)$, giving the partially linearized equations of motion:

$$\frac{1}{4}\,\dot{u}_2 + \frac{1}{2}\cos(\,q_2\,)\,\epsilon\,\dot{u}_1 + J\,\dot{u}_2 = -\,\frac{1}{2}\sin(\,q_2\,) \tag{7.24}$$

$$\begin{aligned}&\frac{5}{4}\,\epsilon\,\dot{u}_1 + J\,\epsilon\,\dot{u}_1 + \frac{1}{2}\,\epsilon\,q_1\,\dot{u}_2\sin(\,q_2\,) + \frac{1}{2}\,\epsilon\,q_1\,{u_2}^2\cos(\,q_2\,)\\ &-\frac{1}{2}\sin(\,q_2\,)\,{u_2}^2 + \frac{1}{2}\cos(\,q_2\,)\,\dot{u}_2 = -\,\frac{3}{2}\,\epsilon\,q_1\end{aligned} \tag{7.25}$$

The resulting 'simplified' equations could be treated as is, or the further step of inserting series in ϵ could be taken. For example one could use the form:

$$q_2 = {q_2}^{(0)} + \epsilon {q_2}^{(1)} + \epsilon^2 {q_2}^{(2)} + \ldots \tag{7.26}$$

with similar expressions for the other variables. Insert these into the approximate equations to derive a hierarchy of perturbation equations. The 'physical' interpretation is that the largest term (called the lowest order term) corresponds to a simple pendulum. Corrective terms account for the modulation of the upper pendulum links movements. Some suggestions for calculations are given in problem 7.2.

7.2.1 The Sophia Series Expansion Package

Sophia provides several functions to assist in the task of deriving approximate equations of motion by the direct method. In cases where the mechanism being treated is beyond Sophia and the general computer algebra systems capabilities, this may be the only way to derive symbolic equations. To see the way the functions are used we reconsider the previous problem using the expansion package. Only input statements are shown. First the statements needed to solve the full problem are given. This is followed by statements needed to derive approximate equations of motion.

The first set of statements set up the natural units. The 'unassignement' statement for the moment of inertia also acts as a reminder. If one wished to choose different natural units, say with $J = 1$ it is a simple matter to edit this statement. In general it is best to insert any specific formula for J at the end of the calculations. The next statement defines the reference frames and moment of inertia tensors, the latter in the specialized manner suitable for planar problems.

```
> m:=1: L:=1: g:=1: J:='J':
> &rot [N,A,3,q1]: &rot [N,B,3,q2]:
> IA := EinertiaDyad(0,0,J,0,0,0,N):
> IB := EinertiaDyad(0,0,J,0,0,0,N):
```

The next two statements define the mass center positions of the bodies, and call for the definition of generalized speeds as generalized velocities, ($\dot{q}_j = u_j$). The last two statements calculate (with simplification and substitution of generalized speeds for coordinate derivatives) the velocity of the mass centers.

```
> rA := A &ev [L/2,0,0]:
> rB := (A &ev [1,0,0]) &++ (B &ev [L/2,0,0]);
> &kde 2:
> vA := &simp (N &to subs(kde,N &fdt rA));
> vB := &simp (N &to subs(kde,N &fdt rB));
```

In natural units with $m = 1$ the mass center velocities and momenta are identical, however it is good practice to define them so that changes of units can be made in the earlier statements specifying the system. The angular velocities and angular momenta are calculated, care being taken to express results in the generalized speed variables by use of the kinematic differential equations and the Maple substitution command.

```
> pA := m &** vA: pB := m &** vB:
> wA := N &to subs(kde,N &aV A);
> wB := N &to subs(kde,N &aV B);
> hA := IA &o wA;
> hB := IB &o wB;
```

The velocity and momentum Kvectors are now formed. This is followed by the formation of the applied force Kvector, first by defining the force and center of mass moments on the individual bodies and finally by gathering them into the Sophia/Maple Kvector structure. In this case the only applied force is due to gravitation, however it should be clear that other forces can easily be introduced at this point.

```
> vK := &KM [vA,wA,vB,wB]: pK := &KM [pA,hA,pB,hB]:
> R1 := N &ev [m*g,0,0]: R2 := N &ev [m*g,0,0]:
>  T1:= N &ev [0,0,0]: T2 := N &ev [0,0,0]:
> RK := &KM [R1,T1,R2,T2]:
```

The velocity Kvector, expressed as a linear function of the generalized speeds, is used to compute a set of two independent tangent Kvectors.

```
> tauK := KMtangents(vK,u,2);
```

The calculation of the full equations of motion is now completed by calculating the dynamic force or rate of change of momentum Kvector. This is used with the Sophia &kane operator and the set of tangent Kvectors, tauK, to calculate the set of 'generalized dynamic forces'. The applied force Kvector is then used to calculate the generalized applied forces. Equating the corresponding members of these lists produces the dynamic part of the equations of motion. The total set of 4 first order equations consists of these and the kinematic differential equations. This completes the derivation of the full equations of motion.

```
> pKt := &Ksimp subs(kde,N &Kfdt pK);
> GDF := tauK &kane pKt;
```

```
> GAF := tauK &kane RK;
> kaneEqns:= combine({GDF[1]=GAF[1],GDF[2]=GAF[2]},trig);
```

The approximation package consists of four basic functions and several modified operators designed to work with approximate frame transformation equations. They all call for an argument which describes the desired scaling. This is entirely up to the user who must define a substitution set that describes the scaling of variables. By retaining the same names for the new variables one can work with previously defined time dependency statements and kinematic differential equations. In the present problem we only partially approximate the variables by leaving q_2 and u_2 in their original forms. The quantities q_1 and u_1 are scaled by the parameter ϵ. Therefore the substitution set for scaling is chosen as:

```
> disturbance := { q1 =epsilon*q1, u1= epsilon*u1};
```

There are three series expansion functions, one each for Evectors, Kvectors and Svectors. The reader should recall that the latter are simply ordered lists of Kvectors. The quantity tauK used to store the tangent Kvectors for the problem is an example of an Svector. They all take three arguments, the vector object, the expansion parameter and the order of the expansion. The latter is the order of the neglected terms. The names of the functions are: EvectSeries, KvectSeries, and SvectSeries. In what follows only the latter two are used. Note that in each case the substitution of the scaled variables is made *before* the expansion, i.e. in the argument to the expansion function.

```
> pKd := KvectSeries(subs(disturbance,pK),epsilon,2);
> RKd := KvectSeries(subs(disturbance,RK),epsilon,2);
```

The calculation of frame based derivatives and the transformation of Evector representations between frames make use of the stored direction cosine matrices. For the derivation of approximate equations it is desirable to also approximate these relations as they are responsible for the large size of the exact expressions that may block an attempt at a full derivation. The next statement tells the system to set up approximate frame relations to the required order. It is called 'SetApproximateRotation', and it takes three arguments, the name of the scaling substitution set, the small parameter name and the order of the expansion. Therefore for the present problem it is given as:

```
> SetApproximateRotation(disturbance,epsilon,2);
```

This statement makes available a set of alternative frame relations which the user can choose to use at any time. The function 'SwitchExactApproximate' takes one argument which is a Maple string, i.e. must include quotes. The string 'E' switches to the exact relations, the string 'A' to the approximate ones. The user may view or operate with any of the exact or approximate direction cosine matrices at any time. They are stored under the global names RF1F2ext and RF1F2apx, where F1 and F2 are frame names and ext indicates exact, apx indicates approximate. Thus the exact and approximate frame relations for frames A and B in the present problem are stored as RABext, RBAext, RABapx, and RBAapx. In the Maple evaluation convention for arrays the actual content of these is obtained with the command op(RF1F2xxx). If one switches to approximate relations all of Sophias operators that require frame transformations will automatically use the approximate relations. In many cases this is not what is wanted, it only being desired that approximate relations are used for specific operations. To this end Sophia has several operators that automatically use the stored approximate relations. They are all named by adding the prefix 'A' to the normal operator name. Thus &Afdt is the approximate frame differentiation operator. It is important in using these to make sure that the SetApproximateRotation command has been entered, otherwise the approximate relations will not be stored in the system. Other approximate operators are &AKfdt, &Akane and &Ato. It should be clear that the user can easily construct other such operators or functions simply by using the SwitchExactApproximate command in a Maple function definition.

Armed with these functions the remainder of the calculation is a straightforward repeat of the exact case. Derivatives are obtained with the approximate operators. To keep expressions simple the series obtained are truncated to the desired order by the appropriate series expansion command. This is needed because multiplication of expressions can introduce higher order terms of no significance. The reason for this is that terms of such higher order have been dropped in previous calculations. Therefore the remaining steps to the approximate equations are given by:

```
> pKtd := KvectSeries(subs(kde,N &AKfdt pKd),epsilon,2);
> tauKd := SvectSeries(subs(disturbance,tauK),epsilon,2);
> GDFd := tauKd &Akane pKtd;
> GAFd := tauKd &Akane RKd;
> kaneEqnsd := GDFd[1]=GAFd[1],GDFd[2]=GAFd[2];
```

This completes the example. The Sophia routines discussed here are summarized in appendix A.

7.3 Impulsive Force Approximations

Short duration intensive forces must frequently be accounted for in the modeling of mechanical systems. Typical examples occur when objects come in contact. This ranges from simple impact to complex situations such as occur with backlash problems in gear systems. The theory of impulse is a classical technique for dealing with these type of problems which can easily be generalized to the Kvector formulation of mechanics. It is an approximation technique which can be formalized along the lines of the discussion in the previous sections of this chapter. The importance of this is that we can develop models which describe impacts in more detail. Results obtained in this way can supply parameters used in the rigid body formulation of an impact problem. It is thus worth a slight diversion to see how the parameters that occur in the typical impact calculation might be calculated as well as measured. Another point is that the measured parameters can be used to deduce the more detailed mechanical behavior. The coefficient of restitution, ν, is a typical example. It is typically defined as the negative of the ratio of the speed of separation to the speed of approach of two bodies which collide, e.g. $\boldsymbol{v}_s = -\nu \boldsymbol{v}_a$. If it is a valid concept it should be measurable as a function of the physical properties of the interacting bodies. It is impossible for a body to change its velocity instantaneously. Hence there must be a time scale during which the bodies remain in contact and large forces are developed.

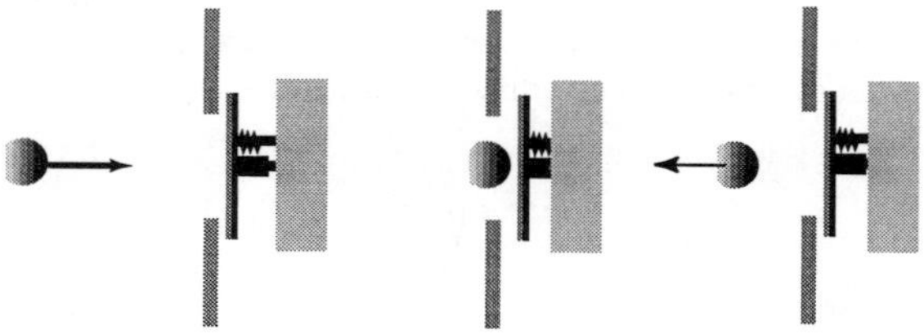

Figure 7.2: Impact on a stiff surface with damping

7.3.1 A Simple Collision Model

Figure 7.2 shows a model of the impact situation which allows for some calculation and perhaps understanding of the process. A particle of mass m moving with a velocity u impacts a surface. The surface is modeled by a massless piston attached to a rigid foundation by a spring-damper. It is arranged so that it can move inward compressing the spring but is stopped in its outward motion. The collision is assumed to occur at the position $q = 0$ and at time $t = 0$. The coordinate q increases in the direction of the initial velocity. Our intuition is that if this is to be a realistic model of observed impact situations the duration of contact between particle and surface should be small. Therefore the spring should be *stiff* so that the time constant of the spring mass system is also small. It is also expected that this is matched by a high damping constant for the same reason. To make this apparent introduce the small parameter ϵ. The spring and damper will be represented by the parameters:

$$k = \frac{1}{\epsilon} k_0, \tag{7.27}$$

$$d = \frac{1}{\epsilon} d_0. \tag{7.28}$$

The asymptotic study of the problem will consider the case where the parameter ϵ becomes small, the other quantities remaining fixed. The equations of motion must cover three cases, the time prior to impact or $t < 0$, the period of contact, $0 < t < t_s$

and the period following impact, $t_s < t$. The time of contact, t_s is to be found. It is also required that both the velocity u and the position q are *continuous* functions of time. The kinematic differential equation $\dot{q} = u$ is valid for all three periods. As no forces are acting on the particle before and after the collision we have for $t < 0$ and $t > t_s$ that $m\dot{u} = 0$, or equivalently that the velocity of the particle is constant. During contact the spring and damper force act, implying that:

$$m\dot{u} + \frac{1}{\epsilon}(k_0 q + d_0 u) = 0. \tag{7.29}$$

Therefore use of the kinematic differential equation shows that during this period q satisfies the damped oscillator equation. In discussing the solution of this equation it is helpful to define several new parameters. These are ω the natural frequency, ω_d the damping frequency and ω_r the reduced frequency:

$$\omega = \sqrt{\frac{k_0}{m}} \tag{7.30}$$

$$\omega_d = \frac{d_0}{2m} \tag{7.31}$$

$$\omega_r = \sqrt{\omega_i^2 - \omega_d^2}. \tag{7.32}$$

Note that ϵ has not been included in these definitions. The reason is that we want to see how the results vary as this parameter tends towards zero. Thus the displacement q during the contact period is given by:

$$q(t) = Ae^{-\omega_d \frac{t}{\epsilon}} \sin \omega_r \frac{t}{\epsilon}, \tag{7.33}$$

where A is a constant to be determined. The initial contact occurs at $t = q = 0$ and the equation for q before contact is $q(t) = u_a t$, where u_a is the initial velocity of approach to the impact. The condition of velocity continuity at $t = 0$ provides an equation for A, thus one finds that $A = \epsilon u_0/\omega_r$. The contact will be broken when the sin function returns to zero, i.e. when:

$$t = t_s = \epsilon \frac{\pi}{\omega_r}. \tag{7.34}$$

The maximum 'penetration' depth of the particle will occur at $t_s/2$ and is given by

$$q_m = \epsilon \frac{u_a}{\omega_r} e^{-\frac{\pi \omega_d}{2\omega_r}}. \tag{7.35}$$

The continuity of velocity condition shows that the wall separation velocity will be the velocity reached by the particle at $t = t_s$. This is given by:

$$\frac{u_s}{u_a} = -e^{-\pi \frac{\omega_d}{\omega_r}}. \tag{7.36}$$

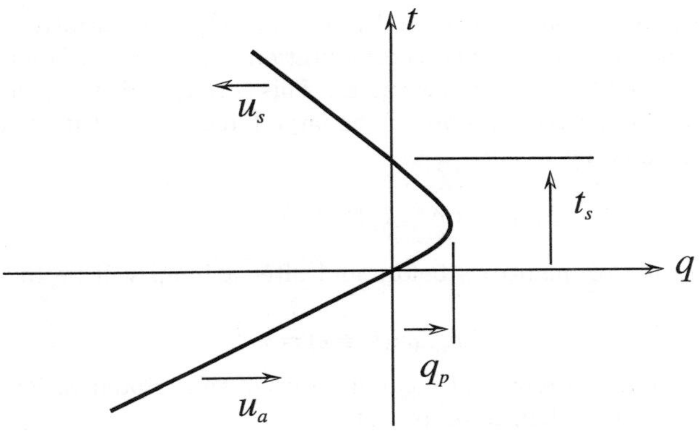

Figure 7.3: Collision path in the q-t plane

The results are summarized graphically in figure 7.3, which shows the path of the particle in the $q - t$ plane. The ratio of the separation speed to the approach speed given in the last equation depends only on the physical properties of the impact surface, that is we may take it as a constant for any given surface. The time of contact and the penetration depth approach zero as ϵ tends to zero. Therefore if ϵ is small we can replace the detailed calculation by the condition that the velocity *jump condition* is given by

$$\Delta u = u_s - u_a = -u_a(\nu + 1), \tag{7.37}$$

where ν, the coefficient of restitution is a function of the impacting bodies material properties. Also taking the second derivative of the displacement shows that the force developed during the impact is proportional to $\frac{1}{\epsilon}$ and the product of the force and duration time is independent of ϵ. While the model chosen is quite simple it does contain the significant physical effects expected in real collisions, i.e. elastic response, contact penetration and energy loss. Perturbation methods, which are outside the scope of this work, could be used to take advantage of the fast time scale that is in effect during collision to calculate 'constants' that describe 'jumps' in velocity components. In the remainder of this section it will be assumed that this has been accomplished by a suitable combination of theory, experiment and physical sense. The theory of *impulse* takes advantage of the force of impact being large and of small duration.

7.3.2 The Impulse Equation

If the system is acted on by large short duration forces it is expected that on the time scale of the forces the configuration will be essentially fixed. Velocities are expected to change significantly as a result of the accelerations associated with the forces. The impact problem of the previous section was an example of this. A useful approximate theory for the velocity changes can be derived in a relatively intuitive manner. First some notation is introduced. The short duration time scale is ignored, hence the velocities are allowed to *jump* to new values. This corresponds to ignoring the ϵ time scale of the collision problem above. If the large forces act at time t_i the jump of a quantity u is denoted by:

$$\Delta_{t_i} u = lim_{\epsilon \to 0}(u(t_i + \epsilon) - u(t_i - \epsilon)). \tag{7.38}$$

The limiting values of quantities before and after a jump will be indicated by the notation:

$$lim_{\epsilon \to 0} u(t_i \pm \epsilon) = \tilde{u}^{(\pm)}. \tag{7.39}$$

Symbols such as q and u represent column arrays of subscripted variables such as q_j. The condition on the configuration is that:

$$\Delta_{t_i} q = 0. \tag{7.40}$$

The value of such fixed quantities as q at the time of the impulsive force will be denoted by a tilde, e.g. $\tilde{q}^{(\pm)} = \tilde{q}$. By definition $\Delta_{t_i} t = 0$. The tangent Kvectors depend on time and configuration and not on speeds, hence:

$$\Delta_{t_i} \boldsymbol{\tau}_j^< = 0, \tag{7.41}$$

or if τ is the column array of tangent Kvectors

$$\Delta_{t_i} \tau = 0. \tag{7.42}$$

The constraint free equations of motion are given by:

$$\dot{\boldsymbol{P}}^< \bullet \tau = \boldsymbol{R}^< \bullet \tau. \tag{7.43}$$

Note that since τ is a column of Kvectors, this is a column of scalars, where each component is a product of the form $\dot{\boldsymbol{P}}^< \bullet \boldsymbol{\tau}_j^<$. Integration of a quantity over the period of the short duration forces is defined by

$$\int_{t_i} f(t) = lim_{\epsilon \to 0} \int_{t_i - \epsilon}^{t_i + \epsilon} f(t) dt. \tag{7.44}$$

Any quantity which has a vanishing jump, such as τ, can be taken outside of the integral. Therefore:

$$\int_{t_i} (\dot{\boldsymbol{P}}^<) \bullet \tilde{\tau} = \int_{t_i} (\boldsymbol{R}^<) \bullet \tilde{\tau}. \tag{7.45}$$

The integral of a derivative, such as on the left side of the above equation, evaluates to the jump of the undifferentiated variable. The integral of the applied force is called the impulse. The notation

$$\tilde{\boldsymbol{R}}^< = \int_{t_i} (\boldsymbol{R}^<) \tag{7.46}$$

is used for this quantity. The basic constraint free dynamic equations for the momentum jump is:

$$\Delta_{t_i} \boldsymbol{P}^{<} \bullet \tilde{\tau} = \tilde{\boldsymbol{R}}^{<} \bullet \tilde{\tau}. \tag{7.47}$$

Because of the basic relation that the velocity is a linear function of the generalized speeds this is actually an equation for the generalized speed jumps, $\Delta_{t_i} u$. To see this explicitly let $\boldsymbol{M}$ indicate a diagonal matrix of dyadic operators. For a rigid body system it is made up of diagonal 2×2 submatrices. These have the form

$$\boldsymbol{M}^{<k} = \begin{bmatrix} m_k \boldsymbol{U} & 0 \\ 0 & \boldsymbol{I}_k \end{bmatrix} \tag{7.48}$$

where m_k is the mass and $\boldsymbol{I}_k$ the center of mass inertia dyad of body k. The ordinary dot product of the elements of $\boldsymbol{M}$ with the center of mass velocity and body frame angular velocity produces the components of the momentum Kvector corresponding to the body, thus:

$$\begin{bmatrix} m_k \boldsymbol{U} & 0 \\ 0 & \boldsymbol{I}_k \end{bmatrix} \cdot \begin{bmatrix} \boldsymbol{v}^{<k} \\ \boldsymbol{\omega}^{<k} \end{bmatrix} = \begin{bmatrix} \boldsymbol{P}^{<k} \\ \boldsymbol{h}^{<k} \end{bmatrix} \tag{7.49}$$

The basic velocity expansion relation, developed in chapters three and five, can be expressed in the form:

$$\boldsymbol{v}^{<} = \tau^T u + \boldsymbol{\tau}_t^{<}, \tag{7.50}$$

where τ^T is a row of the tangent Kvectors. Therefore using the mass operator:

$$\boldsymbol{P}^{<} = \boldsymbol{M} \cdot \boldsymbol{v}^{<} = \boldsymbol{M} \cdot \tau^T u + \boldsymbol{M} \cdot \boldsymbol{\tau}_t^{<}. \tag{7.51}$$

Because the last term in the above is only dependent on the current configuration and time it has a vanishing jump. Inserting this equation into the momentum jump condition gives:

$$\tilde{G}_M(\Delta_{t_i} u) = \tilde{\boldsymbol{R}}^{<} \bullet \tilde{\tau}. \tag{7.52}$$

Where

$$\tilde{G}_M = \tilde{\tau} \bullet (\tilde{\boldsymbol{M}} \cdot \tilde{\tau}^T), \tag{7.53}$$

is the mass metric matrix at the time t_i. Equation 7.52 provides n equations for the jumps $\Delta_{t_i} u$. Unfortunately the impulse $\tilde{\boldsymbol{R}}^{<}$ is often also an unknown. Therefore additional conditions must be specified for the solution of particular problems. Sometimes, as in the collision example, it is possible to obtain equations for the generalized speed jumps from considerations of coefficient of restitution arguments. It should also be appreciated that the impulsive forces are like any other forces, that is when two bodies interact we can expect equal and opposite reaction forces. Another approach is to proceed from experiments which measure the impulse, or from observations as to how the body kinematics develop. Therefore some 'art' may be required to make profitable use of the impulse relations. It is also possible to bring frictional effects into consideration, which when dry friction is involved means that we must find the impulsive constraint forces. The matter can become very difficult and there are basic questions as to the physics of friction during such short duration events. Only a relatively simple example is considered here, that is the completion of the problem of section 7.2 involving the restricted pendulum motion.

7.3.3 The Restricted Double Pendulum

When the upper link of the pendulum strikes the boundary which restricts its motion it is reasonable to expect short duration forces to act on the pendulum system. This will be a repeated event hence it is desirable to derive jump conditions for an arbitrary impact. The configuration of the link which makes contact will always have the same value, i.e. $\tilde{q}_1 = \epsilon$. The angular velocities of the bodies and the angular position of the bottom link will differ for each impact as will the impact time. In any case at such time, t_i the quantities $\tilde{q}_1, \tilde{q}_2, \tilde{u}_1^{(-)}$, and $\tilde{u}_2^{(-)}$ are taken as known. The task is to calculate $\tilde{u}_1^{(+)}$, and $\tilde{u}_2^{(+)}$. Therefore it is necessary to formulate the impulse equation 7.52. This is done by calculation of the mass metric $\tilde{G}_M$ which is a function of the tangent Kvectors at the time of impact. To $O(\epsilon^2)$ inspection of the calculation of section 7.2 shows that:

$$\tilde{\tau_1^<} = \begin{bmatrix} \frac{1}{2}(-\epsilon \boldsymbol{n}_1 + \boldsymbol{n}_2) \\ \boldsymbol{n}_3 \\ -\epsilon \boldsymbol{n}_1 + \boldsymbol{n}_2 \\ \mathbf{o} \end{bmatrix} \tag{7.54}$$

$$\tilde{\tau_2^<} = \begin{bmatrix} \mathbf{o} \\ \mathbf{o} \\ \frac{1}{2}(\tilde{s}_2 \boldsymbol{n}_1 + \tilde{c}_2 \boldsymbol{n}_2) \\ \boldsymbol{n}_3 \end{bmatrix} \tag{7.55}$$

The operator $\tilde{\boldsymbol{M}}$ has diagonal components, using natural units,

$$\tilde{\boldsymbol{M}}_{11} = \boldsymbol{U} \tag{7.56}$$

$$\tilde{\boldsymbol{M}}_{22} = J\boldsymbol{n}_3\boldsymbol{n}_3 \tag{7.57}$$

$$\tilde{\boldsymbol{M}}_{33} = \boldsymbol{U} \tag{7.58}$$

$$\tilde{\boldsymbol{M}}_{44} = J\boldsymbol{n}_3\boldsymbol{n}_3. \tag{7.59}$$

Applying the definition of $\tilde{\boldsymbol{M}}$ it is found that to $O(\epsilon^2)$:

$$\tilde{G}_{M11} = \frac{5}{4} + J, \tag{7.60}$$

$$\tilde{G}_{M12} = \frac{1}{2}(\tilde{c}_2 + \epsilon\tilde{s}_2), \tag{7.61}$$

$$\tilde{G}_{M21} = \frac{1}{2}(\tilde{c}_2 + \epsilon\tilde{s}_2), \tag{7.62}$$

$$\tilde{G}_{M22} = \frac{1}{4} + J. \tag{7.63}$$

To complete the formulation of the impulse equation we must consider the impulse Kvector. At the time if impact the edge of the first bar comes into contact with the boundary, hence we can expect a force and torque about the center of mass of the upper link. It is reasonable to assume that the generated force will be orthogonal to the boundary of the link, that is it will only have a component in the direction of the frame vector $\boldsymbol{a}_2$. If the component (unknown) of this impulse is taken as λ we have that:

$$\tilde{\boldsymbol{R}}^{<1} = \lambda \boldsymbol{a}_2. \tag{7.64}$$

The width of the link in natural units is taken as S. Therefore the position of the impact point relative to the mass center of the link is:

$$\boldsymbol{r}_c = \frac{1}{2}(\boldsymbol{a}_1 + S\boldsymbol{a}_2). \tag{7.65}$$

Therefore the impulsive torque on the link is:

$$\tilde{\boldsymbol{T}}^{<1} = \boldsymbol{r}_c \times (\lambda \boldsymbol{a}_2). \tag{7.66}$$

There is no applied impulse acting directly on the second link, therefore to the considered order:

$$\tilde{\boldsymbol{R}}^{<} = \begin{bmatrix} \lambda \boldsymbol{a}_2 \\ \frac{1}{2}\lambda \boldsymbol{a}_2 \\ \boldsymbol{o} \\ \boldsymbol{o} \end{bmatrix}. \tag{7.67}$$

The impulse relation gives two equations for the jump in generalized speeds, however we have three unknowns $\tilde{u}_1^{(+)}, \tilde{u}_2^{(+)}$, and λ. More assumptions must be made. The most reasonable is to assume a coefficient of restitution argument for the impact, that is the jump in the normal component of the velocity of approach ${}^N\!\boldsymbol{\mathcal{V}}$ at the contact point is related to the normal component of the velocity of separation by the coefficient of restitution ν. Thus

$${}^N\!\boldsymbol{\mathcal{V}} = u_1 \boldsymbol{a}_3 \times (\boldsymbol{a}_1 + \frac{S}{2}\boldsymbol{a}_2). \tag{7.68}$$

The normal component is:

$${}^N\!\boldsymbol{\mathcal{V}} \cdot \boldsymbol{n}_2 = u_1. \tag{7.69}$$

If the separation velocity is equal to the negative of the approach velocity times the coefficient of restitution, it follows that

$$\Delta_{t_i} u_1 = \tilde{u}_1^{(+)} - \tilde{u}_1^{(-)} = -(\nu + 1)\tilde{u}_1^{(-)}. \tag{7.70}$$

The impulse equation, to $O(\epsilon^2)$, is:

$$\begin{bmatrix} \frac{5}{4} + J & \frac{1}{2}(\tilde{c}_2 + \epsilon \tilde{s}_2) \\ \frac{1}{2}(\tilde{c}_2 + \epsilon \tilde{s}_2) & \frac{1}{4} + J \end{bmatrix} \begin{bmatrix} \Delta_{t_i} u_1 \\ \Delta_{t_i} u_2 \end{bmatrix} = \begin{bmatrix} \lambda \\ 0 \end{bmatrix} + O(\epsilon^2), \tag{7.71}$$

which provides all the information needed to calculate the jump in the angular velocity of the second link due to a collision, i.e.

$$\Delta_{t_i} u_2 = \frac{2(\nu+1)}{1+J}(\tilde{c}_2 + \epsilon \tilde{s}_2)\tilde{u}_1^{(-)}. \tag{7.72}$$

It should be realized that while in this case the determination of the jump conditions is relatively simple the need to check for collision events adds considerable complication to numerical simulation studies.

7.4 Problems

•Problem 7.1

Using the method outlined in the discussion of the simple pendulum derive the linearized equation for the case where the base solution is $q_1 = \pi$. Obtain approximate equations that are valid up to $O(\epsilon^4)$. Sketch the equal energy curves in the $q_1 - u_1$ phase plane. Find an expression for the time of traversal of the phase point between two points on the same energy curve. What conclusions can be drawn about the traversal time between an arbitrary point on a curve passing through the upper equilibrium point ($q_1 = \pi, u_1 = 0$) and the equilibrium point. This point is sometimes called a *saddle point*, can you suggest a reason for this name? Generalize your results to systems with damping forces and with more degrees of freedom.

•Problem 7.2

Carry out a numerical study of the simplified double pendulum problem discussed in section 7.2. Compare with numerical solutions to the full equations. You will have to use a test for impact and the jump conditions discussed in section 7.3. Modify the problem by replacing the 'sudden' impact by smooth contact with stiff springs when the limiting angle is reached.

•Problem 7.3

Derive approximate equations of motion for the 'space colony' problem (5.2) under the assumption that the mirror motions are small.

•Problem 7.4

Reconsider the collision problem of section 7.3 by treating the oncoming body as a two particle system. Assume it is constrained to move in a straight line and that the particles are connected by a spring-damper combination. Describe what happens when these have a time scale that is smaller, greater and equal to that of the wall system.

•Problem 7.5 Calculate the velocity changes that occur when the space colony of problem (5.2) is impacted by a rapidly moving mass. Make an estimate for the impulsive forces and consider various collision points and directions on the colony main cylinder.

•Problem 7.6 Develop jump conditions for the example of a double pendulum with an additional restriction placed on the bottom link. This is an extension of the problem presented in section 7.3. Assume that the restriction is the same as for the first link and acts at the lowest point on the double pendulum and develop equations to $O(\epsilon^2)$.

Appendix A

Sophia Command Assistance

A.1 Introduction

This appendix contains examples of a number of basic Sophia commands. The command arguments are typical cases, e.g. if the name of a typical generalized coordinate is called for the example might use q3 for q_3. The Sophia tools are designed to work in a normal manner with Maple, thus they typically define an object which must be assigned to a symbol for storage or further processing. Some of the Sophia tools control the environment by setting global variables. Thus frame information is stored in direction cosine matrices with the naming convention, Rframe1frame2. Also certain sets are defined that record the frame names used or substitution lists used in differentiation operations. The tools can also be used as functions in normal Maple programming.

A.2 Installation of Sophia

The software included with this book is on a disk in PC readable format. The file of Sophia programs is in text format and should be easy to transform to other formats. The software, together with other information and programs, is also available on a server available to internet users. The address is ftp.mech.kth.se. Login as anonymous and use your computer mail address as your password. Look in the directory 'sophia'.

Copy the text file, SoPhIa.txt, into a directory which can be accessed by your Maple software. For example, on a Macintosh, put the file into the Maple folder. Start Maple as usual. Use the Maple 'read' command with the file path and name as argument. On a Macintosh the keyboard combination command-2 will bring up a selection box from which the appropriate file can be selected. Acceptance of the file will cause the path and name to be printed on the console input line after the read command. The Maple input line would be something like, read 'mypath:SoPhIa.txt'; Note the backquotes and the semicolon! Pressing enter will read in the file. Ignore the warnings which are due to reading in Maple's linear algebra package, which is used by Sophia. You can now proceed with your work session. Note that it takes between 30 seconds and several minutes (depending on your system) to read in the routines. To avoid this time delay read in SoPhIa.txt as above, then type save sophia.m This will create a Maple 'm' file which can be read into your workspace in a few seconds. The next time you use Sophia just start your Maple session and read the 'm' file.

The included version of Sophia is for MapleV release 3. Older versions of Maple do not use the 'global' declaration in procedures. If you have an older version of Maple you will have to use a text file to remove all lines in Sophia's procedures that contain a 'global' declaration.

There may be newer information and other material on the distribution disk or in the ftp directory. The SoPhIa.txt file contains a number of procedures that are not documented in this text. Some are help procedures, needed to achieve the objectives of the documented procedures. Others are extensions that are useful for dealing with particular problem situations. Most of the procedures contain some comments on their function.

While it is possible to use Maple to carry out numerical integration of equations of motion derived with Sophia, it is frequently better to export the results to other programs. The ftp sight should be consulted for Sophia extensions that help in this task, for example in export of results to the popular Matlab program or to a 'c' program.

A.3 Setting Frame Information

A common situation that arises is that the frame relation is known between some frames A and B, and between B and C, but not between A and C. If such a transformation is called for, Sophia will automatically detect the intermediate frame B and compute the relations between A and C.

- &rot [A,B,3,q_1]
- chainSimpRot([[A,B,3,q_1],[B,C,1,q_2]])
- dircos(B,A,$\boldsymbol{b}_1 \cdot \boldsymbol{a}_1, \boldsymbol{b}_1 \cdot \boldsymbol{a}_2, \boldsymbol{b}_1 \cdot \boldsymbol{a}_3, \boldsymbol{b}_2 \cdot \boldsymbol{a}_1, \boldsymbol{b}_2 \cdot \boldsymbol{a}_2, \boldsymbol{b}_2 \cdot \boldsymbol{a}_3, \boldsymbol{b}_3 \cdot \boldsymbol{a}_1, \boldsymbol{b}_3 \cdot \boldsymbol{a}_2, \boldsymbol{b}_3 \cdot \boldsymbol{a}_3$)

A.4 Declaring Functional Dependence

- dependsTime($q_1, q_2, q_3, u_1, u_2, u_3$)
- &kde n → *Standard set up in which 2n variables q and u are set to depend on time and kde is assigned so that the generalized speeds are the generalized velocities, i.e. the time derivatives of the generalized coordinates.*
- clearVars() *clears all time dependencies that have been declared*

A.5 Evectors

A.5.1 Construction and Selection

It is important to make sure that the chosen frame name does not conflict with other names in the work space, either defined as part of the particular task or by Maple. For example the letter 'D' is a Maple operator. Frame names need not be a single letter, for example names such as FrameNewton are acceptable if somewhat lengthy. The basic procedure for defining a dyad is also included.

- A &ev $[v_1, v_2, v_3] \rightarrow \mathbf{v} = v_1\mathbf{a}_1 + v_2\mathbf{a}_2 + v_3\mathbf{a}_3$ *This sets up an Evector from the components in frame A.*

- A &> $j \rightarrow \boldsymbol{a}_j$ *This operator creates a base vector for any frame.*

- &vPart v $\rightarrow [v_1, v_2, v_3]$ *extracts the array of components. Note that this is an array, not a list!*

- &fPart eV $\rightarrow$ frame name *extracts the frame name of the representation of the vector*

- v &c j $\rightarrow v_j$ *extracts the jth component in the frame in which the vector is represented.*

- Edyad($d_{11}, d_{12}, d_{13}, d_{21}, d_{22}, d_{23}, d_{31}, d_{32}, d_{33}$,N) $\rightarrow \begin{bmatrix} d_{11} & d_{12} & d_{13} \\ d_{21} & d_{22} & d_{23} \\ d_{31} & d_{32} & d_{33} \end{bmatrix}$ *sets up the representation of a dyad in the given frame*

A.5.2 Changing Representations

- N &to ev *Transforms the representation from whatever frame is current for the object to the stated frame. This works for both Evectors and Edyads.*

- &simp ev *Carries out simplification operations on the components of the Evector or Edyad. This is not done automatically as it can be a costly operation.*

A.5.3 Algebra

The following operations provide the normal algebra of vectors and dyads in three dimensional Euclidian space.

- s &** v $\rightarrow s\boldsymbol{v}$ *Simply the multiplication of a scalar (number) into a vector, i.e. all components of the vector are multiplied by the scalar.*

- v1 &++ v2 → $\boldsymbol{v}_1 + \boldsymbol{v}_2$ *The addition of two vectors. Sophia takes care of transforming the vectors representations to consistent frames before addition. The result is represented in the frame that the last vector was represented in. Multiple additions may be put in the same statement, however it is a good idea to use parenthesis to group other operations.*

- v1 &xx v2 → $\boldsymbol{v}_1 \times \boldsymbol{v}_2$ *This is the normal vector cross product.*

- v1 &o v2 → $\boldsymbol{v}_1 \cdot \boldsymbol{v}_2$ *This is the normal 'dot' product. The object on the left in a product may be an Evector or Edyad.*

- &VtoD v → $Dyad(\boldsymbol{v})$ *Produces the so called Dyad of a Vector. That is the dot product of this object with another vector is equivalent to the cross product of the original vector with the other vector.*

- &DtoV D → $Vect(\frac{1}{2}(D - D^T))$ *This takes the antisymmetric part of an Edyad and converts it to a vector.*

A.5.4 Differentiation

The symbol 't' is reserved for time. Sophia represents time derivatives by appending the appropriate number of t's to the quantity. Sophia requires the declaration of variables which depend on time in order to set up certain substitution lists that are used in carrying out differentiations. The system is setup for first and second order derivatives (all that need to appear in formulating equations of motion).

- &dt s → $\frac{ds}{dt}$ *This is for differentiation of a scalar expression. Quantities such as q1 have derivative q1t.*

  A &fdt v → $\frac{{}^A dv}{dt}$ *This is the important 'frame differentiation operator'.*

- ${}^N\boldsymbol{\omega}^A$ &fdtAV v → $\frac{{}^N dv}{dt}$ *For frame differentiation in terms of a given expression for the angular velocity. This can be useful in the process of looking for a suitable set of kinematic differential equations.*

A.5.5 Kinematic Quantities

The following are to provide assistance in obtaining basic kinematic information from frame information, i.e. it assumes the frames have been specified in terms of suitably declared variables.

- A &aV B → ${}^A\boldsymbol{\omega}^B$ *The angular velocity is computed in terms of the coordinate information.*

- aA(A,B) → ${}^A\boldsymbol{\alpha}^B$ *The angular acceleration is computed in terms of the coordinate information.*

- aA(A,B,wAB) $\rightarrow \frac{^B d}{dt} {}^A\boldsymbol{\omega}^B$ *An alternative computation of the angular acceleration in terms of an expression for the angular velocity, which is invoked as an optional third argument to aA.*

- FFV(N,rB,wNB,rQ) $\rightarrow {}^N\boldsymbol{v}^{B(Q)}$ *Useful when examining rolling constraints, this function gives the frame fixed velocity. The inputs are the base frame, the position of the index point of frame B, the angular velocity of frame B in N and the position of the special point, typically the point of contact between body B and a point in frame N.*

- `[N,B] &ffv [rB,rQ]`

  $\rightarrow {}^N\boldsymbol{v}^{B(Q)}$ *An alternate version of the above which does not require the angular velocity. The left argument list is the base and body frame. The right list are the positions of the index and object points.*

A.5.6 Rigid Body Properties

Tools for determining body properties needed in formulating equations of motion, such as the center of mass and the moment of inertia dyad.

- `B &cm [[m1,r1],[m2,r2],[m3,r3]]`

  $\rightarrow$ center of mass of system of mass points represented in frame B. *Note there must be at least two frames known to the system or an error will occur. It should be realized that this operator is useful for a composite set of bodies. The individual center of mass can be treated as an equivalent point object in the computation of the systems mass center.*

- `m &mpI  r`

  $\rightarrow$ moment of inertia of mass point located at position $\boldsymbol{r}$. *The result is in the frame in which r is given.*

- `B &msI [[m1,r1],[m2,r2],[m3,r3]]`

  $\rightarrow$ moment of inertia of a system of mass points. *The comments above regarding the mass center also applies to this.*

- `[m,Icm] &-> r`

  $\rightarrow$ transfers moment of inertia, Icm is the moment of inertia about center of mass and r is the vector from the cm point to the new point. *Sometimes called the 'parallel axes or Stiener's theorem.*

A.5.7 Equipollent Systems and Screw Transformations

The following are helpful in finding the equivalent force and torque due to a system of forces and torques acting on a rigid body. In setting up the equations of motion it is frequently required to obtain the torque about the mass center or some other index point in each of the involved bodies. The screw object in Sophia is a list of two Evectors. The first object in the list may typically be a torque or velocity, the second a force or angular velocity. For Equipollent Force Systems we generally consider torques and forces.

- A &screw [[r1,R1],[r2,R2],[r3,R3]] → [T,R] *The left argument is the frame in which the result will be represented. The right argument is a list of lists. The latter lists have two members, the point of application of a force and the force. The point of application is an Evector giving the displacement from some index point. The force is also an Evector. The index point is the same for all the forces and the resultant screw object is in terms of that point. The typical case is that the index point is the mass center of a body and the forces are all the applied forces acting on the body at various points.*

- [T,R] &ss rs → [T',R] *The displacement vector rs runs from the point of the screw on the left of the operator to another point. The result is a screw which is based on the new point. Note that this also applies to the pair [v,ω], i.e. provides an expression for the velocities at two points in a rigid body relative to another frame.*

A.6 Redundant Coordinates

This proceedure is for both redundant coordinates and nonholonomic constraints. The first argument is a list of the chosen independent generalized speeds. The second argument is the set of kinematic differential equations for the problem with the redundant coordinates. The third argument is the set of velocity constraint relations. The fourth argument is the name used for the original generalized speeds, the fifth argument is the name used for the transformed generalized speeds. It is implicit that a subset of the new generalized speeds is identically zero. The latter expresses the transformed constraint relation.

- ReducedSpeeds(gslist,kde,vce,u,w) → rtkde,rigst *The output of the procedure is a sequence of two sets. The first set is the reduced transformed kinematic differential equations. The second is the reduced inverse generalized speed transformation. Both of these sets are useful in placing the problem in terms of the reduced generalized speeds and their time derivatives as discussed in chapter 6.*

A.7 KMvectors

The KMvectors are the representations of Kvectors in Maple. They form a vector space and they provide the key to finding constraint free equations of motion. The following are the basic operations for dealing with these objects.

A.7.1 Construction and Selection

- &KM [v1,v2,v3] $\rightarrow \boldsymbol{v}^{<}$ *forms the Kvector from the Evector components. The user need not worry about what representation is used for the latter.*
- &sPartKM vK $\rightarrow K$, *number of Evectors in vK It is important to make sure that in any specific problem one has the right size Kvector!*
- &vPartKM vK $\rightarrow [\boldsymbol{v}_1, \boldsymbol{v}_2, \boldsymbol{v}_3]$ *It is frequently useful to have a list of the component vectors to work with as a Maple object.*

A.7.2 Algebra of KMvectors

- s &*** vK $\rightarrow s\boldsymbol{v}^{<}$ *scaler multiplication*
- vK1 &+++ vK2 $\rightarrow \boldsymbol{v}_1^{<} + \boldsymbol{v}_2^{<}$ *The addition of Kvectors.*
- vK1 & - - - vK2 $\rightarrow \boldsymbol{v}_1^{<} - \boldsymbol{v}_2^{<}$ *Sometimes it is useful to represent the subtraction operation directly.*
- vK &O tau1 $\rightarrow \boldsymbol{v}^{<} \bullet \boldsymbol{\tau}_1^{<}$ *inner product of KMvectors*
- &Ksimp vK $\rightarrow$ *simplification of a KMvector*

A.7.3 Tangent Space Operations

- *Use is made of the basic velocity expansion equation:*

$$\boldsymbol{v}^{<} = \sum_{j=1}^{j=n} u_j \boldsymbol{\tau}_j^{<} + \boldsymbol{\tau}_t^{<}$$

- *The convention is assumed that the kinematic differential equations are stored in a set called kde.*
- *It is also convenient to start all generalized coordinates and speeds with the same variable symbol, e.g. q for generalized coordinates and u for generalized speeds are typical. Thus q1, q2 and q3; u1, u2 and u3.*
- KMtangents(vK,u,3) $\rightarrow [\boldsymbol{\tau}_1^{<}, \boldsymbol{\tau}_2^{<}, \boldsymbol{\tau}_3^{<}]$ *A list of the three independent tangent vectors based on the generalized speeds u1,u2 and u3.*

- vKtime(vK,u,3) → $\boldsymbol{\tau}_t^<$ *Note that this is printed out when using the KMtangents command. It can be used as a rough means of checking for errors*

- tau &kane fK → $[F_1, F_2, F_3]$ *A list of the Generalized Active Forces as defined by Kane*

A.7.4 KMvector Calculus

- N &Kfdt vK → $\frac{^N d}{dt}\boldsymbol{v}^<$ *This carries out the basic frame differentiation operation on all components at once.*

- N &rKfdt rK → *This special form of the time differentiation operator is intended for use on a KM position vector made up of positions of frame origins and possibly frame names. These become angular velocities* $^N\boldsymbol{\omega}^{Fr}$ *for frame Fr.*

A.8 SKvectors

These are lists of KM vectors or 'super' KMvectors which are useful in the theory of transformations of tangent vector base sets developed in chapter 3.

A.8.1 SKvectors Constructors and Selectors

- &SK [tau1,tau2,tau3] → τ

- tau &SKc 2 → $\boldsymbol{\tau}_2^<$

- &SKn tau → 3

A.8.2 SKvector Algebra

- s &**** tau → $s\tau$

- beta &++++ tau → $\beta + \tau$

- beta & - - - - tau → $\beta - \tau$

A.8.3 Linear Operators on SKvectors

- BetaTau &++** tau → $[\boldsymbol{\beta}_1^<, \boldsymbol{\beta}_2^<, \boldsymbol{\beta}_3^<]$ *BetaTau is an ordinary matrix and this gives the linear transformation from the tangent vectors* τ *to* β

A.8.4 Projection

- GK := &Kmetric tau $\rightarrow G_{ij} = \boldsymbol{\tau}_i^< \bullet \boldsymbol{\tau}_j^<$ *The metric coefficients as a matrix for the subspace generated by* τ
- &cv tau $\rightarrow \tau_c$ *The linear transformation using the matrix inverse(G) generates the recipricol bases for the subspace.*
- tauC &<> wK $\rightarrow [c_1, c_2, c_3]$ where

$$\dot{\boldsymbol{\Pi}}_* \bullet \boldsymbol{w}^< = c_1\boldsymbol{\tau}_1^< + c_2\boldsymbol{\tau}_2^< + c_3\boldsymbol{\tau}_3^<$$

- $[c_1, c_2, c_3]$ &SKsum tau $\rightarrow \boldsymbol{\Pi}_* \bullet \boldsymbol{w}^<$
- *Note that this procedure can be used to sum up any list of KM vectors e.g. the velocity expansion gives the velocity KMvector as:*

  ([u1,u2,u3] &SKsum tau) &+++ VKtime(vK,u,3)
- *Note that the generalized speeds are simply the projections of the of the velocity Kvector on the reciprocal basis appropriate to the corresponding tangent Kvectors*

A.8.5 The Direct Kinematic Problem

For holonomic systems a set of kinematic differential equations implies certain relationships between the coefficient matrix and the inhomogeneous column matrix.

- *The General Form of The Kinematic Differential Equation*

$$\dot{q}_i = \sum_{j=1}^{n} Y_{ij}u_j + X_i$$

- *The velocity expansion equation after a transformation from a coordinate basis* τ *to an arbitrary basis* β.

$$\boldsymbol{v}^< = u^T\beta + \boldsymbol{\beta}_t^<$$

  Note that the 'generalized speeds' for the coordinate basis are simply the generalized velocities based on the chosen generalized coordinates $\dot{q}$.

- *The transformation relations are:*

$$W^{\beta\tau} = Y^T$$

$$\beta = W^{\beta\tau}\tau$$

$$\boldsymbol{\beta}_t^< = \boldsymbol{\tau}_t^< + X^T\beta.$$

- *The Maple-Sophia2 implementation of the above is:*

```
> BetaTau := transpose(Y):
> beta := BetaTau \&++** tau:
> betatK := tautK \&+++ ( X &SKsum tau):
```

- CoefficientArray(kdeList,gsList) → coefficient matrix *The arguments are the kinematic differential equations put into ordered form as a list and a list of the generalized speeds. The result is the coefficient matrix that appears in the kinematic differential equations. See section 3.10*

- SourceList(kdeList,gsList) → inhomogeneous terms in the kinematic differential equations *See above and section 3.10*

A.9 Series Expansions

These procedures are tools for producing series expansions in terms of a small parameter. The idea is to do this directly as part of the derivation of the equations of motion. In most cases expression complexity growth will make it impossible to derive approximate equations from the exact equations. This is because of the impossiblility of deriving exact equations using a given set of computer resources. The procedure calls for a user defined set, called *disturbance* below. This set is a group of substitutions that implement a particular variable scaling in terms of some small parameter. The expansions are in terms of the parameter. This must be reflected by the expansion of direction cosine matrices, which is accounted for in the procedure SetApproximateRotation.

- SetApproximateRotationdisturbance,eps,n → *creates approximate direction cosine matrices to order n in the parameter eps. The set, disturbance, must contain a set of scaling relations in terms of the parameter, e.g.* {*q1 = eps * q1, u1= eps * u1*}.

- EvectorSeries(Evec,eps,n) → expansion of the Evector, Evec to order n in eps.

- KvectorSeries(Kvec,eps,n) → expansion of the Kvector, Kvec to order n in eps.

- SvectorSeries(Svec,eps,n) → expansion of the Svector, Svec to order n in eps.

- B &Ato Evec → Evec in frame B to order n.

- A &Afdt Evec → frame based derivative of Evec to order n.

- A &AKfdt Kvec → frame based derivative of Kvec to order n.

- tauK &kane pKt → a list of scalar expressions obtained by taking the fat dot product of each of the members of the set tauK with the Kvector pKt.

A.10 Global Variables

Sophia uses some special global variables to store important information. The user should know about these both to avoid name conflicts and to make other use of the information.

- RfAfB *When a frame relation is defined direction cosine matrices are formed. They are always named RfAfB, where fA and fB are the frame names. If in the course of a computation the system needs a new matrix and has the information to compute it, new names will be assigned as needed. All defined frames are placed in a Global set.*
- SetOfFramesUsed → {{fA,fB},{fA,fC},{fB,fC}} *All the frames known to the system. Each set implies the direction cosine matrices in both directions.*
- kde *In some cases it is assigned by the system, but in general should always be used for the set of kinematic differential equations.*
- toTimeFunction *is used by the system to transform from the time expression form where a derivative is indicated by an appended t, to the functional form where it is indicated by the Maple differentiation expression and functions have explicit time dependencies, for example q1 becomes q1(t), q1t becomes diff(q1(t),t). The user can use subs(toTimeFunction,expr) to convert an expression to the functional form for further processing in Maple, for example in Maples differential equation package.*
- toTimeExpression *The opposite of the above conversion.*

A.11 Comments

Some useful or special considerations when using Sophia in Maple.

- The Maple command *subs(set,expression)* works on fairly general objects, for example KMvectors.
- Always remember the Maple 'last name evaluation rule' for arrays. To get at the actual array one needs to use op(arrayName). arrayName will not evaluate to the array object!
- The toTimeFunction set allows easy conversion of Sophia's time expressions to Maple's functional forms. This is very useful if one wants to use Maple's numerical differential equation solver.
- Consult the Maple help files and manuals to obtain information on how to extract coefficient matrices from sets and lists of linear equations. This is helpful when moving information out of Maple to other applications.

- There is a Sophia package for interacting with Matlab. Parts of this may be obtained by FTP transfer to the server given in the preface to this book.

- Sophia contains many more functions then discussed here or even in the text. Many are used to set up the main functions. The user may find it interesting to explore the procedures for ideas in extending the system.

Appendix B

Annotated List of References and Computer Software

I consider any of these books worth having, even if it is unlikely you will want to read it from cover to cover. Wherever possible I include ISBN numbers to help you in ordering the book, but try to specify a soft cover or international student version to keep cost down. If you are going to use mechanics professionally in your work you will want to own most of these books, but you can start by browsing through a library copy. The comments are personal and should be looked at in that light. Whatever is said it should be realized that mention of a book indicates that I consider it worth looking at and even owning. The computer software represents, at the time of writing, typical modern approaches to numerical mechanics and the use of computer algebra in mechanics.

B.1 Books

- `Angeles, J.`
  `Rational Kinematics`
  `Springer-Verlag 1988`
  `ISBN 0-387-96813-X`

  Modern kinematics makes extensive use of mathematics. This book shows that rigid body kinematics can be put on a firm footing by the use of linear algebra. Theorems and abstraction take priority over computational details.

- `Arnold, V.I.`
  `Mathematical Methods of Classical Mechanics, Second Edition`
  `Springer Verlag 1989`
  `ISBN 0-387-96890-3`

  The acknowledged definitive reference giving the modern mathematicians view of classical analytical mechanics. Concentration is on conservative Hamiltonian systems so expect beautiful mathematics but not much help in dealing with many rigid body systems of the sort found in technical mechanics.

- Bottema and Roth
  Theoretical Kinematics
  Dover Reprint Edition 1990 of 1970 edition
  ISBN 0 486 66346-9

  This book provides a very will written theoretical foundation for the subject of kinematics.

- Drazin, P.G.
  Nonlinear Systems
  Cambridge University Press, 1992
  ISBN 0-521-40668-4

  This is an excellent introduction to the theory of dynamical systems for the engineer or physicist. Drazin is well known for the application of asymptotic methods to stability problems in fluid mechanics. This book gives a useful introduction to tools such as bifurcation theory and Poincaré sections. There is also a very readable discussion of so called chaos theory and strange attractors.

- Huston, R.L.
  Multibody Dynamics
  Butterworth-Heinemann 1990
  ISBN 0-409-90041-9

  The approach of this book is somewhere between the algorithmical style of Kane and the geometric style of the present work. There is extensive background material for setting up numerical forms of the equations of motion.

- Garcia de Jalon and Bayo
  Kinematic and Dynamic Simulation of Multibody Systems
  Springer-Verlag 1994
  ISBN 0-387-94096-0

  This modern work on the formulation of equations of motion for multibody systems concentrates on methods suited for fast numerical simulation of large systems.

- Kane and Levinson
  Dynamics: Theory and Applications
  McGraw-Hill 1985
  ISBN 0-07-037846-0

  This is the standard reference for Kane's approach to mechanics. It covers roughly the same ground as this course but takes an algorithmic rather than a geometrical point of view. Kane's partial velocities are the components of our tangent vectors. The problems are interesting and varied.

- Kane, Likins and Levinson
  Space Craft Dynamics
  McGraw-Hill 1983
  ISBN 0-07-037843-6

  The emphasis in this book is indeed on spacecraft, which calls for extensive discussions of topics such as Euler parameters and gravitational potential theory. Anyone working seriously on problems of modern dynamics should have this book.

- Lanczos, C.
  The Variational Principles of Mechanics
  Toronto University Press 1966 & Dover Books

  Lanczos was an interesting participant of twentieth century science. He discovered the fast fourier transform in the 40's, before the age of digital computers and was a friend and collaborator with among others Einstein. This book gives a very readable and somewhat philosophically oriented account of mechanics based on variational principles. Its a pleasure to read and another one of those books that is nice to have on your bookshelf.

- Landau and Lifshitz
  Classical Mechanics

  This is the first volume in a famous series on the whole of theoretical physics, hence as one might expect the emphasis is on setting the stage for the authors way of thinking about the foundations of physics. Not the best first book on the subject but worth reading at some stage of ones development as a user of mechanics. The latest edition contains an interesting biographical introduction that describes the way Landau worked as a scientist. For the mathematicians point of view read Landau's fellow countryman V.I. Arnold's somewhat pointed comments about this book. They appear as footnotes in Arnold's book, mentioned above.

- Nikravesh, P.E.
  Computer-Aided Analysis of Mechanical Systems
  Prentice-Hall International Editions 1988
  ISBN 0-13-162702-3

  The main focus is on the numerical treatment of two dimensional mechanisms. Basic numerical techniques of linear algebra and differential equations are covered. Cartesian coordinates with added constraint relations provide the main technique used for setting up and solving problems. This is a good source for the standard way engineers use numerical methods for rigid body systems.

• Matzner and Shepley
Classical Mechanics
Prentice-Hall International Editions 1991
ISBN 0-13-138272-1

One of a recent group of mechanics textbooks oriented towards providing the physicist with the modern differential geometry based view of classical mechanics. There is little of use here for solving really complex problems but much in the way of a foundation for theoretical physics.

• Meirovitch, L.
Methods of Analytic Dynamics
McGraw-Hill 1970

One of the best general references on classical mechanics. It covers a wide range of topics fairly well, including the use of quasi-velocities, stability theory and Hamilton-Jacobi theory.

• Moon, F.C.
Chaotic and Fractal Dynamics.
Wiley-Interscience 1992
ISBN 0-471-54571-6

This second version of the book 'Chaotic Vibrations' is a good introduction to the use of current ideas about dynamical systems to practical mechanical problems. The author has been a leader in developing experimental techniques for the study of chaos in mechanical systems.

• Pars, L.A.
A Treatise on Classical Mechanics

A very clear discussion of mechanics based on the various transformations of D'Alembert's principle. Pars gave recognition to Gibbs prior development of the Gibbs Appell equation in the 1870's. The material in the book covers just about all of theoretical mechanics, including stability theory and the three body problem.

• Percival and Richards
Introduction to Dynamics
Cambridge University Press 1982
ISBN 0-521-28149-0

The authors research work is in the dynamics of Hamiltonian systems. This book is intended to introduce the reader to this subject by restricting most of the discussion to one degree of freedom systems. Subjects include action and angle variables and perturbation theory by way of canonical transformations.

- `Roy, A.E.`
  `Orbital Motion, Third Edition`
  `Adam Hilger, 1988`
  `ISBN 0-85274-229-0`

  A good general introduction to celestial mechanics. If your lost in space this is the book you will need.

- `Scheck, F.`
  `Mechanics`
  `Springer-Verlag, 1991`
  `ISBN 0-387-52715-X`

  A physicists book in classical mechanics, including relativity and the modern geometric view.

- `Schiehlen, W. (Editor)`
  `Multibody Systems Handbook`
  `Springer-Verlag 1990`
  `ISBN 0-387-51946-7`

  This is a collection of articles where each author demonstrates his numerical software on two bench mark problems, a seven bar linkage and a robotic arm. The problem is that each author is trying to market his system, however the book does provide a good summary of the state of the art at the time of writing.

- `Wittenburg, J.`
  `Dynamics of Systems of Rigid Bodies`
  `B.G. Teubner Stuttgart 1977`
  `ISBN 3-519-02337-7`

  One of the first general approaches to the problem of formulating equations for complex mechanical systems. The book contains many interesting ideas and examples including applications to biped locomotion.

B.2 Computer Programs

- Interactive Physics II

  The publishers claim that this Macintosh based program is to mechanics what spread sheets are to accountancy. While it is restricted to the treatment of planar mechanisms the claim has a great deal of truth. If you have any interest in mechanics you will want this software which gives you a laboratory in your computer! Using it is like having a Meccano on your computer screen.

- Applied Motion - Mechanica

It is claimed that this full scale design system makes use of Kane's method. It can handle very complex systems and has an interface that will be familiar to users of modern CAD software, but the general nature of the program leads to relatively slow integration times and extensive parameter studies can be painful. The intended industrial customer is expected to think little of spending 20,000 dollars for the software, so don't expect to buy it with your spare lunch money!

- Auto-Lev

  This is Kane's own program, which has its own computer algebra routines to develop equations of motion and to output a FORTRAN code for their integration. The manual itself is an excellent introduction to Kane's methods. Don't expect to get simple analytical forms out of this, but it is excellent in producing useful computer code. The program runs under ms-dos on IBM-PC's or clones. It will run fairly well on a Macintosh using Soft-PC. You will need a FORTRAN compiler to turn the output source code into runable programs. Student editions cost one hundred dollars.

Index

List of Figures